Biosensors and Environmental Health

Biosensors and Environmental Health

Editors

Victor R. Preedy PhD DSc

Professor of Nutritional Biochemistry
School of Medicine
King's College London
and
Professor of Clinical Biochemistry
King's College Hospital
UK

Vinood B. Patel

Department of Biomedical Science
School of Life Sciences
University of Westminster
London
UK

CRC Press
Taylor & Francis Group
Boca Raton London New York

CRC Press is an imprint of the
Taylor & Francis Group, an **informa** business

A SCIENCE PUBLISHERS BOOK

CRC Press
Taylor & Francis Group
6000 Broken Sound Parkway NW, Suite 300
Boca Raton, FL 33487-2742

First issued in paperback 2019

© 2012 Copyright reserved
CRC Press is an imprint of Taylor & Francis Group, an Informa business

Cover Illustrations: Reproduced by kind courtesy of the undermentioned authors:

- Figure No. 1 from Chapter 2 by Javier Ramón-Azcón et al.
- Figure No. 3 from Chapter 3 by Edoardo Puglisi et al.
- Figure No. 3 from Chapter 4 by Georges Istamboulie et al.
- Figure No. 2 from Chapter 6 by Pavlina Sobrova et al.

No claim to original U.S. Government works

ISBN-13: 978-1-57808-735-8 (hbk)
ISBN-13: 978-0-367-38098-4 (pbk)

Library of Congress Cataloging-in-Publication Data

Biosensors and environmental health / editors, Victor R. Preedy, Vinood Patel. -- 1st ed.
       p. cm.
  Includes bibliographical references and index.
  ISBN 978-1-57808-735-8 (hardback)
  1. Biosensors. 2. Environmental health. 3. Environmental toxicology. 4. Pollution--Measurement. I. Preedy, Victor R. II. Patel, Vinood B.
  R857.B54.B5523 2012
  610.28--dc23
                                                    2012012971

Visit the Taylor & Francis Web site at
http://www.taylorandfrancis.com

and the CRC Press Web site at
http://www.crcpress.com

Preface

Biosensors have a simplistic concept but a great deal of sophistication in design, manufacture and application. They essentially have a biological component within them and are used to detect, monitor or quantify substances. They use a variety of physical platforms and technologies. The biological components may include enzymes, membranes and cells or any other naturally occurring biological product. Some have artificial biological components such as modified molecules or polymers. Biosensors may be used to detect single or groups of molecules and have wide applicability to the life sciences. In this book we aim to disseminate the information on biosensors in a readable way by having unique sections for the novice and expert alike. This enables the reader to transfer their knowledge base from one discipline to another or from one academic level to another. In this book we focus on environmental issues. Chapters in **Biosensors and Environmental Health** have an abstract, key facts, applications to other areas of health and disease and a "mini-dictionary" of key terms and summary points. The book describes new methods, prototypes and applications. For example coverage includes: personal toxicity testing, soil and risk assessment, pesticide, insecticides, parasites, nitrate, endocrine disruptors, heavy metals, food contamination, whole cell bioreporters, bacterial biosensors, antibody-based biosensors, enzymatic, amperometric and electrochemical aspects, quorum sensing, DNA-biosensors, cantilever biosensors, bioluminescence and other methods and applications.

Contributors to **Biosensors and Environmental Health** are all either international or national experts, leading authorities or are carrying out ground breaking and innovative work on their subject. The book is essential reading for environmental scientists, toxicologists, medical doctors, health care professionals, pathologists, biologists, biochemists, chemists and physicists, general practitioners as well as those interested in disease and sciences in general.

The Editors

Contents

List of Contributors

Vojtech Adam
Mendel University in Brno, Department of Chemistry and Biochemistry, Zemedelska 1, 61300 Brno, Czech Republic.
E-mail: vojtech.adam@mendelu.cz

Erol Akyilmaz
Ege University Faculty of Science Biochemistry Department 35100 Bornova/ Izmir, Turkey.
E-mail: erol.akyilmaz@ege.edu.tr

Silvana Andreescu
206 Science Center, Clarkson University, PO Box 5810, Potsdam, NY 13699-5810 USA.
E-mail: eandrees@clarkson.edu

Maria Daniela Angione
Department of Chemistry, University of Bari, Via Orabona 4, I-70126 Bari, Italy.
E-mail: angione@chimica.uniba.it

Mariarita Arenella
Dipartimento di Scienze delle Produzioni Vegetali, del Suolo e dell'Ambiente Agroforestale, Università di Firenze, Piazza delle Cascine 18, 50144 Firenze, Italy.
E-mail: mariarita.arenella@unifi.it

Umberto Bencivenga
Istituto of Genetica and Biofisica del CNR, Napoli, Italia.
E-mail: benciven@igb.cnr.it

S. Buchinger
German Federal Institute of Hydrology (BfG), Division of Qualitative Hydrology, Am Mainzer Tor 1, 56068 Koblenz, 56068 Germany.
E-mail: Buchinger@bafg.de

Xinxia Cai
State Key Laboratory of Transducer Technology, Institute of Electronics, Chinese Academy of Sciences, Beijing 100190, P.R. China.
E-mail: xxcai@mail.ie.ac.cn

Swati Choudhary
Department of Biochemistry, Molecular Biology and Biophysics, University of Minnesota, 140 Gortner Laboratory, St. Paul, MN 55198, USA.
E-mail: swati@umn.edu

Nicola Cioffi
Department of Chemistry, University of Bari, Via Orabona 4, I-70126 Bari, Italy.
E-mail: cioffi@chimica.uniba.it

Serafina Cotrone
Department of Chemistry, University of Bari, Via Orabona 4, I-70126 Bari, Italy.
E-mail: cotrone@chimica.uniba.it

Liu Deng
State Key Laboratory of Electroanalytical Chemistry, Changchun Institute of Applied Chemistry, Chinese Academy of Sciences, Changchun, Jilin, 130022, P. R. China.
E-mail: dengliu@ciac.jl.cn

Shaojun Dong
State Key Laboratory of Electroanalytical Chemistry, Changchun Institute of Applied Chemistry, Chinese Academy of Sciences, Changchun, Jilin, 130022, P. R. China.
E-mail: dongsj@ciac.jl.cn

Gabriele Favero
Department of Chemistry and Drug Technologies, Sapienza University of Rome, P.le Aldo Moro, 5 – 00185 Roma, Italy.
E-mail: gabriele.favero@uniroma1.it

Zhixian Gao
Tianjin Key Laboratory of Risk Assessment and Control Technology for Environment and Food Safety, Tianjin Institute of Health and Environmental Medicine, No. 1 Dali Road, Heping District, Tianjin, China.
E-mail: Gaozhx@163.com

Maria Hepel
Department of Chemistry, State University of New York at Potsdam, Potsdam, New York 13676, USA.
E-mail: hepelmr@potsdam.edu

Wei E. Huang
Kroto Research Institute, Department of Civil and Structural Engineering, North Campus, University of Sheffield, Broad Lane, Sheffield S3 7HQ, UK.
E-mail: w.huang@shef.ac.uk

Jaromir Hubalek
University of Technology in Brno, Department of Microelectronics.
E-mail: hubalek@feec.vutbr.cz

Georges Istamboulie
IMAGES EA 4218, Biosensors group, Building S, Université de Perpignan Via Domitia, 52 Av Paul Alduy 66860 Perpignan cedex – France.
E-mail: georges.istamboulie@univ-perp.fr

Stephen L. Kaattari
Virginia Institute of Marine Science, Department of Environmental and Aquatic Animal Health, College of William & Mary, PO Box 1346, Gloucestor Point, Virginia 23062-1346 USA.
E-mail: kaattari@vims.edu

Rene Kizek
Mendel University in Brno, Department of Chemistry and Biochemistry, Zemedelska 1, 61300 Brno, Czech Republic.
E-mail: kizek@sci.muni.cz

Maria Lepore
Dipartimento di Medicina Sperimentale, Seconda Università di Napoli, Napoli, Italia.
E-mail: maria.lepore@unina2.it

Guanghe Li
Department of Environmental Science & Technology, Tsinghua University, Beijing 100084, P.R. China.
E-mail: ligh@mail.tsinghua.edu.cn

Nan Liu
Tianjin Key Laboratory of Risk Assessment and Control Technology for Environment and Food Safety, Tianjin Institute of Health and Environmental Medicine, No. 1 Dali Road, Heping District, Tianjin, China.
E-mail: LNQ555@126.com

Xiaohong Liu
Graduate School of Chinese Academy of Sciences, Beijing 100080, P.R. China.
E-mail: xinqi_0120@163.com

Jinping Luo
State Key Laboratory of Transducer Technology, Institute of Electronics, Chinese Academy of Sciences, Bejing 100190, P.R. China.
E-mail: jpluo@mail.ie.ac.cn

Maria Magliulo
Department of Chemistry, University of Bari, Via Orabona 4, I-70126 Bari, Italy.
E-mail: magliulo@chimica.uniba.it

Antonia Mallardi
CNR-IPCF, Istituto per i Processi Chimico-Fisici, Via Orabona 4, I-70126 Bari, Italy.
E-mail: a.mallardi@ba.ipcf.cnr.it

Jean-Louis Marty
IMAGES EA 4218, Biosensors group, Building S, Université de Perpignan Via Domitia, 52 Av Paul Alduy 66860 Perpignan cedex – France.
E-mail: jlmarty@univ-perp.fr

Franco Mazzei
Department of Chemistry and Drug Technologies, Sapienza University of Rome, P.le Aldo Moro, 5 – 00185 Roma, Italy.
E-mail: franco.mazzei@uniroma1.it

Damiano Gustavo Mita
Dipartimento di Medicina Sperimentale, Seconda Università di Napoli, Napoli, Italia.
E-mail: mita@igb.cnr.it

Fumio Mizutani
Graduate School of Material Science, University of Hyogo, 3-2-1 Kouto, Kamigori-cho, Ako-gun, Hyogo 678-1297, Japan.
E-mail: mizutani@sci.u-hyogo.ac.jp

Raj Mutharasan
Department of Chemical and Biological Engineering, Drexel University, Philadelphia, PA 19104, USA.
E-mail: mutharasan@drexel.edu

Thierry Noguer
IMAGES EA 4218, Biosensors group, Building S, Université de Perpignan Via Domitia, 52 Av Paul Alduy 66860 Perpignan cedex – France.
E-mail: noguer@univ-perp.fr

Anna Nowicka
Department of Chemistry, State University of New York at Potsdam, Potsdam, New York 13676, USA.
Permanent address: Department of Chemistry, University of Warsaw, PL-02-093 Warsaw, Poland.
E-mail: anowicka@chem.uw.edu.pl

Gerardo Palazzo
Department of Chemistry, University of Bari, Via Orabona 4, I-70126 Bari, Italy.
E-mail: palazzo@chimica.uniba.it

Rosa Pilolli
Department of Chemistry, University of Bari, Via Orabona 4, I-70126 Bari, Italy.
E-mail: pilolli@chimica.uniba.it

Karen Polizzi
Division of Molecular Biosciences & Centre for Synthetic Biology and Innovation, Imperial College London, London SW7 2AZ, UK.
E-mail: k.polizzi@imperial.ac.uk

Marianna Portaccio
Dipartimento di Medicina Sperimentale, Seconda Università di Napoli, Napoli, Italia.
E-mail: marianna.portaccio@unina2.it

Victor R. Preedy
Diabetes and Nutritional Sciences, School of Medicine, Kings College London, Franklin Wilkins Buildings, 150 Stamford Street, London SE1 9NU, UK.
E-mail: victor.preedy@kcl.ac.uk

Edoardo Puglisi
Istituto di Microbiologia, Università Cattolica del Sacro Cuore, Via Emilia Parmense 84, 29122 Piacenza, Italy.
E-mail: edoardo.puglisi@unicatt.it

Rajkumar Rajendram
Locum Consultant Physician, Department of Medicine, John Radcliffe Hospital, Oxford OX3 9DU, UK.
E-mail: rajkumarrajendram@doctors.org.uk

Javier Ramón-Azcón
Graduate School of Material Science, University of Hyogo, 3-2-1 Kouto, Kamigori-cho, Ako-gun, Hyogo 678-1297, Japan.
E-mail: ramonazconjavier@gmail.com

G. Reifferscheid
German Federal Institute of Hydrology (BfG), Division of Qualitative Hydrology, Am Mainzer Tor 1, 56068 Koblenz, 56068 Germany.
E-mail: Reiffersceid@bafg.de

Giancarlo Renella
Dipartimento di Scienze delle Produzioni Vegetali, del Suolo e dell'Ambiente Agroforestale, Università di Firenze, Piazza delle Cascine 18, 50144 Firenze, Italy.
E-mail: giancarlo.renella@unifi.it

Claudia Schmidt-Dannert
Department of Biochemistry, Molecular Biology and Biophysics, University of Minnesota, 140 Gortner Laboratory, St. Paul, MN 55198, USA.
E-mail: schmi232@umn.edu

Li Shang
Institute of Applied Physics and Center for Functional Nanostructures (CFN), Karlsruhe Institute of Technology (KIT), Wolfgang-Gaede-Strasse 1, 76131 Karlsruhe, Germany.
E-mail: lishang208@gmail.com

Pavlina Sobrova
Mendel University in Brno, Department of Chemistry and Biochemistry, Zemedelska 1, 61300 Brno, Czech Republic.
E-mail:pavlina.sobrova@seznam.cz

Yizhi Song
Department of Environmental Science & Technology, Tsinghua University, Beijing 100084, P.R. China.
E-mail: song.ezhi@gmail.com

Candace R. Spier
Virginia Institute of Marine Science, Department of Environmental and Aquatic Animal Health, College of William & Mary, PO Box 1346, Gloucestor Point, Virginia 23062-1346 USA.
E-mail: cspier@vims.edu

Magdalena Stobiecka
Department of Chemistry, State University of New York at Potsdam, Potsdam, New York 13676, USA.
E-mail: stobiemk@potsdam.edu

Qing Tian
State Key Laboratory of Transducer Technology, Institute of Electronics, Chinese Academy of Sciences, Beijing 100190, P.R. China.
E-mail: tianqing@ustc.edu

Luisa Torsi
Department of Chemistry, University of Bari, Via Orabona 4, I-70126 Bari, Italy.
E-mail: torsi@chimica.uniba.it

Marco Trevisan
Istituto di Chimica Agraria ed Ambientale, Università Cattolica del Sacro Cuore, Via Emilia Parmense 84, 29122 Piacenza, Italy.
E-mail: marco.trevisan@unicatt.it

Libuse Trnkova
Masaryk University, Brno, Department of Chemistry.
E-mail: libuse@chemi.muni.cz

Daniela Di Tuoro
Dipartimento di Medicina Sperimentale, Seconda Universita di Napoli, Napoli, Italia.
E-mail: daniela.dituoro@libero.it

Michael A. Unger
Virginia Institute of Marine Science, Department of Environmental and Aquatic Animal Health, College of William & Mary, PO Box 1346, Gloucestor Point, Virginia 23062-1346 USA.
E-mail: munger@vims.edu

Mona Wells
Helmholtz Centre for Environmental Research, Department of Environmental Microbiology, Permoserstraβe 15, Leipzig, 04318, Germany.
E-mail: mona.well@ufz.de

Sen Xu
Department of Chemical and Biological Engineering, Drexel University, Philadelphia, PA 19104, USA.
E-mail: xuzhonghou@gmail.com

Tomoyuki Yasukawa
Graduate School of Material Science, University of Hyogo, 3-2-1 Kouto, Kamigori-cho, Ako-gun, Hyogo 678-1297, Japan.
E-mail: yasu@sci.u-hyogo.ac.jp

Weiwei Yue
State Key Laboratory of Transducer Technology, Institute of Electronics, Chinese Academy of Sciences, Beijing 100190, P.R. China.
E-mail: physics_yue@163.com

Dayi Zhang
Kroto Research Institute, Department of Civil and Structural,
Engineering, North Campus, University of Sheffield, Broad Lane,
Sheffield S3 7HQ, UK.
E-mail: d.zhang@shef.ac.uk

Eko Ge Zhang
Fetal and Maternal Medicine, Imperial College NHS Trust, London SW7
2AZ, UK.
E-mail: GeEko.Zhang@Imperial.nhs.uk

Immunochips for Personal Toxicity Testing

Zhixian Gao,[1,a] Nan Liu[1,b] and Rajkumar Rajendram[2]

ABSTRACT

People are exposed to many chemicals in the course of day-to-day life. Measurement of the exposure of the environment and its inhabitants to pollutants is a useful estimate of the toxic effects of environmental pollution on health. The microarray is a sensitive and precise device which can be used to obtain this information from complex biological samples. Microarrays can thus be used to assess the "health" of the environment or an individual person. The use of microarrays allows complex, automated, high-throughput processes to be performed in small devices. The "immunochip" is a one of the formats of protein microchip based on the molecular specific immunological recognition of antigens (Ags) by antibodies (Abs) immobilized on a certain surface that together respond in a concentration-dependent manner. Recent work from our laboratory demonstrated that immunochip technology can simultaneously detect at least five different chemicals. This chapter discusses the various types of immunochips available and their application in personal toxicity testing. The definition, main features,

[1]Tianjin Key Laboratory of Risk Assessment and Control Technology for Environment and Food Safety, Tianjin Institute of Health and Environmental Medicine, No. 1 Dali Road, Heping District, Tianjin, China.
[a]E-mail: Gaozhx@163.com
[b]E-mail: LNQ555@126.com
[2]Locum Consultant Physician, Department of Medicine, John Radcliffe Hospital, Oxford OX3 9DU, England.
E-mail: rajkumarrajendram@doctors.org.uk

List of abbreviations after the text.

and probes of "conventional" immunochips, "Lab-on a chip" and "suspension arrays" are included. "Lab-on a chip" integrates several laboratory processes including preparation, incubation, detection and analysis on a single microchip. This eliminates the need for several different pieces of laboratory equipment to prepare and analyze a biological sample. The "high-throughput suspension" array is a novel method for multi-analysis of veterinary drugs. It is easy to use, very sensitive and inexpensive. However, immunochips are difficult to use in the field. A high-quality Ab with good bioactivity and specificity is the key reagent in the production of immunochips. Although further investigation is required, the potential advantages of immunochip technology for the detection of chemicals for environmental assessments are of great interest.

INTRODUCTION

Millions of chemicals have been created as civilization has advanced and industry and agriculture have developed. Modernization has simultaneously facilitated human existence and wrought havoc on the environment. Pollution is a global phenomenon of major concern.

The air quality of large- and medium-sized cities is poor worldwide. Air pollution affects many countries but is the worst in the developing world. Soil is commonly contaminated with solvents, hydrocarbons derived from petrochemicals, heavy metals and pesticides. Organic pollutants are a major problem affecting water quality. Eutrophication occurs naturally as bodies of water age, but the process is accelerated by pollution. Many lakes in China are in the intermediate or advanced stages of eutrophication. Pollution from industry, agriculture and domestic life damages the ecosystem and is a major hazard to human health. Air pollution increases excess mortality rates especially in the developing countries. The incidence of gastrointestinal tract tumors is increased with consumption of contaminated water. Many environmental contaminants, pollutants and toxins interfere with immune defense, immune signal transduction, and induce hypoimmunity and do harm to population health.

The general population is exposed to many chemicals in the course of day-to-day life. Measurement of the exposure of the environment and its inhabitants to pollutants is a useful estimate of the toxic effects of environmental pollution on health. "Safe" levels of exposure for many chemicals have yet to be determined. Newly synthesized compounds are released regularly and data on their intermediate products and by-products are scanty. However, relationships between levels of exposure and adverse effects on health remain unclear. So, even when guidance is available, standards for safe or acceptable concentrations or levels of exposure often vary between countries. Assessment of the toxicity of chemicals *in vivo* is urgently in need of further research.

Personal toxicity testing is used to detect environmental toxins/pollutants/contaminants and/or their metabolites and biomarkers. Chromatogram (Koblížek et al. 2002), liquid chromatography with mass spectrometry (LC-MS) technology (Ramos et al. 2003) and enzyme-linked immunoadsorbent assay (ELISA) (Estévez et al. 2006; Mart'ianov et al. 2005) are widely employed for the detection of chemicals. However, these techniques require meticulous sample preparation and are complex to perform, so are expensive and time consuming. These shortcomings cannot be easily overcome.

The use of recently developed scientific techniques such as proteomics, metabolomics and metallomics could enhance personal toxicity testing and environmental assessments. The research, development and use of these and other new techniques, could improve monitoring of environmental pollution and the effects on human health. The recent production of microchips which can perform these techniques has enabled production of small, automated analytical systems for this purpose.

Microarray or biochip technology allows the simultaneous monitoring of several biological processes in a single experiment. These sensitive and precise, automated techniques can obtain vast amounts of data from each biological sample. Small, high-throughput systems utilizing microarrays can be used to detect and analyze samples containing bacteria (Song and Dinh 2004), viruses (Zhou et al. 2005), veterinary drug residues (Du et al. 2005) and tumor biomarkers (Ghobrial et al. 2005; Huang et al. 2004; Miller et al. 2003; Ghobrial et al. 2005; Huang et al. 2004).

Biochips have several uses in the agricultural, veterinary, healthcare, and medical sectors. DNA microarrays, which contain single strand nucleotide probes bound to a solid substrate, are used to identify DNA or RNA. DNA arrays function on the principle of nucleic acid hybridization on a surface-immobilized template. Protein arrays (protein microchips) contain various proteins probes imprinted on solid surface and can detect interactions between different proteins (Chiem and Harrison 1998; Wildt et al. 2000).

The immunochip is a protein microchip based on the specific recognition of Ags by Abs immobilized on a solid planar surface (e.g. glass) that responds in a concentration-dependent manner to a chemical target. Although plastic, gold and silicon have been used, and several novel surfaces have been developed including porous polyacrylamide gel pad slides (Arenkov et al. 2000) and microwells (Zhu et al. 2000), the most common surface used for immobilization of bio-recognition agents is the glass microscope slide.

When complementary Ags or Abs react with probes on the array, the resulting array image or fluorescent intensity (FI) can be captured by laser scanner and analyzed by software. Microarray images are necessarily of high resolution and are therefore large (typically at least 1500 × 3500 pixels). Multifunctional data analysis software such as GenePix Pro, ScanAlyze and BlueFuse can be used to analyze FI data.

Several modifications of the conventional immunochip platform have been described. These include the "lab-on a chip" (Du et al. 2005), a hydrogel-based immunochip (Rubin et al. 2005) and the suspension array (Connolly et al. 2010; Kalogianni et al. 2007) which immobilizes bio-active materials on a fluorescent coded bead/microsphere.

High-output, automated, sensitive, analysis of several chemicals can be performed simultaneously. The potential scope for use of the immunochip is therefore remarkable. Although immunochip platforms are fairly versatile, configurations are limited as chips must be designed to detect a specific group of compounds. Abs with good bioactivity and specificity are the best probes currently available. The use of high titres of Abs with high specificity; especially monoclonal Ab (mAb), can significantly increase the accuracy and sensitivity of the immunochip. Phage display technology and ribosome display technology are tools for the generation of high-quality Abs. However, acquiring the key reagent, i.e. the specific Ab required, is still a limiting factor in the production of immunochips.

The Luminex (xMAP®) suspension array is another type of immunochip. It is a microarray on a microsphere surface which enables greater efficiency and output than a glass slide (planar, solid microarray). This system is based on internally color-coded microspheres with surface linked Abs, receptors, or oligonucleotides. Beads containing one of 100 specific dye sets can be differentiated by flow-cytometry. Binding of the surface label indicates analyte binding and can also be detected using a second, shorter-wavelength dye and a dual laser detector (Dasso et al. 2002).

It is inexpensive, versatile and considered as an open platform that has been widely employed for multiplex analysis of ribosomal RNA (rRNA), microRNA, cytokines and single-nucleotide polymorphisms (SNP); selecting and screening of mAbs; detection and testing for Abs, bacterial toxins, polysaccharides and autoantigens in serum, cerebrospinal fluid, dried blood spot specimens, and stool samples (Liu et al. 2009).

Binding to the solid support of ELISA or other microarrays denatures or dries bioactive materials. However, due to robust multiplexing, the Luminex suspension array can obtain more data more quickly from samples in the aqueous phase. Analysis of samples in the aqueous phase maintains their bioactivity for probes. Suspension array technology also offers several other advantages over traditional methods. It is versatile, flexible, accurate, reproducible and high-throughput.

APPLICATION TO HEALTH AND DISEASE

The aim of personal toxicity testing and environmental assessments is to determine the effect of pollution on health. Traditionally the diagnosis of disease was based on a single diagnostic test. However, modern clinical practice, bases the diagnosis of disease on the synthesis of data from several sources. Similarly environmental medicine and environmental health assessments can now use biomarkers and cytokines as well as the detection of toxins to determine the effects of pollution. The accuracy and validity of environmental health assessments are improved simply by taking more factors into account.

Immunochip based analytical techniques have a wide range of potential uses for personal toxicity testing and environmental assessments. Immunochips are already used extensively in many similar fields. For example, clinical diagnostics, biomedicine and pharmaceutical analysis use immunochips to detect disease related protein changes. Immunochips have also been used for the assessment of food and drink. Potential uses for immunochips include analyses of nutrients, organic toxins (e.g. bacterial toxins, mycotoxins and hormones) and inorganic toxins (e.g. pesticides, and heavy metals). These chemicals pollute the environment and cause disease. In comparison to other diagnostic techniques, those which use immunochips can provide large amounts of data in simple, rapid, automated, and relatively inexpensive processes.

KEY FACTS ON THE USE OF IMMUNOCHIPS FOR PERSONAL TOXICITY TESTING

- Many environmental contaminants, including polychlorinated biphenyl compounds, formaldehyde, heavy metals, organ phosphorus pesticide, tobacco and smog affect white blood cell function and immune signal transduction. This impairs cellular and non-cellular immunity and reduces population health.
- Personal toxicity testing requires rapid, accurate detection of a diverse range of chemicals which are often present in small traces and analysis of the interactions of these toxins with health and disease. This remains a formidable challenge.
- Analytical techniques which use microarray or biochip technology allow sensitive and precise monitoring of multiple processes in biological samples in a single experiment.
- Immunochips have a wide range of potential uses and can provide large amounts of data in simple, rapid, automated, and relatively inexpensive processes.

- Abs with good bioactivity and specificity are the best probes currently available and are the key reagents in the production of immunochips.
- The use of high titres of Abs with high specificity; especially monoclonal Abs, significantly increases the accuracy and sensitivity of the immunochip.
- Phage display technology and ribosome display technology are other alternative ideal tools for the generation of high-quality Abs.

DEFINITIONS OF KEY TERMS, GENES, CHEMICALS AND PATHWAYS

- Endocrine disruptor chemical (EDC) is the external agent that interferes with hormonal function *in vivo*. Any stage of hormone production and activity can be affected. For example by preventing hormones synthesis, binding directly to hormone receptors, or interfering with the natural breakdown of hormones. These agents can impair endocrine function *in vivo* and are toxic to human health and the environment.
- Immunochip technology is based on the specific, concentration-dependent, immunological recognition of Ags by immobilized Abs. The most common surface used for immobilization of Abs is a glass slide.
- Suspension array is another kind of immunochip and is simply a transfer of the microarray format from a glass slide (planar and solid microarray) to a high-throughput and efficient microsphere format. This system is based on internally color-coded microspheres with surface linking chemistry to accommodate Abs, receptors, or oligonucleotides. Beads containing one of 100 specific dye sets can be differentiated using flow-cytometry. Binding of the surface label indicates analyte binding. This can be detected using a second, shorter-wavelength dye and a dual laser detector (Dasso et al. 2002).
- "Lab-on-chips" integrate several laboratory processes on a single microchip. These procedures include preparation, incubation, detection and analysis.
- Phage display technology is widely used to display Ab libraries on the surface of filamentous bacterio-phages. The libraries allow the selection of Abs with high specificity and affinity for any Ags. A phagemid-based Ab display library is made by cloning Ab genes into a phagemid vector at the 5'end of the g III within the

phagemid vector. This is followed by transformation of *Escherichia coli* (Hoogenboom et al. 1991).

*   Ribosome display is a powerful *in vitro* display technology. It exploits cell-free translation to generate a selection unit comprising a "stalled" ribosome linking a protein to its encoding mRNA (McCafferty et al. 1994). This pro-karyotic cell-free translation system can develop high affinity Abs.

THE USE OF CONVENTIONAL IMMUNOCHIPS FOR PERSONAL TOXICITY TESTING

Based on the molecular weight (MW) and the structure of the target molecules, chemicals or toxins can be divided into macromolecules and "small molecules". The immunochips use patterned Abs/Ags and sandwich immunoassays for the detection of their complimentary Ags/Abs. Most toxins are small molecules. As there is generally only one binding site in the structure of small molecules, direct or indirect competitive immunoassays are used to detect toxins in the immunochip.

Sapsford et al. (2006) employed the Naval Research Laboratory (NRL) array biosensor, an indirect competitive immunochip, to detect aflatoxin B_1 (AFB_1) in corn and nut products. The NeutrAvidin slides patterned with biotinylated AFB_1 were designed so that the orientation of the flow channels of the Poly(dimethyl)siloxane (PDMS) flow cells was perpendicular to the strips of immobilized biotinylated molecules (Fig. 1.1; Sapsford et al. 2006).

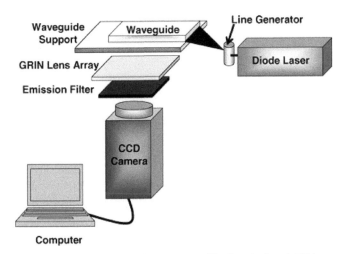

Figure 1.1. Schematic of the NRL array biosensor used by Sapsford et al. 2006.

After washing the PDMS channels, spiked food sample, containing "free"AFB$_1$ and the Cy5-labeled monoclonal mouse anti-AFB$_1$, was applied to each channel. AFB$_1$-spiked foods were extracted with methanol and Cy5-anti-AFB$_1$ then added. The mixture of the extracted sample and Ab was passed over a waveguide surface patterned with immobilized AFB$_1$. The resulting fluorescence signal decreased as the concentration of AFB$_1$ in the sample increased. The limit of detection for AFB$_1$ in buffer (0.3 ng/mL) increased to between 1.5 and 5.1 ng/g and 0.6 and 1.4 ng/g when measured in various corn and nut products, respectively.

In a recent work, we explored the feasibility of using immunochips to detect five different chemicals (Gao et al. 2009) simultaneously. We used atrazine (Ar), nonylphenol (NP) and 17-beta estradiol (E$_2$), paraverine and chloramphenicol. Atrazine (Ar), nonylphenol (NP) and 17-beta estradiol (E$_2$) are endocrine disruptor chemicals (EDCs) which are harmful to human health and the ecological environment (Cooper et al. 2000; Han et al. 2004; Spearow et al. 1999). Effects include heteroplasia (Kavlock et al. 1996), metabolic disorders (Friedmann, 2002), changes in sexual characteristics (Hayes et al. 2002) and development of some tumors (Choi et al. 2004). Papaverine (Pap) is an isoquinoline alkaloid, derived from poppies which can cause addiction and chronic poisoning. Chloramphenicol (CAP) may cause aplastic anemia.

Different concentrations of Ar, NP, E$_2$, Pap and CAP were used and standard curves were produced for each of the five chemicals (Fig. 1.2). The equations for these standard curves are shown in Table 1.1.

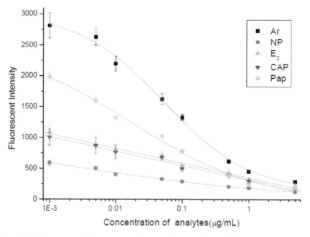

Figure 1.2. Standard curves of fluorescent intensity produced from immunochip analysis of five different chemicals (Liu et al. 2009).

Table 1.1. Equations for standard curves. Comparison of real values and those found by immunochip detection.

Item	Standard curve	R^2	Real (ng/mL)	Found (n=3, $x \pm s$, ng/mL)	RSD(%)
CAP	$y_{CAP}=317.6746-231.6716x$	0.9933	10.00	10.39 ± 0.84	3.9
NP	$y_{NP}=-9.15721+1171.6080/(1+(x/0.00123)^{0.2455})$	0.9917	10.00	9.41 ± 0.31	5.9
Pap	$y_{Pap}=-49.5254+2557.6292/(1+(x/0.0183)^{0.4544})$	0.9928	10.00	10.98 ± 0.53	9.8
E_2	$y_{E_2}=351.0196-236.9911x$	0.9914	10.00	9.42 ± 0.40	5.8
Ar	$y_{Ar}=131.3843+2873.1845/(1+(x/0.0549)^{0.6803})$	0.9956	10.00	10.89 ± 0.66	8.9

Our observations demonstrated that simultaneous quantitative assessment of these five chemicals can be performed by the immunochip. Fluorescent intensity decreased as concentrations of the added standard chemicals increased. The detection ranges were 0.001–5 µg/mL with logistic and linear correlation. The determination coefficients-R^2 were all greater than 0.99 implying good correlation between the target chemicals and the FIs. Concentrations of any of the five chemicals within the detection range could quantify the standard curves. To simplify the experimental procedure the concentrations were increased in eight steps. Concentrations under 0.001 µg/mL and over 5 µg/mL, may deviate from the standard curve. Due to inter-chip variation and human error, validity, reliability, and stability are still major problems when immunochip technology is used for multi-analysis. For this reason, it is necessary to plot standard curves with simultaneous assessment of real samples when new multi-analyses are performed with immunochips.

To simplify and accelerate multianalytical procedures, each aldehyde glass slide was divided into 10 relatively small units. Each small unit was made up of a 6 × 4 array. From left to right, 3% BSA (negative control), mAbs of CAP, Pap, E_2, NP and Ar were spotted as probes in turn and repeated four times. The five corresponding Ag conjugates were successively added homogeneously to each of the 10 units on the slide. After spotting, the slide was deposited and incubated in an enclosed box for 2 hr at 37°C. The slide was then washed with PBST and water before being blocked with 2% BSA. After incubation in an enclosed box for another 2 hr the slide was washed again with PBST and water. The addition sequences of Ag conjugates are shown in Fig. 1.3. The slide was then incubated in an enclosed box for another 2 hr at 37°C for specific Ag-Ab competitive reactions. The GenePix™ 4000B scanner and GenePix™ Pro 4.0 software were used to analyze the results.

The specificity of the immunochip was assessed by expose to a mixture of Ags. Fig. 1.3A, demonstrates four mAbs of CAP, NP, E_2 and Ar with their complementary Ag conjugates spotting on a slide showing strong

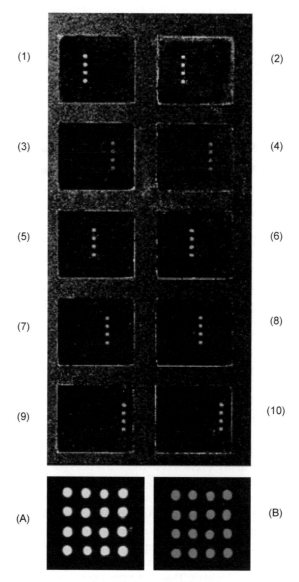

Figure 1.3. Multi-analysis of five different chemicals. Abs addition: In one unit, from left to right: 3% BSA (as the negative control), monoclonal antibodies of CAP, Pap, E_2, NP and Ar were spotted as probes; addition of the 10 units was homogeneous. Antigen conjugates addition: (1) and (2) units: CAP-BSA-Cy3; (3) and (4) units: NP-OVA-Cy3; (5) and (6) units: Pap-OVA-Cy3; (7) and (8) units: E_2-OVA-Cy3; (9) and (10) units: Ar-OVA-Cy3. **(A)** Four mAbs of CAP, NP, E_2 and Ar with their complementary Ag conjugates spotting on slide. **(B)** CAP, NP, E_2 and Ar with the same concentration of 0.01µg/mL mixture were added on the chip (Gao et al. 2009).

Color image of this figure appears in the color plate section at the end of the book.

FIs. After addition of the mixture of Ags including CAP, NP, E_2 and Ar at the same concentration (0.01μg/mL), the FIs significantly decreased (Fig. 1.3B). However, FIs were not reduced significantly after addition of erythromycin, gentamicin, BPA, diethylstilbestrol and 17-α-ethinylestradiol onto the immunochip. This observation suggested that these chemicals did not significantly cross-react on the immunochip. Phage display technology and ribosome display technology are other tools for the generation of high-quality Abs. Antibodies with higher titres and specificity, especially mAb, can significantly increase the accuracy and sensitivity of the immunochip.

Our observations indicate that immunochips can simultaneously detect five different chemicals. This offers exciting opportunities for detection of small molecules in online environmental and food hygiene assessments. However, the sensitivity is too low to detect traces of chemicals and further studies are required to determine whether more target chemicals can be integrated onto a single microchip. Improvements in the production of immunochip technology and access to high-quality Abs, are important for the rapid, sensitive, and high-throughput detection of chemicals, toxins and pollutants in food, water and the environment.

Kloth et al. (2009) developed a regenerable (i.e. reusable) immunochip for the rapid determination of 13 different antibiotics in raw milk. The newly developed hapten microarrays which use an indirect competitive chemiluminescence microarray immunoassay (CL-MIA) are designed to analyze 13 different antibiotics in milk within 6 min. Antigen immobilization was performed by the contact spotter system BioOdyssey Calligrapher miniarrayer (Bio-Rad, München, Germany) using a TeleChem Stealth SNS 9 microspotting solid pin. Printing conditions were set via the Bio-Rad Calligrapher Software. The relative humidity was 35% and the air temperature inside the spotting chamber was 21°C. Two 15 × 5 clusters were set on one microarray (grid spacing 1100 mm; cluster distance 11.75 mm).

To regenerate the immunchip for reuse the high affinity Abs that bound Ag must be removed from the chip surface. This was based on epoxy-activated PEG chip surfaces, onto which microspotted antibiotic derivatives like sulfonamides, β-lactams, aminoglycosides, fluorquinolones and polyketides are coupled directly without further use of linking agents. Using the chip reader platform MCR 3, this Ag solid phase is stable for at least 50 consecutive analyses. Fig. 1.4A shows the scheme of the direct covalent coupling of different types of antibiotics and the assessment of samples of milk.

Figure 1.4B shows the two different CCD exposures of microarrays detecting 13 different antibiotics and the reference substance DNT with five replicates in each row. Normal milk and a sample of milk containing cloxacillin which tested positive were measured on the same biochip after one regeneration cycle. The concentration of cloxacillin was high enough

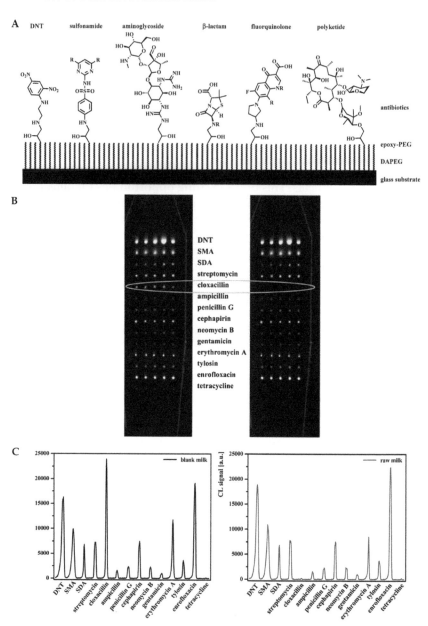

Figure 1.4. Direct covalent coupling of different antibiotics without further use of linking agents and the assessment of samples of milk. **(A)** Direct covalent coupling of different types of antibiotics without use of linking agents; **(B)** Measurements of milk samples: CCD exposures of blank milk without any added antibiotics (left) and of a raw milk sample contaminated with cloxacillin (right); **(C)** CL signal variation profiles of both analyses: blank milk (black) and raw milk (gray) (Kloth et al. 2009).

to reduce the signal intensity to background level. The signal intensities of all other analytes were reproducible (Fig. 1.4B; average CL signal deviation: 5.3%). The concentration of cloxacillin in this sample was estimated to be about 370 mg/L which is significantly greater than the MRL (30 mg/L). This is similar to the semi-quantitative result obtained by a microbial inhibitor test (cloxacillin: >220 mg/L). With this test, Kloth et al. 2009 also found all samples of milk containing one analyte and an inhibitor could be identified and quantified by the MCR 3. Similar residue levels were found in contaminated samples analyzed with both methods. Furthermore, milk samples which tested negative were identified correctly. The new microarray system offers a means for rapid, routine, identification and quantification of antibiotics in milk and will therefore aid the food industry to maintain quality and safety levels.

LAB-ON-CHIP FOR PERSONAL TOXICITY TESTING

The micrototal analysis system (μ-TAS) or "lab-on-a-chip" has been popular since it was introduced around 10 yr ago (Lee and Lee 2004). The ability to tailor-make an integrated system for a specific immunoassay is very useful. The lab-on-a-chip integrates several laboratory processes including all preparation procedures, incubation, detection and analysis on a single chip. It eliminates the need for several different pieces of laboratory equipment to prepare and perform a single assay. Technology from semi-conductor industries was adapted to design and fabricate a myriad of interconnecting micro-sized channels, chambers and reactors, required to produce the micrototal analysis system (Lim and Zhang 2007). Like microfluidic circuits, the micrototal analysis system offers the possibility of developing small, easy-to-use, fully integrated, automated devices for analysis of chemicals. The lab-on-a-chip should include fluidic systems, like channels, pumps, and valves, able to perform tasks of separation, transfer of liquids, purification, amplification etc., as well as bioarrays and the array-readers.

Lab-on-chip technology greatly exceeds the ability of conventional bioanalytical techniques to detection of environmental toxins. Other advantages include automated analysis, miniaturization, multiplex analysis, use of minute samples, reduced reagent consumption. However, perhaps most importantly, lab-on-chip technology facilitates integration of systems which share characteristics in respect to their production, bio-recognition interfaces, and signal enhancement and transition processes.

One approach to the production of a whole-cell lab-on-chip integrated system was described by Rabner et al. 2006. This disposable plastic biochip is prepared with a 4 × 4 micro-lab (mLab) chamber array of bioluminescent *E. coli* reporter cells that responds to a predetermined class of chemical agents and microfluidic channels for liquids translocation. The device includes electro-optics for signal acquisition with motorized read out calibration accessories, hydropneumatic modules for water sample translocation into biochip mLabs and electronics for control and communication with the host computer. This prototype is sensitive to broad classes of water-borne toxins including naladixic acid (a model genotoxic agent), botulinum toxin, and acetylcholine esterase inhibitors.

Many immunochips incorporate nanotechnology. This does not mean that the devices are manufactured on nanoscale dimensions, but refers to the use of, for example, molecular monolayers of a material inside the structure, or the immobilization of individual molecules (proteins, DNA) inside the channels. However, some microfabricated fluidic devices are smaller than 1µm, for example, the carbon nanotube shown in Fig. 1.5.

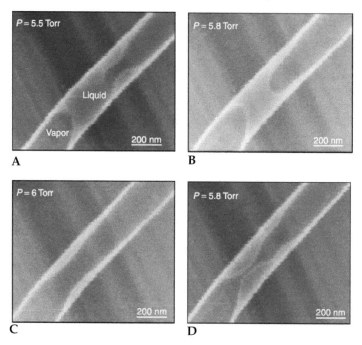

Figure 1.5. Sequence of environmental scanning electron microscopy (ESEM) images. These were obtained when partial pressure of water in the ESEM chamber was gradually raised in a controlled manner, while observing a single open carbon nanotube filled with water **(A–C)**. Note the liquid-volume recovery during subsequent pressure decrease **(C–D)** (Rabner et al. 2006).

SUSPENSION ARRAY FOR PERSONAL TOXICITY TESTING

We assessed the ability of suspension array technology to simultaneously assess three different veterinary drugs, chloramphenicol (CAP), clenbuterol

(CL) and 17-beta-estradiol (E_2) (Liu et al. 2009). The high-throughput suspension array is a novel method for multi-analysis of veterinary drugs. It is easy to use, inexpensive and very sensitive. Three different conjugate-coupled beads were mixed in the same proportions. The optimized mAbs were then added into the 96-well plate. Meanwhile, CAP, CL and E_2 were diluted into eight concentration gradients as 5×, 5× and 2×, respectively. Mixed 6000 beads were added to each well and the concentrations of CAP, CL and E_2 were adjusted to 0, 50, 250, 1250, 6250, 31250, 156250, 781250 ng L^{-1}; 0, 56, 280, 1400, 7000, 35000, 175000, 875000 ng L^{-1} and 0, 1, 3, 9, 27, 81, 243, 729 µg L^{-1}, respectively in the total volume of 50 µL per well in a plate. The plate was then spun for 1 hr at medium speed at 37°C for Ag-Ab competitive reaction. Then, SA-PE was added and the plate was spun for 15 min at medium speed at 37°C. 100 microspheres were read out well by well by Bio-Plex™ suspension array to obtain the MFIs. Based on the MFIs for each target, standard curves could be plotted.

There are negative logistic correlations between MFIs and the concentrations of the veterinary drugs, and all determination coefficients-R^2 were greater than 0.989 implying good correlation. The detection ranges are 40–6.25 × 10^5 ng L^{-1}, 50–7.81 × 10^5 ng L^{-1} and 1 × 10^3–7.29 × 10^5 ng L^{-1} for CAP, CL and E_2, respectively (Fig. 1.6).

Within these ranges, any concentration of the three veterinary drugs can be quantified using the standard curves. The sensitivity means the minimum

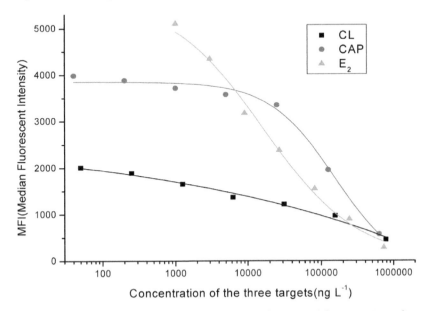

Figure 1.6. Standard curves produced by simultaneous detection of three veterinary drugs by suspension array technology (Liu et al. 2009).

detectable concentrations (Min DC) in suspension array detection, and MFIs of the blank controls or background for the three veterinary drugs are very close to that of the Min DC. On statistical analysis (Table 1.2) there were no significant differences between the MFIs of the blank controls and the groups of the Min DC. Consequently, the Min DC are the lowest detection limits (LDLs) for CAP, CL and E_2, i.e. 40, 50 and 1000 ng L^{-1}.

Table 1.3 shows that the resulting MFIs had no significant differences from blank control groups after addition of different concentrations of salbutamol, ractopamine, gentamicin, erythromycin or 17-alpha-estradiol, P >0.05 indicating no significant alteration by or cross-reaction with other chemicals. Salbutamol, ractopamine and 17-alpha-estradiol are structural analogues of CL and E_2. The specificity of the suspension array is dependent on the specificity of the mAbs.

As competitive ELISA is an established and widely used method of quantifying small molecules, it was employed to assess real samples to confirm the feasibility in comparison with suspension array technology. The results from the two methods are shown in Table 1.3 and the relative standard deviations (RSDs) were between 8.09–17.03% and 9.19–17.74% from the real values. These are relatively small detection ranges. However,

Table 1.2 Standard curves for CAP, CL and E_2 detection and MFI for blank control and Min DC.

Item	Standard curve	R^2	MFI ($\bar{x} \pm s$, n=3) Blank Control	MinDC
CAP	$Y_{CAP}=-250.323 + 4103.517/$ $[1+(x/30121.243)^{0.980}]$	0.994	2247.167 ± 90.423^a	2071.667 ± 85.0431
CL	$Y_{CL}=-263012.682 + 265751.765/$ $[1 + (x/6.063)^{0.115}]$	0.989	4516.833 ± 177.340^b	4195.833 ± 185.581
E_2	$Y_{E_2}=8.368 + 5651.421/$ $[1+(x/16409.422)^{0.679}]$	0.995	5249.000 ± 201.765^c	5110.833 ± 134.612

Compared with blank control, a, b and c in CAP, CL and E_2 group respectively: $P > 0.05$

Table 1.3 MFIs for different groups of chemicals.

Item	Concentration (ng L^{-1})	MFI ($\bar{x} \pm s$, n=3) 0ng L^{-1} (blank control)	50 ng L^{-1}	1250 ng L^{-1}	31.25 µg L^{-1}
Salbutamol		1750.17 ± 92.59^a	1691.82 ± 25.59	1727.20 ± 42.06	1529.48 ± 7.26
Ractopamine		1750.17 ± 92.59^a	1749.90 ± 51.65	1618.16 ± 76.21	1609.67 ± 61.94
Gentamicin		3662.58 ± 75.47^b	3719.23 ± 74.68	3849.20 ± 56.60	3970.16 ± 82.41
Erythromycin		3662.58 ± 75.47^b	3527.83 ± 52.53	3673.25 ± 63.37	3485.62 ± 47.75
17 α-estradiol		5520.50 ± 127.08^c	5460.80 ± 118.95	5510.83 ± 339.50	5497.94 ± 219.81

Compared with 0 ng L^{-1} group (Negative control), a, b and c: $P > 0.05$.

the detection ranges of the suspension array are broader and more sensitive than ELISA, especially for the detection of the three veterinary drugs. Moreover, as multiplex analysis is possible with the suspension array, it will be a very useful application.

As a booming technology, suspension array has revealed great developing prospects and potentials not only for the research and detection of macromolecules (protein and nucleic acid) detection, but also for providing a novel pathway for analysis and assessment of small molecules such as pesticides, veterinary drugs and toxins. With the increasing availability of commercial kits multiplex analysis of these small molecules by suspension array could be applied extensively.

SUMMARY POINTS

- Immunochips provide large amount of information with rapid, timely, simple, low cost, wide range of usage and advantages of automatization.
- High-quality Ab with good bioactivity and specificity is the key reagent in the production of immunochips.
- Suspension array has great potential for use in personal toxicity testing not only for the research and detection of macromolecules (protein and nucleic acid), but also for providing a novel method of analysis and assessment of small molecules such as pesticides, veterinary drugs and toxins.
- All forms of immunochips are difficult to use for field detection and require further development and improvement.
- The online environmental and health inspection by suspension array for a large numbers of chemicals, pollutants and toxins requires further investigation and development.

ACKNOWLEDGEMENTS

The authors gratefully acknowledge the financial support of the National Science and Technology Supporting Program of China (No. 2009BADB9B03-Z05), National High Technology R&D Program of China (No. 2010AA06Z302) and National Nature Science Foundation of China (No. 81030052, 30771810 and 30800915).

ABBREVIATIONS

AFB_1	:	aflatoxin B_1
Ags	:	antigens

Abs	:	antibodies
Ar	:	atrazine
CAP	:	chloramphenicol
CL-MIA	:	competitive chemiluminescence microarray immunoassay
E_2	:	17-beta estradiol
EDC	:	endocrine disruptor chemical
ELISA	:	enzyme-linked immunoadsorbent assay
FI	:	fluorescent intensity
LC-MS	:	liquid chromatographic method with mass spectrometry
LDLs	:	lowest detection limits
MFI	:	median fluorescent intensity
Min DC	:	minimum detectable concentrations
MW	:	molecular weight
mAbs	:	monoclonal antibodies
GalNAc	:	N-acetyl galactosamine
Neu5Ac	:	N-acetyl neuraminic acid
NP	:	nonylphenol
Pap	:	papaverine
PDMS	:	Poly (dimethyl) siloxane
PBS	:	phosphate-buffered saline
RSDs	:	relative standard deviation
RT	:	room temperature
SEM	:	scanning electron microscopy
SEB	:	staphylococcal enterotoxin B
SNR	:	signal-to-noise ratio
SA-PE	:	streptavidin-R-phycoerythrin

REFERENCES

Arenkov P, A Kukhtin, A Gemmell, S Voloshchuk, V Chupeeva and A Mirzabekov. 2000. Protein microchips: use for immunoassay and enzymatic reactions. Anal Biochem 278: 123–131.

Chiem NH, and DJ Harrison. 1998. Microchip systems for immunoassay: an integrated immunoreactor with electrophoretic separation for serum theophylline determination. Clin Chem 44: 591–598.

Choi SM, SD Yoo and BM Lee. 2004. Toxicological characteristics of endocrine-disrupting chemicals: developmental toxicity, carcinogenicity, and mutagenicity. J Toxicol Environ Health B Crit Rev 7: 1–32.

Connolly AR, R Palanisamy and M Trau. 2010. Quantitative considerations for suspension array assays. J Biotechnol 145: 17–22.

Cooper RL, TE Stoker, L Tyrey, JM Goldman and WK McElroy. 2000. Atrazine disrupts the hypothalamic control of pituitary-ovarian function. Toxicol Sci 53: 297– 307.

Dasso J, J Lee, H Bacha and RG Mage. 2002. A comparison of ELISA and flow microsphere-based assays for quantification of immunoglobulins. J Immunol Methods 263: 23–33.

Du H, M Wu, W Yang, G Yuan, Y Sun, Y Lu, S Zhao, Q Du, J Wang, S Yang, M Pan, Y Lu, S Wang and J Cheng. 2005. Development of miniaturized competitive immunoassays on a protein chip as a screening tool for drugs. Clin Chem 51: 368–375.

Estévez MC, M Kreuzer, SB Francisco and MP Marco. 2006. Analysis of Nonylphenol: Advances and Improvements in the Immunochemical Determination Using Antibodies raised against the Technical Mixture and Hydrophilic Immunoreagents. Environ Sci Technol 40: 559–568.

Friedmann AS. 2002. Atrazine inhibition of testosterone production in rat males following peripubertal exposure. Reprod Toxicol 16: 275–279.

Gao ZX, N Liu, QL Cao, L Zhang, SQ Wang, W Yao and FH Chao. 2009. Immunochip for the detection of five kinds of chemicals: Atrazine, nonylphenol, 17-beta estradiol, paraverine and chloramphenicol. Biosens Bioelectron 24: 1445–1450.

Ghobrial IM, DJ McCormick, SH Kaufmann, AA Leontovich, DA Loegering, NT Dai, KL Krajnik, MJ Stenson, MF Melhem, AJ Novak, SM Ansell and TE Witzig. 2005. Proteomic analysis of mantle-cell lymphoma by protein microarray. Blood 105: 3722–3730.

Han XD, ZG Tu, Y Gong, SN Shen, XY Wang, LN Kang, YY Hou and JX Chen.2004. The toxic effects of nonylphenol on the reproductive system of male rats. Reprod Toxicol 19: 215–221.

Hayes TB, A Collins, M Lee, M Mendoza, N Noriega, AA Stuart and AH Vonk. 2002. Demasculinized frogs after exposure to the herbicide atrazine at low ecologically relevant doses. Proc Natl Acad Sci 99: 5476–5480.

Hoogenboom H, A Griffiths, K Johnson, D Chisswell, P Hudson and G Winter. 1991. Multi-subunit proteins on the surfaces of filamentous phage: methodologies for displaying antibody (Fab) heavy and light chains. Nucleic Acids Res 19: 4133–4137.

Huang R, Y Lin, Q Shi, L Flowers, S Ramachandran, IR Horowitz, S Parthasarathy and RP Huang. 2004. Enhanced Protein Profiling Arrays with ELISA-Based Amplification for High-Throughput Molecular Changes of Tumor Patients' Plasma. Clin Cancer Res 10: 598–609.

Issaragicil S, DW Kaufman, T Anderson, K Chansung, T Thamprasit, J Sirijirachai, A Piankijagum, Y Porapakkham, S Vannasaeng, PE Leaverton, S Shapiro and NS Young. 1997. Low Drug Attributability of Aplastic Anemia in Thailand Blood 89: 4034–4039.

Kalogianni DP, DS Elenis, TK Christopoulos and PC Ioannou. 2007. Multiplex quantitative competitive polymerase chain reaction based on a multianalyte hybridization assay performed on spectrally encoded microspheres. Anal Chem 79: 6655–6661.

Kavlock RJ, GP Daston, C DeRosa, CP Fenner, LE Gray, S Kaattari, G Lucier, M Luster, MJ Mac, C Maczka, R Miller, J Moore, R Rolland, G Scott, DM Sheehan, T Sinks and HA Tilson. 1996. Research needs for the risk assessment of health and environmental effects of endocrine disruptors: A Report of the U.S. EPA sponsored workshop. Environ Health Perspect 104 (Suppl 4): 715–740.

Kloth K, RJ Maria, D Andrea, D Richard, M Erwin, N Reinhard and S Michael. 2009. A regenerable immunochip for the rapid determination of 13 different antibioties in raw milk. Analyst 134: 1433–1439.

Koblížek M, J Malý, J Masojídek, J Komenda, T Kučera, MT Giardi, AK Mattoo and R Pilloton. 2002. A biosensor for the detection of triazine and phenylurea herbicides designed using Photosystem II coupled to a screen printed electrode. Biotechnol Bioeng 78: 110–116.

Kuo JS. 2003. Interfacing Chip-Based Nanofluidic-Systems to Surface Desorption Mass Spectrometry, JIN Reports, Pacific Northwest National Laboratory. http://www.pnl.gov/nano/institute/2003reports/.

Lee SJ and SY Lee. 2004. Micro total analysis system (μ-TAS) in biotechnology. Appl Microbiol Biotechnol 64: 289–299.

Lim CT and Y Zhang. 2007. Bead-based microfluidic immunoassays: The next generation. Biosens Bioelectron 22: 1197–1204.

Liu N, P Su, ZX Gao, MX Zhu, ZH Yang, XJ Pan, YJ Fang and FH Chao. 2009. Simultaneous detection for three kinds of veterinary drugs: chloramphenicol, clenbuterol and 17-beta-estradiol by high-throughput suspension array technology. Anal Chim Acta 632: 128–134.

Mart'ianov AA, BB Dzantiev, AV Zherdev, SA Eremin, R Cespedes, M Petrovic and D Barcelo. 2005. Immunoenzyme assay of nonylphenol: Study of selectivity and detection of alkylphenolic non-ionic surfactants in water samples. Talanta 65: 367–374.

Mary TM, G Stuart, P Maurice, WOB Thomas, S Thomas, A Jennifer, GL Richard, C Bill and SV Kodumudi. 2003. Multiplexed liquid arrays for simultaneous detection of simulants of biological warfare agents. Anal Chem 75: 1924–1930.

McCafferty JK, J Fitzgerald, J Earnshaw, DJ Chiswell, J Link, R Smith and J Kenten. 1994. Selection and rapid purification of murine antibody fragments that bind a transition-state analog by phage display. Appl Biochem Biotechnol 47: 157–173.

Miller JC, H Zhou, J Kweke, R Cavallo, J Burke, EB Butler, BS Teh and BB Haab. 2003. Antibody microarray profiling of human prostate cancer sera: antibody screening and identification of potential biomarkers. Proteomics 3: 56–63.

Rabner A, S Belkin, R Rozen and Y Shacham. 2006. Whole-cell luminescence biosensor-based lab-on-chip integrated system for water toxicity analysis. Proceedings of the SPIE-The International Society for Optical Engineering. 6112: 611205/611201–611205/611210.

Ramos M, P Munoz, A Aranda, I Rodriguez, R Diaz and J Blanca. 2003. Determination of chloramphenicol residues in shrimps by liquid chromatography-mass spectrometry. J Chromatogr B 791: 31–38.

Raymond EB, LS Deborah, PS Jerome, AM Barbara, AFS Cynthia, S Vera, SC Evelen, S Karen, EF Alison, PQ Conraduinn and ES John. 2004. Comparison of a multiplexed fluorescent covalent microsphere immunoassay and an enzyme-linked immunosorbent assay for measurement of human immunoglobulin G antibodies to anthrax toxins. Clin Diagn Lab Immunol 11: 50–55.

Rubin Y, VI Dyukov, EI Dementiev, AA Stomakhin, VA Nesmeyanov, EV Grishin and AS Zasedatelev. 2005. Quantitative immunoassay of biotoxins on hydrogel-based protein microchips. Anal Biochem 340: 317–329.

Sapsford KE, RT Chris, F Stephanie, HM Martin, ELMichael, MM Chris and CSL Lisa. 2006. Indirect competitive immunoassay for detection of aflatoxin B_1 (AFB_1) in corn and nut products. Biosens Bioelectron 21: 2298–2305.

Song JM and TV Dinh. 2004. Miniature biochip system for detection of *Escherichia coli* O157:H7 based on antibody-immobilized capillary reactors and enzyme-linked immunosorbent assay. Anal Chim Acta 507: 115–121.

Spearow JL, P Doemeny, R Sera, R Leffler and M Barkley. 1999. Genetic Variation in Susceptibility to Endocrine Disruption by Estrogen in Mice Science 285: 1259–1261.

Wildt RMT, CR Mundy, BD Gorick and IM Tomlinson. 2000. Antibody arrays for high-throughput screening of antibody-antigen interactions. Nat Biotechnol 18: 989–994.

Zhou Y, R Yang, S Tao, Z Li, Q Zhang, H Gao, Z Zhang, J Du, P Zhu, L Ren, L Zhang, D Wang, L Guo, Y Wang, Y Guo, Y Zhang, C Zhao, C Wang, D Jiang, Y Liu, H Yang, L Rong, Y Zhao, S An, ZH Li, X Fan, J Wang, Y Cheng, O Liu, Z Zheng, H Zhu H, JF Klemic, S Chang, P Bertone, A Casamayor, KG Klemic, D Smith, M Gerstein, MA Reed and M Snyder. 2000. Analysis of yeast protein kinases using protein chips. Nat Genet 26: 283–289.

Zuo, Q Shan, L Ruan, Z Lü, T Hung and J Cheng. 2005. The design and application of DNA chips for early detection of SARS-CoV from clinical samples. J Clin Virol 33: 123–131.

Detection of Pesticide Residues Using Biosensors

Javier Ramón-Azcón,[1,a] Tomoyuki Yasukawa[1,b] and Fumio Mizutani[1,c]

ABSTRACT

Pesticides are substances used in food production in order to minimize or prevent damage caused by pests. Thus, unlike other groups of chemicals, pesticides are intentionally released into the environment, and there is a high risk of these chemicals appearing in the food chain. Unfortunately, the exponential increase in the demand for food in the world today makes it impossible to eliminate them from food production. Therefore, consumers and governments consider pesticide regulation and control a very relevant issue for the economy as well as human health. Often expensive and instrumental single-analyte methods are applied by regulatory and industrial laboratories. There is an urgent need for validated screening tools that are not only simple, inexpensive, and rapid but also show multiplex capabilities by detecting simultaneously as many contaminants as possible. In recent years, many efforts have been made to develop new analytical techniques integrating biorecognition elements and detection components in order to obtain small devices with the ability to carry out direct, selective, and continuous measures for one or several analytes present in samples. In

[1]Graduate School of Material Science, University of Hyogo, 3-2-1 Kouto, Kamigori-cho, Ako-gun, Hyogo 678-1297, Japan.
[a]E-mail: ramonazconjavier@gmail.com
[b]E-mail: yasu@sci.u-hyogo.ac.jp
[c]E-mail: mizutani@sci.u-hyogo.ac.jp

List of abbreviations after the text.

this context, biosensors can fulfill these requirements. Biosensors offer a good alternative to conventional methodologies in pesticide analysis due to their high sensitivity and selectivity.

INTRODUCTION

Today, environmental contamination is a problem recognized worldwide. A significant portion of environmental pollution is caused by the application of pesticides in agriculture, horticulture, and forestry. Pesticide is a term used in a broad sense for a chemical, synthetic or natural, which is used for the control of insects, fungi, bacteria, weeds, nematodes, rodents, and other pests. A large number of these compounds and/or their degradation products are highly toxic, and they have negative effects not only on the ecosystem but also on human health. Surveillance of the environment and food for pesticide residues has become essential in recent years for the prevention of risks to the population. It is necessary to control the presence of pesticides in the environment and at the same time to assess the risk to human health due to the presence of these chemicals in workplaces and in food in order to prevent adverse effects.

To control environmental contamination and to protect the population from it, governmental agencies have established several directives. In 1976, European Union (EU) set a "black list" of 132 dangerous substances (based on their toxicity, stability, and bioaccumulation) that should be monitored in water (Directive 76/464/CE). With the aim to protect the health of the general population, the EU has established a value of 0.1 µg/L as the maximum individual concentration and 0.5 µg/L as the total concentration of pesticides and related products in drinking water (Directive 80/778/CEE). In the United States of America (USA), the Environmental Protection Agency (EPA) has established a maximum level for each pesticide or its transformation products according to their toxicity. Similarly, for the protection of the public against the toxic effects of pesticides, regulatory agencies in many countries have established standards specifying the residue levels of each pesticide in various foodstuffs. Thus, the World Health Organization (WHO) has evaluated and reviewed the acceptable daily intakes (ADI) of pesticides (Lu 1995). An example of the degree of exposure to the European population suffers from pollution is reflected in the detection of more than 41 toxic chemicals in the blood tests to 39 Member of the European Parliament (MEP) in 2004.

Control of the environment and food must be assessed by reliable, fast, and sensitive analytical techniques. Chromatographical systems are conventional analytical techniques characterized by high precision and sensitivity. Nevertheless, these sophisticated techniques need experienced personnel and costly instruments and are not easily adoptable for field analysis. For all these reasons, there is considerable delay between sample collection and data display, thus resulting in loss of money for the food industry and a possible risk to the population. A simple and advantageous alternative is the use of biosensors.

In the early 1950s, potentiometric detection was adopted for pesticide detection, and in the middle of the 1980s, it was used for the construction of the first integrated biosensor for the detection of pesticides based on inhibition of acetylcholinesterase (AChE). In the following decades, important advances have been achieved in the field of biosensors with new elements of recognition and new systems of transduction. The advances in nanotechnology and microelectronics in recent years have been particularly important for this field. However, the commercialization of biosensor technology in the environmental and food industries has significantly lagged behind the research output. In clinical diagnostics, commercial biosensors are well established, and an important number of companies produce them. It is not easy to explain the slow transfer of technology within research and industry, but it could be attributed to cost considerations and some key technical barriers such as stability, detection sensitivity, and reliability. Furthermore, the level of acceptation for governmental agencies of standard analytical techniques is low because of the lack of well-established methodologies of validation. Table 2.1 lists EPA requirements for biosensors that may be used in field assay applications (Rogers and Lin 1992). Medical applications overshadow the other applications, but there are some companies that work in environmental and food control. Table 2.2 summarizes companies with biosensors specially designed for the environmental or agro-alimentary industries.

Table 2.1. General requirements for biosensors in environmental field applications.

Requirement	Specific range
Cost	US$1–15 per analysis
Portable	Can be carried by one person; no external power requirements
Fieldable	Easily transported in a van or truck; limited external power required
Assay time	1–60 min
Personnel training	Can be operated by minimally trained personnel after 1–2 hr training period
Matrix	Minimal preparation for ground water, soil extract, blood, urine, or saliva
Sensitivity	Parts per million/parts per billion
Dynamic range	At least two orders of magnitude
Specificity	Enzymes/receptors: specific to one or more groups of related compounds Ab: specific to one compound or one group of closely related compounds

Adapted from Rogers and Lin (Rogers and Lin 1992) with permission from Elsevier.

Table 2.2. List of companies commercializing biosensors for the agricultural and food industries.

Company name	Country	URL address	Biosensor name
Abtech Scientific	USA	www.abtechsci.com	ToxSen™
Affinity Sensors	United Kingdom	www.affinity-sensors.com	IASys plus™ e IASys, Auto+Advantage™
Ambri Limited	Australia	www.ambri.com	AMBRI Biosensor
Applied Biosystems	USA	www.appliedbiosystems.com	8500 Affinity Chip Analyzer
Biacore AB	Sweden	www.biacore.com	Biacore®Q
BioFutura Srl	Italy	www.biofutura.com	PerBacco 2000 y 2002
Biomerieux	France	www.biomerieux.com	Vitek™ Bactometer™
Biosensor Systems Desing	USA	www.biosd.com	OptiSense Technology™
Biosensores S.L.	Spain	www.biosensores.com	Politox
Biotrace	USA	www.biotrace.com	Uni-lite® Bev-Trace™
Chemel AB	Sweden	www.chemel.com	SIRE®
Innovative Biosensors, Inc	USA	www.innovativebiosensors.com	CANARY™
Molecular Devices	Switzerland	www.moleculardevices.com	Threshold® System
MicrooVacuum	Hungary	www.microvacuum.com	OWLS 210
Nippon Laser Electronics	Japan	www.nle-lab.co.jp	SPR-670M, SPR-MACSNANOSENSOR
Reichert Analytical Instruments	USA	www.reichertai.com	Reichert SR 7000™
Research International	USA	www.resrchintl.com	Analyte 2000™ FAST 6000™ Raptor™
Texas Instruments Inc.	USA	www.ti.com	Spreeta™
Universal Sensors	USA	intel.ucc.ie/sensors/universal	ABD 3000 Biosensor Assay System
Yellow Springs Instruments Co	USA	www.ysi.com	YSI 2700 SELECT™ Biochemical Analyzer

Unpublished table.

PESTICIDES BIOSENSOR CLASSIFICATION

Biosensors can be classified according to the type of recognition element (catalytic or affinity-based biosensor) used or the transduction system (optical, electrochemical, piezoelectric and nanomechanical) (see Fig. 2.2).

Catalytic Biosensors: Enzymatic Biosensors

Enzymatic biosensors for the detection of pesticides in food and environmental samples have been extensively described in several reviews (Cock et al. 2009; Manco et al. 2009). Enzymatic biosensors, which are based on the use of enzymes, use either of the two principles: enzyme inhibition or hydrolysis of the pesticide.

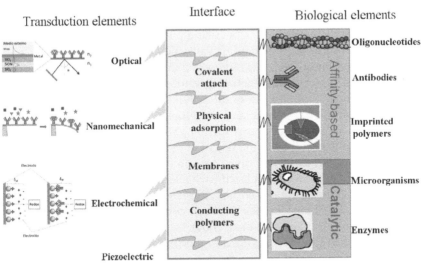

Figure 2.1. Biosensors classification. Classification of the biosensors depends on the bioactive element of recognition or the transduction system. The immobilization of the biological element onto a transducer is a key step in optimizing the analytical performance of a biosensor in terms of response, sensitivity, stability, and reusability. The immobilization strategies most generally employed are physical or chemical methods. Figure unpublished.

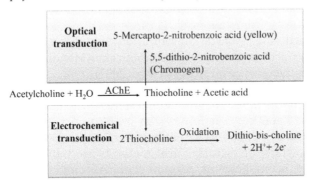

Figure 2.2. Enzymatic formation of thiocholine and transduction systems. Unpublished scheme.

Enzyme inhibition-based biosensors

In the case of inhibition, enzymes used for the detection of pesticides are inhibited by the pesticide, and the extent of inhibition is correlated to the concentration of the analyte. Other enzymatic methods such as the organophosphorus hydrolase assay use the analyte as a substrate, thus giving the result that a positive signal is generated through the production of hydrolysis products rather than merely the inhibition of the enzyme. Although the most widely used transduction system by enzymatic biosensors has mainly been electrochemical, it is possible to find some examples of enzymatic biosensors using piezoelectric transduction (Abad et al. 1998) or optical transduction (Vamvakaki and Chaniotakis 2007).

Inhibition-based methods have some disadvantages and can be prone to false positives because handling and storage can cause the loss of enzyme activity (Shimomura et al. 2009). Furthermore, many pesticides irreversibly inhibit enzymes, and therefore regeneration of the sensor is required after each sample; this further extends the testing time. Finally, the main problem found in the use of enzymatic biosensors is the lack of specificity and selectivity in the detection of pesticides (Luque de Castro and Herrera 2003). Despite these drawbacks enzyme-based biosensors are effective tools that can be used in general screening methods and further investigation in chemometrics has been conducted for the differentiation of mixtures of pesticides (Ni et al. 2004).

Most pesticide biosensors are designed based on the inhibitory property of enzymes. AChE and butyrylcholinesterase (BChE) are widely used in the development of pesticide biosensors. Inhibition leads to a decrease in activity, which is indirectly proportional to the amount of inhibitors or pesticides in the sample. The other often-employed enzymes in pesticide biosensors are acetolactate synthase, acid phosphatase, alkaline phosphatase (AP), and tyrosinase. Pesticides can also inhibit the activity of luciferase, which is a major enzyme in bioluminescence reactions. By employing fireflies, luciferase pesticide concentrations have been determined on the basis of the fact that the pesticide concentration is indirectly proportional to the bioluminescence (Trajkovska et al. 2005).

Organophosphorus pesticides (OPP) and carbamate pesticides are potent inhibitors of the enzyme cholinesterase. The thiocholine produced during the catalytic reaction can be monitored using spectrometric, amperometric (see Fig. 2.2), or potentiometric methods, but as it has been previously stated, electrochemical methods are more common. As an example of the applications, Arduini et al. (Arduini et al. 2006) analyzed different pesticides with AChE and BChE enzymes. AChE-based biosensors have higher sensitivities toward aldicarb (50% inhibition with 50 μg/L) and carbaryl (50% inhibition with 85 μg/L) while BChE biosensors have higher affinities for diethyl-p-nitrophenyl phosphate (50% inhibition with 4 μg/L) and chlorpyrifos-methyl oxon (50% inhibition with 1 μg/L). The limits of detection (LOD) were 12 and 25 ppb (μg/L) for aldicarb and carbaryl, respectively, using the AChE enzyme and 2 and 0.5 ppb (μg/L) for paraoxon and chlorpyrifos-methyl oxon, respectively. BChE and AChE biosensors were then tested using both wastewater and river water samples, and no inhibition of the signals was observed, thus obtaining good recovery percentages with spiked samples.

Catalysis-based biosensors

Inhibition-based biosensors are poor in selectivity and are rather slow and tedious since the analysis involves multiple reaction steps such as measurement of initial enzyme activity, incubation with an inhibitor, measurement of residual activity, and regeneration and washing. Biosensors based on direct pesticide hydrolysis are more straightforward. The enzyme organophosphorus hydrolase (OPH) hydrolyzes esters in a number of OPP and insecticides (e.g. paraoxon, parathion, coumaphos, diazinon) (Jeffrey et al. 1987). According to one example found in the literature, Mulchandani et al. (Mulchandani et al. 1999) purified OPH from recombinant *E. coli* and immobilized it on a pH electrode to develop a potentiometric biosensor by catalyzing the hydrolysis of OPP (parathion, paraoxon, and methyl parathion) to release protons, the concentrations of which were proportional to the amounts of hydrolyzed substrates.

Affinity Based Biosensors: Immunosensors for Pesticides

Immunosensors are based on the immunological reaction derived from the binding of the antibody (Ab) to the corresponding antigen (Ag). This reaction is reversible and is stabilized by electrostatic forces, hydrogen bonds, and Vander Waals interactions. The formed complex has an affinity constant (k_a) that can achieve values on the order of 10^{10} M^{-1}. In immunosensors procedures the quantification of pesticides molecules is performed under competitive conditions. The general strategy of competitive assays is based on the competition between the free Ag (analyte) and a fixed amount of labeled Ag for a limited amount (low concentration) of Ab. At the end of the reaction the amount of labeled Ag and subsequently the free Ag is determined. The labels used to quantify the immunoreaction can be of a different nature. A wide variety of antibody biosensors reported for different pesticides in food and environmental applications exists and are summarized and discussed in several reviews (González-Martínez et al. 2007; Bojorge Ramírez et al. 2009).

Electrochemical immunosensors

Because of its simplicity, electrochemical transduction is the oldest and most common method used in biosensors (for recent review see (Rivas et al. 2007)). Researchers can determine the level of pesticides by measuring the change in potential, current, conductance, or impedance caused by the immunoreaction.

Amperometric biosensors are based on the measurement of the current generated by oxidation/reduction of redox species at the electrode surface, which is maintained at an appropriate electric potential. The current observed has a linear relationship with the concentration of the electroactive species. For the simultaneous analysis of several samples using only one device, Skládal and Kaláb (Skládal and Kaláb 1995) developed a multichannel immunosensor. The 2,4-D molecule conjugated to horseradish peroxidase was used as a tracer, which was determined amperometrically using hydrogen peroxide and hydroquinone as substrates. 2,4-Dichlorophenoxyacetic acid (2,4-D) is widely used in amounts of 10^5 tons per year as a herbicide for the control of broadleaf weeds. "Agent Orange," which was used extensively during the Vietnam war for defoliation, is composed of a 50:50 mixture of n-butyl esters of 2,4-dichlorophenoxyacetic acid (2,4-D) and 2,4,5-trichlorophenoxyacetic acid (2,4,5-T) and can still be detected in human tissues, soil as well as in the biosphere of Vietnam. A similar immunosensor coupled with an enzyme-linked immunosorbent assay (ELISA) microtiter plate has been also reported (Deng and Yang 2007), and a detection limit of 0.072 µg/L was achieved. The advantages of the presented electrochemical detector were high stability and sample throughput, a low detection limit, the ability to be repeatedly used without the need for regeneration.

Recently, several papers have been published combining microparticles with electrochemical amperometric detection. The use of microbeads greatly improves the performance of the immunological reaction, minimizing the matrix effect due to improved washing and separation steps. This strategy has been used for the detection of atrazine herbicide in orange juice (Zacco et al. 2006). The Ab is immobilized on the surface of magnetic beads. The immunological reaction for the detection of atrazine is based on a direct competitive assay using a peroxidise tracer as the enzymatic label (see Fig. 2.3). The LOD obtained in orange juices was 0.025 µg/L, which is below the maximum residue level (MRL) established by actual European legislation (0.1 µg/L) in oranges. Furthermore, in the case of orange juice, preliminary experiments performed with the magneto-ELISA demonstrated that nonspecific interferences from a matrix can be easily eliminated by a very simple sample pre-treatment, which consists of simply adjusting the pH to 7.5 (the original pH of the sample was 3.5), filtering the sample through a 0.2 µm filter, and diluting the filtrate five times with buffer.

Electrochemical impedance spectroscopy (EIS) is being rapidly developed because of the possibility to directly record information on biorecognition events occurring at the electrode surfaces and inducing capacitance and resistance changes (Katz and Willner 2003), allowing

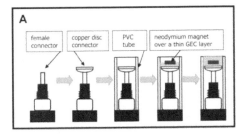

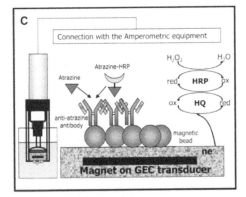

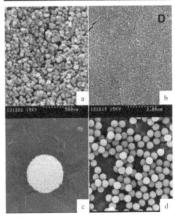

Figure 2.3. Schematic representation of the electrochemical magnetoimmunosensing strategy for the detection of atrazine. **(A)** Preparation of the magneto graphite-epoxy composite (m-GEC) electrode. **(B)** After the immunoreaction, the antibody (Ab) modified magnetic beads were captured for the m-GEC electrode. **(C)** Chemical reactions occurring at the m-GEC surface polarized at –0.150 V (vs Ag/AgCl) upon the addition of H_2O_2 in the presence of a mediator (hydroquinone). **(D)** Scanning electron microphotographs of carboxylated magnetic particles (MP-COOH) **(A, B)** on the surface of sensors taken at 0.5 and 2 µm of resolution, respectively and tosylated magnetic beads (MB-tosyl) **(C, D)** taken at 2 and 10 µm of resolution, respectively. An identical acceleration voltage (15 kV) was used in all cases. This figure is from Zacco et al. (Zacco et al. 2006) with permission from ACS.

the development of label-free biosensing devices. EIS in connection with immunochemical methods was tested for the direct determination of the herbicide 2,4-D (Navratilova and Skladal 2004). The changes in the impedance parameters (ϕ_{max} and Z_{min}) due to immunocomplex formation, which served as a parameter characterizing changes on the sensing surfaces, were evaluated. It was possible to measure the response to 2,4-D in a concentration range from 45 nM to 450 µM. In this context, interdigitated microelectrodes (IDµE) have recently received enormous attention because their sensitivity is higher than that of conventional electrodes (Berggren et al. 2001; Navratilova and Skladal 2004). Using thin Au/Cr (~200-nm thickness) IDµEs (3.85-µm thick and with electrode gaps of 6.8 µm) on a Pyrex 7740 glass substrate, researchers have recently reported the detection of atrazine without the use of any label with a limit of detection of 0.04 µg/L

(Ramón-Azcón et al. 2008) (see Fig. 2.4). The sensor has been evaluated to assess its potential to analyze pesticide residues in a complex sample matrix, such as red wine. An atrazine hapten-bovine serum albumin (BSA) conjugate was covalently immobilized within the microelectrodes on the glass substrate.

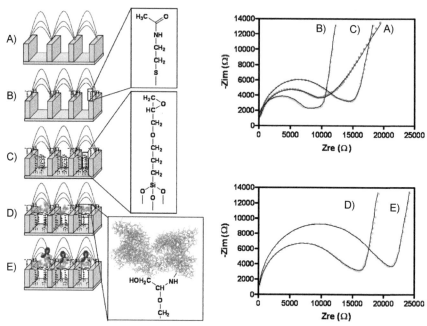

Figure 2.4. Scheme showing steps used to prepare the immunosensor surfaces and antibody (Ab) binding. Nyquist plots of impedance corresponding to **(A)** ID m E; **(B)** Step I: N-acetylcysteamine, gold protection; **(C)** Step II: functionalization of the Pyrex substrate with (3-glycidoxypropyl)trimethoxysilane; **(D)** Step III: coating atrazine hapten-bovine serum albumin (BSA) conjugate, covalent immobilization (1 μg/mL); and **(E)** Step IV: specific Ab produced versus atrazine herbicide, incubation step (1 μg/mL). Symbols represent the experimental data. Solid curves represent the computer fitting data with the parameters calculated by the commercially available software Zplot/Zview (Scibner Associates). Parts of this figure are reprinted from Ramón-Azcón et al. (Ramón-Azcón et al. 2008) with permission from Elsevier.

Optical biosensors

Optical transducers are based on various technologies involving optical phenomena, which are the result of an interaction between the analyte and receptor. This group may be further subdivided according to the optical properties applied in sensing (i.e. absorbance, reflectance, luminescence, fluorescence, refractive index and SPR, and light scattering).

An evanescent wave (EW) is a near field standing wave with an intensity that decays exponentially with increase in distance from the boundary at which the wave was formed. When biomolecules are located in the evanescent field, they absorb energy, leading to attenuation in the reflected guide of the waveguide. In attempts to improve detectability, many researchers reported that immunosensors combine this principle with the use of labeled molecules that are able to re-emit the absorbed evanescent photons at a longer wavelength as fluorescence. This phenomenon is known as total internal reflection fluorescence (TIRF). As an example, a TIRF immunosensor was shown to allow the detection of a multitude of analytes in one single test cycle (Klotz et al. 1998). Calibration curves obtained for 2,4-D and simazine had detection limits of 0.035 and 0.026 µg/L respectively. One limiting factor on the ability to simultaneously perform more than one assay on the same transducer was the availability of low cross-reactant Ab combined with a high affinity between the antibody and the analyte. Similarly, the River Analyzer (RIANA) is also a highly sensitive, fully automated biosensor able to rapidly and simultaneously detect multiple organic targets (Mallat et al. 2001; Rodriguez-Mozaz et al. 2006). Thus, this system was used to measure two herbicides, atrazine and isoproturon, in raw river water and in water obtained after each treatment step in the waterworks. The analysis of these compounds could be performed in one unique run and in a very short period (one measurement cycle, including the regeneration step, took 15 min), and the LOD reached the legislation requirements (0.1 µg/L, as set in the EU drinking water directive 2000/60/EC as the maximum admissible concentration for individual pesticides). The performance of the immunosensor method developed was evaluated against a method based on solid phase extraction, followed by liquid chromatography-mass spectrometry (LC-MS). In conclusion, the chromatographic method was superior in terms of linearity, sensitivity, and accuracy, and the biosensor method in terms of repeatability, speed, cost, and automation.

Other EW immunosensor approaches such as grafting couplers (Grego et al. 2008) and Mach–Zehnder Interferometers (MZI) (Prieto et al. 2003) have been investigated in order to obtain possible measurements of pesticides without the use of fluorescent labels. An optical waveguide lightmode spectroscopy (OWLS)-based biosensor is a recently developed device in the field of integrated optics, and exploits the science of light guided in structures that are smaller than the wavelength of light. This technique can be applied for the detection of the herbicide trifluralin (Székács et al. 2003). Within the immobilized Ab-based immunosensors, this method allowed the detection of trifluralin only above 100 µg/L due to the small molecular size of Ag, while the immobilized Ag-based OWLS system allowed the detection of trifluralin in the concentration range of 2×10^{-7} to 3×10^{-5}

μg/L. Trifluralin concentrations detected by the indirect OWLS sensor were correlated with those detected by ELISA and gas chromatography-mass spectrometry (GC-MS) methods. On an MZI, the propagating light is split into two arms, one with the appropriate sensing layer and the other acting as a reference. Schipper et al. (Schipper et al. 1997) discussed the feasibility of an evanescent wave interferometer immunosensor for pesticide detection. The interferometer immunosensor demonstrated a high resolution (2×10^{-3} nm). An increase in both the receptor layer binding capacity and net response rate should result in the required increase in the sensitivity of the MZI, making the direct detection of pesticides feasible.

Fluorescent techniques have also been combined with nano/ microparticles with magnetic properties to be manipulated by magneto forces. Microparticles can also be manipulated in a flowing stream by dielectrophoresis (DEP) (Yasukawa et al. 2007). In this context, several publications have been presented to detect pesticides in a faster and easier manner. In a first approach (Ramón-Azcón et al. 2009), polystyrene microparticles (6-μm diameters) modified with bovine serum albumin conjugated with atrazine (atrazine-BSA) were manipulated and captured by DEP forces using three dimensional microelectrodes. The performance of this n-DEP immunosensing technique was evaluated using wine samples. The immunodevice showed a limit of detection of 0.11 μg/L for atrazine in buffer samples and 6.8 μg/L in pretreated wine samples; these detection limits are less than the MRL established by the European Community for residues of this herbicide in wine (50 μg/L). Subsequent works have improved the immunosystem, allowing the simultaneous measurement of two pesticides (Ramón-Azcón et al. 2010). Simultaneous detection of atrazine and bromopropylate was investigated using a DEP device with two channels. Since the formation of the immunocomplexes was accelerated significantly by n-DEP, a period as short as 5 min was sufficient to detect the atrazine in orange juice. The LOD in orange juice prepared by four times dilution with PBS were 4.0 and 1.5 μg/L for atrazine and bromopropylate, respectively. Finally, to improve the sensitivity, as described (Ramón-Azcón et al. 2011) a combination of immunological recognition events, based on the particle surface, and complementary recognition events of single-stranded deoxyribonucleic acid (ssDNA) oligomers between the channels and particles. As can be seen in Fig. 2.5, microparticles were functionalized with two specific oligomers and two pesticides Ag. These particles were incubated together with their characteristic antibody and pesticide and introduced in the channels in which specific double strand capture occurs.

Plasmon resonance is an evanescent electromagnetic field generated at the surface of a metal conductor (usually Ag or Au) when excited by the impact of light of an appropriate wavelength at a particular angle (α).

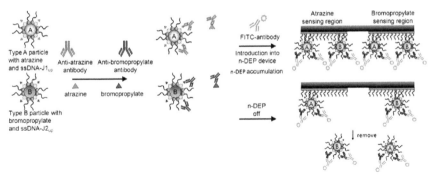

Figure 2.5. The principle of measurement for a multi-analyte system. In this device, two different sets of microparticles: one modified with atrazine-conjugated bovine serum albumin and single-stranded deoxyribonucleic acid 1_{up} (ssDNA-1_{up}) and the other with bromopropylate-conjugated aminodextran and ssDNA-2up were used. The microparticles were incubated in a mixture of analyte-specific antibody (Ab) and analyte at different concentrations to trap the unreacted Ab prior to being labeled with Ab conjugated with a fluorescence molecule. A suspension containing both types of microparticles was introduced into a negative dielectrophoretic (n-DEP) device consisting of an interdigitated microarray (IDA) electrode and two channels modified with ssDNA-1down and ssDNA-2down, which are complementary to ssDNA-1_{up} and ssDNA-2_{up}, respectively. The n-DEP force generated by applying an AC voltage to the IDA electrode displaced the microparticles toward the encoded areas, causing them to rapidly accumulate on the upper surfaces. Hybridization allowed it to distinguish the microparticles and sense multiple analytes by spatial recognition in the DNA-encoded areas. This figure is from Ramon-Azcon et al. (Ramón-Azcón et al. 2011) with permission from ACS.

Since distinct SPR prototypes (Biacore, IASys, etc.) have appeared in the market, a significant number of applications of this principle for pesticides have been reported in recent years. Monitoring of the pesticide DDT in water samples was performed using SPR immunosensors (Mauriz et al. 2007). The DDT acronym is derived from an old and imprecise name, Dichloro Diphenyl Trichloroethane. DDT is a chlorinated compound with insecticidal properties that has been used worldwide for controlling insect pests. However, it is highly hydrophobic with great stability to physical, chemical, and biological degradation, resulting in the accumulation of its residues in animal and human tissues as well as in the environment. The analyte derivative was covalently immobilized onto a gold-coated sensing surface and competed with free insecticide for binding to the Ab. This portable immunosensor based on SPR technology could provide a highly sensitive detection of DDT at nanogram per liter levels (15 ng/L). The regeneration of the immunosurface was accomplished throughout 270 assay cycles without degradation of the covalently immobilized molecule. Other examples of single and multi-analyte assays for simultaneous detection of different pesticides by SPR were reported by the same research group, and the SPR biosensor portable platform (β-SPR) is already commercialized by the company SENSIA, S.L. (Spain). In another example, atrazine was measured in water using SPR transduction, and a study of the matrix effects of different types of wastewater was also performed (Farré et al. 2007).

Nanomechanical immunosensors

Microcantilevers (MCL) translate molecular recognition of biomolecules into nanomechanical motion that is commonly coupled to an optical or piezoresistive read out detector system (Alvarez et al. 2008). Microcantilever sensors rely on their deflection to indicate sensing. Thus, molecular adsorption onto the sensing element shifts the resonance frequency and changes its surface forces (surface stress). One example is the use of a competitive immunosystem to measure DDT in a buffer solution (Alvarez et al. 2003). A synthetic hapten of DDT conjugated with BSA was covalently immobilized on the gold-coated side of a cantilever using thiol self-assembled monolayers, and the cantilever was exposed to a mixed solution of the monoclonal Ab and DDT. This biosensor could achieve subnanomolar sensitivity (10 nM) without the need for labeling with fluorescent and radioactive molecules.

Piezoelectric immunosensors

Piezoelectrics are materials that may be brought into resonance by the application of an external alternating electric field. (for review see (O'Sullivan et al. 1999)). Piezoelectric immunosensors may adopt two modes: (1) Bulk acoustic (BA) devices in which adsorption of the analytes occurs on the coated surface of a piezoelectric crystal connected to an oscillator circuit and (2) surface acoustic wave (SAW) devices in which an acoustic wave moves just at the surface of the crystal. For BA devices, resonance occurs over the entire mass of the crystal. Several immunosensors have been developed for the assay of 2,4-D. For example, Horácek and Skladal (Horácek and Skladal 1997) obtained a biosensor with an LOD of 0.24 ng/mL for free 2,4-D. Piezoelectric immunosensors for the determination of atrazine (Pribyl et al. 2003) were developed, and direct and competitive determination systems were studied and compared. In the competitive format, the mixture of object pesticide and specific Ab was pre-incubated in solution and passed through the flow cell with the piezoelectric crystal modified with the atrazine hapten. The competitive assay provided a much lower limit of detection (0.025 ng/mL) compared with the direct determination (1.5 ng/mL), but the protocol for the direct determination is faster and can be accomplished in 10 min.

SUMMARY

- A significant portion of environmental pollution is caused by the application of pesticides in agriculture, horticulture, and forestry. Much interest has been focused on these compounds not only because of their possible adverse effects but also for the great amounts that are produced worldwide.
- Biosensors are a good tool for pesticides control that could be used in various formats for specific purposes, ranging from quantitative laboratory tests to simple "yes/no" screening tests that are field-portable. These advantages can be exploited in monitoring programs in which great number of samples need to be analyzed. Thus, as effective screening techniques, biosensors are complementary in effect to the standard analytical techniques.
- Commercialization of biosensor technology in the environmental and food industries has significantly lagged behind the research output. It is not easy to explain the slow transfer of technology within research and industry, but it could be attributed to cost considerations and some key technical barriers such as stability, detection sensitivity, and reliability.
- Enzymatic biosensors, which are based on the use of enzymes, use either of the two principles: enzyme inhibition or hydrolysis of the pesticide. In the case of inhibition, enzymes used for the detection of pesticides are inhibited by the pesticide, and the extent of inhibition is correlated to the concentration of the analyte. Compared with enzyme inhibition, biosensors based on direct pesticide hydrolysis are more straightforward, because inhibition-based biosensors are poor in selectivity and are rather slow and tedious since the analysis involves multiple reaction steps such as measurement of initial enzyme activity, incubation with an inhibitor, measurement of residual activity, and regeneration and washing.
- Immunosensors are biosensors based on the immunological reaction derived from the binding of the Ab to the corresponding Ag. In immunosensors procedures the quantification of pesticides molecules is performed under competitive conditions. The general strategy of competitive assays is based on the competition between the free Ag (analyte) and a fixed amount of labeled Ag for a limited amount (low concentration) of Ab. At the end of the reaction the amount of labeled Ag and subsequently the free Ag is determined.
- Recently, several immunodevices have been published combining microparticles with different transduction systems for pesticide

analysis. The use of microbeads greatly improves the performance of the immunological reaction, minimizing the matrix effect due to improved washing and separation steps. Optical as well as electrochemical transduction has been used in combination with microparticles, and microparticles movement has been manipulated using magneto force (with magnetic particles) or DEP force (e.g. polystyrene, gold, carbon nanotubes).

CONCLUSIONS

Biosensors offer a good alternative to conventional methodologies in pesticide analysis due to their high sensitivity and selectivity. They have two especially useful properties: (i) They can be carried out for use in the field and (ii) they can work with complete automation and produce results after a short period. Thus, they are a good alternative to the conventional techniques in specific places where they can reach easily, for example in the field or in the hands of the initial manufacturer. Furthermore, they also represent a good complementary tool that can be used as screening methods.

The future of biosensors lies in the combination of knowledge from different scientific disciplines. Collaboration and interchange of expertise between analytical and immune chemists and technologists are needed to achieve new objectives. Many new micro/nanotechnological advances present new and interesting possibilities that should be investigated to find new transduction systems to convert biological specific recognition phenomena into electrical or optical signals. Finally, the extensive research in this branch of science should not be exported to industry without performing rigorous validation studies with environmental samples. For this last purpose, rigorous legislation with well-established protocols is necessary for the validation of new biosensor technology.

KEY FACTS

- Pesticides arrive intentionally to the environment and their use and release should be controlled.
- The commercialization of biosensor technology in the environmental and food industries has significantly lagged behind the research output.
- Enzymatic biosensors, which are based on the use of enzymes, use either of the two principles: enzyme inhibition or hydrolysis of the pesticide.
- Immunosensors are based on the immunological reaction derived from the binding of the Ab to the corresponding Ag.

- Because of its simplicity, electrochemical transduction is the oldest and most common method used in biosensors for pesticides detection.
- Immunosensors in combination with microparticles have achieved important goals to control pesticides residues in the agroalimentary industry.

DEFINITIONS

- *Antibody*: Ab are globular proteins generated by the immune system as a defense against foreign agents (Ag).
- *Dielectrophoresis*: electrokinetic phenomenon produced by the interaction of an induced polarization with a spatially inhomogeneous electric field.
- *Impedance spectroscopy*: analytical tool capable of characterizing the evolution of the impedance in a system for a certain range of frequencies.
- *Interdigitated microelectrodes*: device constructed with two metal electrodes formed by a large number of digits that are interspersed to increase the length/area relation.
- *Maximum residue level*: MRL are defined as the maximum concentration of pesticide residue likely to occur in or on food and feeding stuffs after the use of pesticides according to Good Agricultural Practice (GAP).
- *Pesticide*: substance used to kill, control, repel, or mitigate any pest. Insecticides, fungicides, rodenticides, herbicides, and germicides are all pesticides.

ACKNOWLEDGMENTS

This work has been supported by the Japan Society for the Promotion of Science (JSPS)—Postdoctoral Fellowships for Foreign researchers.

ABBREVIATIONS

2,4-D	:	2,4-Dichlorophenoxyacetic Acid
2,4,5-T	:	2,4,5-Trichlorophenoxyacetic Acid
Ab	:	Antibody
AChE	:	Acetylcholinesterase
ADI	:	Acceptable Daily Intakes
Ag	:	Antigen
AP	:	Alkaline Phosphatase

BA	:	Bulk Acoustic
BSA	:	Bovine Serum Albumin
BChE	:	Butyrylcholinesterase
DDT	:	Diphenyl Trichloroethane
DEP	:	Dielectrophoresis
DNA	:	Deoxyribonucleic Acid
DPV	:	Differential Pulse Voltammetry
EIS	:	Electrochemical Impedance Spectroscopy
ELISA	:	Enzyme-Linked Immunosorbent Assay
EPA	:	Environmental Protection Agency
EU	:	European Union
EW	:	Evanescent Wave
GAP	:	Good Agricultural Practice
IDμD	:	Interdigitated Microelectrodes
IC_{50}	:	Concentration at 50% of Signal Inhibition
HRP	:	Horseradish Peroxidase
GC-MS	:	Gas Chromatography-Mass Spectrometry
LC-MS	:	Liquid Chromatography-Mass Spectrometry
LOD	:	Limit of Detection
MCL	:	Microcantilever
MEP	:	Member of the European Parliament
MRL	:	Maximum Residue Level
MW	:	Molecular Weight
MZI	:	March–Zehnder Interferometers
OPH	:	Organophosphorus Hydrolase
OPP	:	Organophosphorus Pesticides
OWLS	:	Optical Waveguide Lightmode Spectroscopy
RIANA	:	River Analyzer
SAW	:	Surface Acoustic Wave
SPR	:	Surface Plasmon Resonance
TIR	:	Total Internal Refraction
TIRF	:	Total Internal Reflection Fluorescence
USA	:	United States of America
WHO	:	World Health Organization

REFERENCES

Abad JM, F Pariente, L Hernández, HD Abruña and E Lorenzo. 1998. Determination of Organophosphorus and Carbamate Pesticides Using a Piezoelectric Biosensor. Anal Chem. 70: 2848–2855.

Alvarez M, A Calle, J Tamayo, LM Lechuga, A Abad and A Montoya. 2003. Development of nanomechanical biosensors for detection of the pesticide DDT. Biosens Bioelectron 18: 649–653.

Alvarez M, K Zinoviev, M Moreno, LM Lechuga, SL Frances and T Chris Rowe. 2008. Cantilever Biosensors. pp. 419–452. *In*: F Ligler and C Taitt. [eds.] Optical Biosensors. 2nd edn. Elsevier, Amsterdam, The Nethererlands.

Arduini F, F Ricci, CS Tuta, D Moscone, A Amine and G Palleschi. 2006. Detection of carbamic and organophosphorous pesticides in water samples using a cholinesterase biosensor based on Prussian Blue-modified screen-printed electrode. Anal Chim Acta 580: 155–162.

Berggren C, B Bjarnason and G Johansson. 2001. Capacitive biosensors. Electroanal 13: 173–180.

Bojorge Ramírez N, AM Salgado and B Valdman. 2009. The evolution and developments of immunosensors for health and environmental monitoring: problems and perspectives. Braz J Chem Eng 26: 227–249.

Cock L, Z Arenas, A Ana and Aponte. 2009. Use of enzymatic biosensors as quality indices: A synopsis of present and future trends in the food industry. Chilean J Agric Res 69: 270–280.

Deng A-P and H Yang. 2007. A multichannel electrochemical detector coupled with an ELISA microtiter plate for the immunoassay of 2,4-dichlorophenoxyacetic acid. Sensor Actuat B-Chem 124: 202–208.

Farré M, E Martínez, J Ramón, A Navarro, J Radjenovic, E Mauriz, L Lechuga, MP Marco and D Barceló. 2007. Part per trillion determination of atrazine in natural water samples by surface plasmon resonance immunosensor. Anal Bioanal Chem 388: 207–214.

González-Martínez M, R Puchades and A Maquieira. 2007. Optical immunosensors for environmental monitoring: how far have we come? Anal Bioanal Chem 387: 205–18.

Grego S, JR McDaniel and BR Stoner. 2008. Wavelength interrogation of grating-based optical biosensors in the input coupler configuration. Sensor. Actuat B-Chem 131: 347–355.

Horácek J and P Skládal. 1997. Improved direct piezoelectric biosensors operating in liquid solution for the competitive label-free immunoassay of 2,4-dichlorophenoxyacetic acid. Anal Chim Acta 347: 43–50.

Karns JS, MT Muldoon, WW Mulbry, MK Derbyshire, and PC. Kearney. Use of Microorganisms and Microbial Systems in the Degradation of Pesticides. pp. 156–170. *In*: HM LeBaron, RO Mumma, RC Honeycutt, JH Duesing, JF Phillips and MJ Haas [eds.] 1987. Biotechnology in Agricultural Chemistry, American Chemical Society, USA.

Katz E and I Willner. 2003. Probing Biomolecular interactions at Conductive and Semiconductive Surfaces by Impedance Spectroscopy: Routes to Impedimetric Immunosensorsm, DNA-Sensors, and Enzyme Biosensors Electroanal 15: 913–947.

Klotz A, A Brecht, C Barzen, G Gauglitz, RD Harris, GR Quigley, JS Wilkinson and R A Abuknesha. 1998. Immunofluorescence sensor for water analysis. Sensor. Actuat B-Chem 51: 181–187.

Lu FC. 1995. A review of the acceptable daily intakes of pesticides assessed by WHO. Regul. Toxicol Pharmacol 21: 352–364.

Luque de Castro MD and MC Herrera. 2003. Enzyme inhibition-based biosensors and biosensing systems: questionable analytical devices. Biosens Bioelectron 18: 279–294.

Mallat E, C Barzen, R Abuknesha, G Gauglitz and D Barceló. 2001. Part per trillion level determination of isoproturon in certified and estuarine water samples with a direct optical immunosensor. Anal Chim Acta 426: 209–216.

Manco G, R Nucci and F Febbraio. 2009. Use of esterase activities for the detection of chemical neurotoxic agents. Protein Pept Lett 16: 1225–34.

Mauriz E, A Calle, JJ Manclús, A Montoya, A Hildebrandt, D Barceló and LM Lechuga. 2007. Optical immunosensor for fast and sensitive detection of DDT and related compounds in river water samples. Biosens Bioelectron 22: 1410–1418.

Mulchandani A, P Mulchandani, W Chen, J Wang and L Chen. 1999. Amperometric Thick-Film Strip Electrodes for Monitoring Organophosphate Nerve Agents Based on Immobilized Organophosphorus Hydrolase. Anal Chem 71: 2246–2249.

Navratilova I and P Skladal. 2004. The immunosensors for measurement of 2,4-dichlorophenoxyacetic acid based on electrochemical impedance spectroscopy. Bioelectrochem 62: 11–18.

Ni Y, C Huang and S Kokot. 2004. Application of multivariate calibration and artificial neural networks to simultaneous kinetic-spectrophotometric determination of carbamate pesticides. Chemom Intell Lab 71: 177–193.

O'Sullivan CK, R Vaughan and GG Guilbault. 1999. Piezoelectric Immunosensors—Theory and Applications. Anal Letters 32: 2353–2377.

Pribyl J, M Hepel, J Halámek and P Skládal. 2003. Development of piezoelectric immunosensors for competitive and direct determination of atrazine. Sensor Actuat B-Chem 91: 333–341.

Prieto F, B Sepúlveda, A Calle, A Llobera, C Domínguez and LM Lechuga. 2003. Integrated Mach-Zehnder interferometer based on ARROW structures for biosensor applications. Sensor Actuat B-Chem 92: 151–158.

Ramón-Azcón J, E Valera, Á Rodríguez, A Barranco, B Alfaro, F Sanchez-Baeza and MP Marco. 2008. An impedimetric immunosensor based on interdigitated microelectrodes (ID[mu] E) for the determination of atrazine residues in food samples. Biosens Bioelectron 23: 1367–1373.

Ramón-Azcón J, R Kunikata, FJ Sanchez, MP Marco, H Shiku, T Yasukawa and T Matsue. 2009. Detection of pesticide residues using an immunodevice based on negative dielectrophoresis. Biosens Bioelectron 24: 1592–1597.

Ramón-Azcón J, T Yasukawa, HJ Lee, T Matsue, F Sánchez-Baeza, M-P Marco and F Mizutani. 2010. Competitive multi-immunosensing of pesticides based on the particle manipulation with negative dielectrophoresis. Biosens Bioelectron 25: 1928–1933.

Ramón-Azcón J, T Yasukawa and F Mizutani. 2011. Sensitive and Spatially Multiplexed Detection System Based on Dielectrophoretic Manipulation of DNA-Encoded Particles Used as Immunoreactions Platform. Anal Chem 83: 1053–1060.

Rivas GA, MD Rubianes, MC Rodríguez, NF Ferreyra, GL Luque, ML Pedano, SA Miscoria and C Parrado. 2007. Carbon nanotubes for electrochemical biosensing. Talanta 74: 291–307.

Rodriguez-Mozaz S, MJL de Alda and D Barceló. 2006. Fast and simultaneous monitoring of organic pollutants in a drinking water treatment plant by a multi-analyte biosensor followed by LC-MS validation. Talanta 69: 377–384.

Rogers KR and JN Lin. 1992. Biosensors for environmental monitoring. Biosens Bioelectron 7: 317–321.

Schipper EF, AM Brugman, C Dominguez, LM Lechuga, RPH Kooyman and J Greve. 1997. The realization of an integrated Mach-Zehnder waveguide immunosensor in silicon technology. Sensor Actuat B-Chem 40: 147–153.

Shimomura T, T Itoh, T Sumiya, F Mizukami and M Ono. 2009. Amperometric biosensor based on enzymes immobilized in hybrid mesoporous membranes for the determination of acetylcholine. Enzyme Microb Tech 45: 443–448.

Skládal P and T Kaláb. 1995. A multichannel immunochemical sensor for determination of 2,4-dichlorophenoxyacetic acid. Anal Chim Acta 316: 73–78.

Székács A, N Trummer, N Adányi, M Váradi and I Szendro. 2003. Development of a non-labeled immunosensor for the herbicide trifluralin via optical waveguide lightmode spectroscopic detection. Anal Chim Acta 487: 31–42.

Trajkovska S, K Tosheska, JJ Aaron, F Spirovski and Z Zdravkovski. 2005. Bioluminescence determination of enzyme activity of firefly luciferase in the presence of pesticides. Luminescence 20: 192–196.

Vamvakaki V, and NA Chaniotakis. 2007. Pesticide detection with a liposome-based nano-biosensor. Biosens Bioelectron 22: 2848–2853.

Yasukawa T, M Suzuki, T Sekiya, H Shiku and T Matsue. 2007. Flow sandwich-type immunoassay in microfluidic devices based on negative dielectrophoresis. Biosens Bioelectron 22: 2730–2736.

Zacco E, MI Pividori and S Alegret. 2006. Electrochemical magnetoimmunosensing strategy for the detection of pesticides residues. Anal Chem 78: 1780–1788.

Biosensors for Ecotoxicity of Xenobiotics: A Focus on Soil and Risk Assessment

Edoardo Puglisi,[1] Mariarita Arenella,[2,a]
Giancarlo Renella[2,b] and Marco Trevisan[3,*]

ABSTRACT

In this chapter we discuss the potential application of biosensors for the ecotoxicologal risk assessment of xenobiotics in soil environments. In the first part we introduce basic concepts of ecotoxicological risk assessment, ecotoxicology, environmental fate of xenobiotics, with particular focus on bioavailability processes and their importance in exposure and ecotoxicity of xenobiotics in soils. Specifically, all the main processes controlling the environmental fate of xenobiotics in soils are explained.

In the second part we introduce some principles of the functioning of different classes of whole cell biosensors and chemical sensors for

[1]Istituto di Microbiologia, Università Cattolica del Sacro Cuore, Via Emilia Parmense 84, 29122 Piacenza, Italy; E-mail: edoardo.puglisi@unicatt.it
[2]Dipartimento di Scienze delle Produzioni Vegetali, del Suolo e dell'Ambiente Agroforestale, Università di Firenze, Piazza delle Cascine 18, 50144 Firenze, Italy.
[a]E-mail: mariarita.arenella@unifi.it
[b]E-mail: giancarlo.renella@unifi.it
[3]Istituto di Chimica Agraria ed Ambientale, Università Cattolica del Sacro Cuore, Via Emilia Parmense 84, 29122 Piacenza, Italy; E-mail: marco.trevisan@unicatt.it
*Corresponding author

List of abbreviations after the text.

xenobiotics of ecotoxicological interests and present some applications and their contribution to improve the ecoxicological risk assessment of xenobiotics in soils. We show how biosensors can be applied together with chemical assessment of bioavailability of xenobiotics in soil to assess ecotoxicological risk. The application of biosensors to improve the classical risk assessment paradigm is presented, in order to show how this technology can be used to assess the ecotoxicological risk of xenobiotics in soils. Examples of biosensors with potential application for ecotoxicological risk assessment are also presented and discussed. The relevance of these studies of environmental health is also acknowledged.

We conclude that biosensors technology holds relevant potential for ecotoxicological risk assessment of xenobiotics in soils.

INTRODUCTION—PRINCIPLES OF ECOTOXICOLOGICAL RISK ASSESSMENT

Risk assessment for xenobiotics is currently the main regulatory decision-support process behind the measures to prevent, assess and manage environmental pollution. Risk assessment can be defined as a scientific procedure aimed at identifying the hazards correlated with pollutants and the risks related with their use/presence in the environment. It is possible to distinguish between a human health and an ecotoxicological risk assessment: in both cases the concern is related to toxic substances, but the focus is moved from human to ecological receptors. The US National Academy of Science already defined about 30 years ago a risk assessment "paradigm", that is applicable for both human or ecological receptors, and it is made up of four components (NRC 1983): i) hazard identification, ii) exposure assessment, iii) dose-response assessment and iv) risk characterization (Fig. 3.1). In the first step, all available information about the toxicity (or ecotoxicity) of the chemical of concern are gathered; exposure is aimed at identifying the magnitude of the releases as well as the possible pathways and potential exposures for human and ecological receptors; dose-response assessments are divided in the evaluation of the observable data range and the extrapolation of ranges to toxicological (or ecotoxicological) endpoints; risk characterization finally integrates the information of the previous components in order to assess the potential or existing risk of an adverse effect (Newman and Unger 2003).

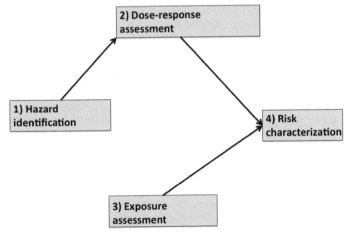

Figure 3.1. The risk assessment paradigm scheme. Modified from NRC (1983). This figure shows that different steps must be undertaken in order to characterize and assess risk of a xenobiotic. First, the hazard (i.e. the toxicity or ecotoxicity) of the xenobiotic must be identified. This hazard is then quantified by a dose-response assessment: increasing doses of the xenobiotic are applied and the response measurement. Finally, exposure is assessed by quantifying the fraction of xenobiotic that is reaching the receptor.

ECOTOXICITY AND ENVIRONMENTAL FATE OF XENOBIOTICS

The term ecotoxicology was firstly used in 1969 by Truhaut (see Truhaut 1977), and aimed at extending toxicology, i.e. the science of studying the effects of poisons on an organism, to an ecosystem level. Ecotoxicology thus takes some approaches used in toxicology (e.g. dose-response assessments, effective doses and other end-points) and extends them to an ecosystem level. As outlined by Moriarty (1983), it would be a mistake to think of ecotoxicology as a simple translation of toxicology, where the only difference is the test species (e.g. a water flea rather than a lab rat): ecotoxicology is concerned with ecosystems health, and thus takes into account effects at population or community level, and considers not only death or carcinogenicity, but also effects that could have a deep impact on ecosystems, such as developmental or endocrine effects. Several definitions of ecotoxicology were proposed in time: a recent and synthetic one defines it as "the study of harmful effects of chemicals on ecosystems" (Walker et al. 2001).

Given the definition above, ecotoxicology is quite a multidisciplinary science, aimed at assessing the distribution and the effects of contaminants in the ecosystems. Knowledge about the environmental distribution of a contaminant is thus at the basis of any ecotoxicological risk assessment, including those based on the use of biosensors. Any ecosystem, either aquatic or terrestrial, is studied by ecotoxicology: here we will focus on xenobiotics in soil ecosystems, especially on agricultural soils.

First of all, it is necessary to make some definition and distinction among terms that are often used as synonyms by non-experts. In this chapter we focus on xenobiotics: the term derives from the Greek, and literary means "stranger for life". Xenobiotics are thus compounds that are not produced in nature and not normally considered constitutive components of a specific biological system (Rand and Petrocelli 1985). Xenobiotics are not to be confounded with contaminants and pollutants: a contaminant is a compound with natural or synthetic origin that is present in a given environment at concentrations higher than the natural ones, while pollutant indicates a compound that is present at concentrations able to cause deleterious effects on living organisms. This implies that a xenobiotic is always a contaminant, but it becomes a pollutant of ecotoxicological concern only when present at concentrations exerting toxic effects.

In order to show how biosensors can be applied to assess the ecotoxicity of xenobiotics, some definitions of environmental fate are also necessary. Once released in the environment, a xenobiotic undergoes a series of processes of transport and, eventually, transformation. These processes depend on the physico-chemical properties of the pollutant and the conditions of the surrounding ecosystem. Focusing on the soil environment, xenobiotics can migrate from it to the surrounding compartments: air (volatilization), water (leaching in the case of groundwater, lateral drainage and runoff and transport with sediment eroded in the case of surface water), or living organisms (uptake processes) (Fig. 3.2). From an ecotoxicological point of view, the uptake by living organisms (whether animals, plants or microorganisms) is a process of major concern, as it represents the necessary (but not sufficient) conditions to have ecotoxicological effects. Xenobiotics with low vapour pressures, high Kow (octanol-water partition coefficients, a measure of chemicals hydrophobicity), and low degradability have a low tendency to escape (i.e. fugacity) towards air and water compartments, and are thus the ones of major interest for soil environments.

Another relevant concept in ecotoxicology is that of bioavailability, which is the capacity of a xenobiotic to be actively or passively absorbed by living organisms. The peculiar problem of assessing soil xenobiotics bioavailability is related to the fact that the soil environment is structured

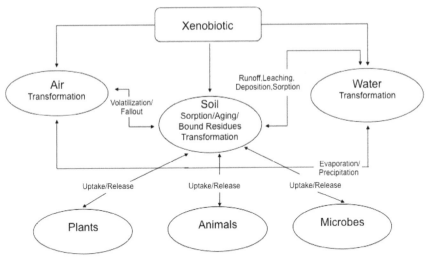

Figure 3.2. Processes controlling the distribution of a xenobiotic in the soil and in the surrounding environments. Once released in the environment, a xenobiotic undergoes a number of processes controlling its fate. In soil, it can be sorbed, bound or transformed in other chemical forms. It can also move to the air through volatilization or to the water through runoff or leaching. A portion of what is remaining in the soils, termed the bioavailable fraction, can finally be uptaken by biological receptors (plants, animals or microbes). This bioavailable fraction is the one that can potentially pose a risk. Unpublished material of the authors.

in aggregates and pores of various dimensions (Ø µm-cm) and constituted by reactive minerals and organic substances. Once entering into the soil, xenobiotics can indeed be trapped into micropores or be adsorbed onto organic and inorganic particles (Fig. 3.3). Therefore, normally the xenobiotic bioavailable fraction does not correspond to its total concentration, and may not be entirely available for uptake by ecological receptors. Furthermore, entrapment and sorption processes generally increases with time, and the aging phenomenon has to be taken into account for a meaningful definition of xenobiotic bioavailability in soil. Due to the large variability of the soil properties, the bioavailable fraction differs in various types of soils and xenobiotics, but it also depends on the ecological considered receptor.

The xenobiotic availability in soil has been traditionally estimated by chemical extraction methods: a summary of these methods is presented in Table 3.1. In this chapter we show how biosensor technology can integrate the information given by chemical methods with more biologically meaningful information on xenobiotic bioavailability, in order to provide an improved and integrated approach to ecotoxicological risk assessment in soils.

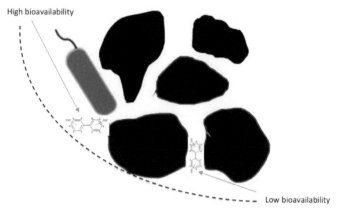

Figure 3.3. Schematization of bioavailability processes in soil environments. Xenobiotics can be trapped within soil particles or within the organic matter, and their biovailability for ecological receptors such as bacteria is reduced. Unpublished material of the authors.

Color image of this figure appears in the color plate section at the end of the book.

Table 3.1. Examples of chemical extraction methods for the assessment of the bioavailable fraction of organic xenobiotics in soils.

Extraction method	Description	Reference(s)
Cyclodextrins	Modified cyclodextrins (e.g. hydroxypropyl-β-cyclodextrin, HPCD) are used to entrap xenobiotics such as PAHs or PCBs, thus increasing their solubility.	Semple et al. 2003 Puglisi et al. 2007
Hydrophobic resins	Resins such as XAD or Tenax are used to desorb and quantify xenobiotics labile fractions	Northcott and Jones 2001
Mild solvents	Use of mild solvents such as buthanol to increase the solubility of hydrophobic xenobiotics	Kelsey et al. 1997
Persulphate oxidation	The fraction of PAHs oxidized by persulphate is used as an estimate of the bioavailable fraction	Cuypers et al. 2000
Supercritical fluid extraction	The bioavailable fraction is extracted by putting samples in pressure and temperature controlled conditions in chambers flushed with supercritical CO_2	Björklund et al. 1999

Different extraction methods for the quantification of the bioavailable fraction have been developed. These methods can be coupled with biosensors in order to achieve an integrated ecotoxicological risk assessment.

BIOSENSORS WITH FOCUS ON SOIL APPLICATIONS

A biosensor can be defined as a measurement device or system composed of a biological sensing component, which recognizes a chemical or physical change, coupled to a transducing element that produces a measurable signal in response to the environmental change (Daunert et al. 2000) (Fig. 3.4). The sensing component classifies the biosensors in one of the three basic types such as: molecular, tissue and cellular (Pancrazio et al. 1999). The sensing elements of molecular biosensors can be subcellular components or macromolecules such as nucleic acids, lipid bilayers, enzymes, ion channels or antibodies. Tissue based biosensors are derived from intact tissue, whereas the whole-cell biosensors are usually bacterial cells producing a measurable signal or product.

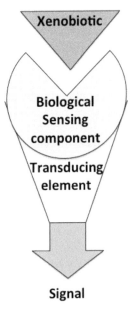

Signal

Figure 3.4. Schematic representation of a biosensor for detection of xenobiotics. A biosensor is made up of different components assuring its functioning: a biological sensing component (e.g. a microorganism, a tissue, an enzyme or an antibody) recognizes a specific xenobiotic or class of xenobiotics, and through a transducing element produces a signal (e.g. light or fluorescence) easily measurable. Unpublished material of the authors.

Whole Cell Biosensors

Bacteria have long being used as models for exploring the dose-dependent toxicity and mutagenicity of specific analytes in drug tests and monitoring of environmental contamination. The first use of a bacterial strain dates back to the Microtox test which was introduced in the early 1980s and it is based on the natural bioluminescence of the marine bacterium *Vibrio fischeri* to assess samples toxicity (Bulich and Isenberg 1981). The basic assumption for this test is that bioluminescence is an energy- and cofactor-demanding process and any factor reducing the bioluminescence indicates that bacteria have been exposed to a toxicant. Microtox is a typical example of the so-called "light-off" or non-specific biosensor: similar "light-off" biosensor have been developed by inserting *lux* genes into enteric bacterial strains such as *Escherichia coli* or soil borne bacteria such as *Pseudomonas fluorescens* (Hamin-Hanjani et al. 1993). The lack of specificity is the major limitation of these 'light off' biosensors, which may also produce false positive responses to organic xenobiotics if they are metabolized and produce cellular reducing potential leading to an increase of bioluminescence.

A major step forward in biosensor technology has been made by the production of the first specific biosensor, constructed by the insertion of a promoterless *lux* gene from *Vibrio fisherii* into the soil borne bacterium *Pseudomonas fluorescens* under the control of the promoter *nahG* gene, specifically activated by salycilate, the metabolite produced after naphthalene uptake (King et al. 1990). This work demonstrated the possibility of stabilizing a genetic construct and transmitting it to the following generations in ecological representative microorganism. It is also the first example of a specific "light-on" biosensor, where the luminescence is emitted only in response to the assimilation of specific chemicals or classes of chemicals. Since then, a large number of biosensors have been created for analysis of the bioavailability of the major classes of organic xenobiotics, but also for trace elements, nutrients and ecological interactions, with important applications for soil systems.

"Light-on" biosensors are characterized by two main elements: a reporter gene and a promoter (or responsive element). The concept of the reporter gene is relatively old in basic biology and biochemistry, and it can be defined as a gene whose phenotypic expression can be easily detected. Reporter genes traditionally used in microbiology and biochemistry are *lacZ*, coding for β-galactosidase, or *xylE* coding for catechol 2,3-dioxygenase, but they are not useful for monitoring microbial responses in soil, because their products cannot be easily recovered or targeted against the high background in soil samples. Differently, bioluminescence-based reporter genes, expressing light emitting molecules, such as the promoterless *lux* (bacterial luciferase), *luc* (firefly luciferase) or *gfp* (green fluorescent protein encoding gene) are more suitable to study activity and location of microorganisms in the soil environment.

Four genes have been successfully used to construct whole cell reporter strains, mainly due to the possibility to detect their products in the soil environment. The *luxCDABE* gene cassette codes for the luciferase enzyme, which emits constant light when supplied with oxygen and intracellular energy, whilst bioluminescence is reduced by uncoupling factors, loss of membrane integrity, or direct enzyme inhibition. The *gfp* (Table 3.2), isolated from the jellyfish *Aequorea victoria*, encodes for the green fluorescent protein

Table 3.2. Key features of fluorescent proteins.

1.	The first fluorescent protein discovered was the green fluorescent protein (GFP), a component of the bioluminescent organs of the jellyfish *Aequoria victoria*.
2.	Three scientists (Osamu Shimomoura, Martin Chalfie and Roger Y. Tsien) received in 2008 the Nobel Prize in Chemistry for the discovery of GFP and development of GFP-related applications.
3.	Other fluoresecent proteins, such as the yellow and the red fluorescent proteins (YFP and RFP) have been also engineered.
4.	Scientists can now clone fluorescent proteins and insert them into living cells of different organisms and visualize the location and the dynamics of the gene product using fluorescence microscopy.
5.	Fluorescent based biosensors are constructed by inserting the fluorescence genes under the control of promoters activated by specific compounds.
6.	The advantage of this approach for soil studies is that fluorescent proteins respond fast, and have no interferences with soil constituents.
7.	It is possible to insert more fluorescent proteins each under the control of a different promoter in a single organism, thus obtaining multiple-specific biosensors.

Fluorescent proteins have very important applications in science, as acknowledged by the Nobel Prize in Chemistry awarded in 2008 to their three discoverers. In the case of biosensors application for soils, fluorescent proteins present several advantages as compared to other reporter systems.

(GFP), a protein of 27 kDa size that converts by chemiluminescence the blue light (395 nm) of the Ca^{2+}-photoprotein (aequorin) into green light (510 nm). The chromogenic part of the protein is a tripeptide, which requires oxygen for maturation. The cDNA of the *gfp* was firstly cloned by Prasher et al. (1992) and afterwards introduced in intact cells and/or organelles by Inouye and Tsuji (1994). The main advantages of the gfp-based whole cell biosensors are that no substrates are required to perform the assays, GFP protein is stable in a broad range of pH (6–12) and resists to different proteases. The latter two protein characteristics are particularly important for soil applications, because soil pH may vary in different soils and because the soil holds an intense proteolytic activity either due to the biontic or abiontic proteases. The disadvantage of GFP is that the formation of the GFP active form requires some hours and that under UV light excitation several organic and inorganic soil constituents are also fluorescent. However, reporters with mutant unstable GFP having a rapid intracellular turnover (Andersen et al. 1998) allows the construction of gfp-based reporter bacteria, which respond faster and no interferences by soil constituents in the fluorescence measurements have been reported.

When a specific biosensor for organic xenobiotics is required, the most useful promoter genes are usually the ones involved in degradative pathways. Specific biosensors have been developed employing genes involved in the degradation of xenobiotic compounds such as phenols (Shingler and Moore 1994), middle-chain alkanes (Sticher et al. 1997), salicylates (King et al. 1990), chlorobenzoates from PCBs degradation (Boldt et al. 2004), and dioxins (Garrison et al. 1996).

Chemosensors

Chemical sensors (chemosensors) consist of two main components: a receptor and a detector. Receptors include enzymes, antibodies, and lipid layers, and are responsible for the selectivity of the sensor. The detector is not selective and acts as transducer, because it translates the physical or chemical changes of the receptor into an electrical signal. Detectors include transduction platforms having electrochemical (potentiometric, amperometric, impedance), piezoelectric, thermal or optical (reflectrometry, interferometry, optical waveguide lightmode spectroscopy, total internal reflection fluorescence, surface plasmon resonance) properties.

The chemosensors for environmental monitoring based on the use of piezoelectric detectors are among the most commonly used. Piezo-electric crystals (e.g. quartz) vibrate under the influence of an electric field, and the variations of the resonant frequency of an oscillating piezoelectric crystal in relation to the mass deposited on the crystal surface are used as an index of interactions between the receptor and the analyte. The main differences in the piezoelectric chemosensors concern the physical dimensions of the quartz plate and the thickness of the deposited electrode.

The piezoelectric DNA-based biosensors are constructed by immobilizing double stranded DNA (dsDNA), and are then placed in contact with the environmental liquid phase or extracts, allowing the contact between DNA and eventual environmental pollutants. As for the other types of sensors, in the analysis of soil and soil solution the extraction conditions are a critical step. A DNA-based biosensor for the qualitative/semiquantitative detection of genotoxic effects of various aromatic xenobiotics in soil was presented by Bagni et al. (2005), and its response in soils polluted by benzene, naphthalene and anthracene derivatives were in agreement with standard plant and animal toxicity, and the comet tests. This paper also well illustrated the different responses obtained using different extractants and extraction conditions.

Supramolecular Sensors

Supramolecular chemistry is the study of the complex multimolecular species formed by aggregation of relatively simpler molecules (Ganjali et al. 2006). Supramolecular chemistry has been used for constructing sensitive membranes, selective for particular analytes, by matching the size and binding properties of the species to enhance the sensor selectivity. Ceresa et al. (2001) used ion-selective membranes based on the Pb-selective ionophore thioacetic acid dimethylamide, in this way they established that potentiometric polymeric membrane electrodes based on electrically neutral ionophores are useful analytical tools for heavy metal ion determinations in drinking water at nanomolar total concentrations.

APPLICATION OF BIOSENSORS FOR ECOTOXICOLOGICAL RISK ASSESSMENT OF XENOBIOTICS IN SOILS

The potential application of biosensors in different stages of the basic risk assessment procedure is graphically illustrated in Fig. 3.5. As shown, the hazard can be identified through the application of a range of different biosensors, by selecting those with wider applicability and more ecological relevance. Wide applicability and ecological relevance are important pre-requisites to allow the biosensors to be accepted as tools in risk assessment future and legislation. From this point of view, owing to the biological complexity, responses of whole cell biosensors offer a more meaningful picture of the exposure and toxicity of xenobiotics than of chemosensors, which are more suitable for detection, particularly specific biosensors which can be applied for a deeper screening of toxicity, and to obtain dose-response curves and ecotoxicological end-points such as EC_{50}s and NOECs. Biosensors can also be very useful to determine the exposure level, since they provide a direct measure of the amount of xenobiotic that is actually able to cross the cell membrane. This approach also shares the possibility of selecting ecologically representative microbial strains as hosts: compared to

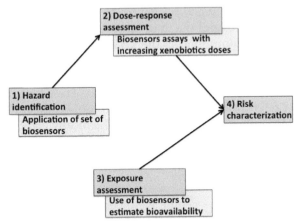

Figure 3.5. Modification of the risk assessment paradigm scheme to take into account the possible contribution of biosensors in soil ecotoxicological risk assessment. **Biosensors can be used to improve the risk paradigm in the case of ecotoxicological risk assessment of xenobiotics in soils.** 1) Different biosensors can be used as a screening tool to identify the major classes of xenobiotics present in a contaminated site; 2) specific biosensors are then used to carry out dose response assessments and obtain ecotoxicological endpoints; 3) biosensors can also be used as tools to estimate the bioavailable fraction of xenobiotics; 4) all the information is used to produce a risk characterization. Unpublished material of the authors.

classical biosensors tools such as the marine bacterium *Vibrio fisherii*, more representative strains (e.g. *Pseudomonas fluorescens* for soil) should be chosen. As a final step, all obtained information is elaborated to produce a risk assessment (Fig. 3.5). Biosensors of course do replace all tools necessary for risk assessment: their role is complementary, by providing information, such as ecotoxicity or real exposure, not obtainable with classical tools such as chemical analyses and environmental modelling. Chemical methods are also very useful to improve the application of biosensors. It is indeed difficult to apply biosensors directly on soil. *In situ* application of biosensors has been successfully achieved for nutrients status report (Puglisi et al. 2008), but for xenobiotics the situation is more difficult because of the quenching phenomena disturbing the expression of the signal. A common approach is thus to extract the xenobiotic from the soil, and then expose the biosensors to the extracts. If this is done for an ecotoxicological risk assessment, the bioavailable and not the total extractable fraction is relevant. At present several extraction methods have been developed and reviews on the topic published (Semple et al. 2003). Examples of some of these methods are reported in Table 3.1.

An example of the application of biosensors coupled to chemical methods for the extraction of the bioavailable fraction, has been given in the case of dioxins. Dioxins and furans are extremely toxic chemicals, formed as a by-product of industrial processes involving Cl such as waste incineration, chemical and pesticide manufacturing, with proven adverse effects on human and animal health in various parts of the world. Dioxins are characterized by a long persistence in the environment. The most known toxic compound is the 2,3,7,8-tetrachlorodibenzo-p-dioxin (TCDD), and the toxicity of the dioxin family as well as other chemicals with comparable effects (e.g. PCB) are measured in relation to TCDD. Kurosawa et al. (2006) developed an immunosensor with a modified anti-TCDD (2,3,7,8-tetrachlorodibenzo-p-dioxin) monoclonal antibody as the molecular recognition component on the chemically activated surface of the electrode and a high sensitive bisphenol-A detection method using a signal amplification protocol to solve the problem of low signal emission. The immunosensor method described by Kurosawa has demonstrated its effectiveness as an alternative screening method for environmental monitoring because these results were compared with results obtained through environmental monitoring methods such as GC/MS and ELISA. Another biosensor system for dioxin compounds has been developed by modification of rat hepatoma cells with insertion of luciferase gene under the control of the aryl hydrocarbon receptor: it is known as DR-CALUX (Dioxin Response Chemical Activated Luciferase eXpression) and it is widely used to assess the presence and toxicity of dioxin and dioxin-like xenobiotics in food or environmental matrixes (Garrison et al. 1996; Besselink et al. 2004).

As compared to the chemical specific immunosensor, the DR-CALUX is more "mechanism" specific: in fact, the light emission is proportional not only to the total concentration of dioxins but also to their relative toxicity. This offers several advantages for risk assessment, although the above mentioned limitations should be born in mind when the total rather than the bioavailable fraction is used.

To evaluate this potential limitation, Puglisi et al. (2007) conducted a study coupling non-exhaustive extraction techniques with resins and cyclodextrins for bioavailability assessment with the DR-CALUX assay, in an ecotoxicological risk assessment of dredged sediments historically contaminated with PCBs and dioxins. Results showed that after a long time a fraction varying from 38 to 70% of the total xenobiotic contents was bioavailable, and that the coupling of non-exhaustive extraction techniques and DR-CALUX bioanalyses lead to a lower risk estimated for as compared to commonly adopted exhaustive extraction techniques.

Biosensors can also be used for ecotoxicological risk assessment of other xenobiotics such as BTEX compounds, organochlorinateds compound and antibiotics. The mono-aromatic volatile compounds like benzene, toluene, ethylbenzene, xylenes, collectively termed BTEX are emitted from various human and industrial activities and undergo variable partitioning and distribution between the solid and liquid phases of water, soil, and vegetation. While the emission of the BTEX has led to environmental concern for their contribution to global warming, caused by their chemical reactivity and their potential to reduce the ozone in the troposphere, the BTEX, as well as other volatile xenobiotics, also have harmful effects to human health even at lower concentrations. In fact, they can affect different organs such as nervous systems, liver, kidney, reproductive apparatus and cause asthma in children, particularly in the urban environment (Hinwood et al. 2007).

Whole cell biosensors responding to bioavailable benzene, toluene and xylene (BTEX) in soil have been constructed by inserting the *luxCDBAE* or *luc* genes in different plasmidial genes involved in the catabolic pathways of such compounds, using *Pseudomonas putida* strains (Burlage and Kuo 1994) or *Escherichia coli*. The insertion of different genes improve the sensitivity towards specific BTEX compounds because their degradation to pyruvate or acetaldehyde is controlled by several intermediates and enzymes. To date, there is no comprehensive comparison between the BTEX chemosensor responses and the models developed for assessing their chemical fate in the environment.

Available whole cell biosensors for detection of chlorinated and organic polychlorinated pollutants are based on a *Pseudomonas fluorescens* 10586s, *Burkholderia sp. Rasc* strain or equipped with a complete *luxCDABE* cassette on the same plasmid pUCD607 (Palmer et al. 1998; Boyd et al. 2001), or on *Pseudomonas fluorescens* 8866 and *Pseudomonas putida* (Weitz et al. 2001) with chromosomally integrated *luxCDABE* responding to soil pollution by different chlorobenzene derivatives. However, the latter two strains also responded to heavy metals such as Zn and Cu (Daunert et al. 2000). Whole cell biosensors for detection of polychlorinated biphenils (PCBs) have been obtained by fusions of *luxCDBAE* cassette with different genes of plasmidial operons of *Ralstonia eutropha* strains, coding for the degradative enzymes regulated by the presence of monochlorinated or polychlorinated biphenyls (Layton et al. 1998). Two gfp-based whole cell biosensors sensitive to 2,4-dichlorophenoxyacetic acid have been constructed by the fusion of gfp with tfdCI gene in *Ralstonia eutropha* (Füchslin et al. 2003) and by the fusion of gfp with orf0-bphA1 in *Pseudomonas fluorescens* (Boldt et al. 2004). A luc-based *Arthrobacter chlorophenolicus* strain responding to 4-chlorophenol in soil has been successfully used by Elvang et al. (2001) for monitoring soil remediation.

Another class of whole cell biosensors includes strains capable of detecting groups of substances functionally similar such as antibiotics (Bahl et al. 2004) and endocrine disruptors (Desbrow et al. 1998). Such bioreporters have been constructed inserting reporter genes in DNA coding for co-regulation mechanisms, as structurally similar compounds may activate common intracellular reactions. An early example of this type of whole cell biosensors can be considered the *Pseudomonas putida* TVA8 bioreporter, in which the chromosomal *tod* genes were fused with a complete *luxCDABE* cassette, controlled by the Ptod promoter. Thus emitted bioluminescence in response to benzene, toluene and xylene isomers, and phenol (Applegate et al. 1998). These type of biosensors may be useful to study the fate of different xenobiotics in the rhizosphere, as they can be partially or totally degraded or transformed in different substances in soil. However, none of these bioreporters have been used for studying soils. In this class of whole cell biosensors, multiple stress-responsive bioreporters may also be included which are particularly suited to study the rhizosphere and soil conditions, as bacterial physiology under natural conditions (e.g. exposure to fluctuating temperature or osmolarity, variable moisture content, presence of radicals) are controlled by the expression of global regulatory metabolic genes (regulons). Stress-responsive bioreporters respond to a broad range of conditions and can be useful to characterize a soil environment in terms of general cytotoxicity. Due to the extended knowledge on such systems in enteric bacteria, *Escherichia coli* and *Salmonella typhimurium* have been used in early works, e.g. linking the lux reporter system to general stress promoters, such as the heat shock promoters dnaK, protein-damage sensitive grpE, oxidative-damage sensitive katG or membrane-damage sensitive fabA promoters (Bechor et al. 2002). More recently, lux constructs soil-borne *Pseudomonas* species have been genetically engineered for bioreporters of genotoxicity in environmental samples based on the SOS response involving DNA repair (Rabbow et al. 2002).

APPLICATIONS TO AREAS OF HEALTH AND DISEASE

The examples reported below show how an integrated use of biosensors and chemical extraction techniques is very useful for an ecotoxicological risk assessment of xenobiotics in soils. Being an ecotoxicological approach, the main focus is on ecological receptors in soils, primarily bacteria. It should be however taken into account that most soil xenobiotics considered here (e.g. chlorinated pollutants or BTEX compounds) also hold relevance for human health, and that soil is very often the primary route through which these compounds can enter the food chain and finally reach human receptors. For these reasons, we believe that an ecotoxicological risk assessment in soils is also very important to prevent xenobiotics-related human health issues.

Recent advances in both metabolic pathways engineering capacities, and detection technologies along with discovery of new mutants, driven primarily by studies in medicine, pharmacology and drug discovery, let us also hypothesize that more sensitive and specific whole-cell biosensors will be available for application to soil studies with relevance for human health too.

KEY TERMS

- **Aging**: reduction of a xenobiotic' bioavailability in time.
- **Total xenobiotic fraction**: the total mass of xenobiotic that can be extracted from soil with the most-exhaustive methods available.
- **Bioavailable contaminant fraction**: the fraction of xenobiotic accessible for assimilation and possible toxicity to biological receptors. It is the relevant fraction from an ecotoxicological point of view.
- **Ecotoxicological risk assessment**: procedure to assess the risk posed by environmental pollutants to representative ecological receptors. As every risk assessment, is the resultant of a hazard and an exposure assessment.
- **Ecotoxicity**: the property of a chemical to cause harmful effects on ecosystems.
- **Contaminant**: a substance present in greater than natural concentration as a result of human activity.
- **Pollutant**: a substance that occurs in the environment at least in part as a result of man's activities, and which has a deleterious effects on living organisms (Moriarty 1983).

- **Xenobiotic**: a foreign chemical or material not produced in nature and not normally considered a constitutive component of a specific biological system (Rand and Petrocelli 1985).

KEY FACTS ABOUT THE USE OF BIOSENSORS FOR THE ECOTOXICOLOGICAL RISK ASSESSMENTS OF XENOBIOTICS IN SOILS

- Environmental contamination by xenobiotics can potentially result in adverse effects for both human and ecological receptors.
- Ecotoxicological risk assessment is a procedure where information about the ecotoxicity of a xenobiotic is linked to information about the degree of exposure of ecological receptors to the xenobiotic.
- In the soil environment, ecological receptors are exposed to a fraction of the total xenobiotics mass: the bioavailable fraction.
- Biosensors are useful tools for ecotoxicological risk assessment of xenobiotics: they can provide information on exposure, ecotoxicity or on both.
- A correct ecotoxicological risk assessment of xenobiotics in soils can be achieved by coupling chemical methods for the assessment of the bioavailable fraction of the xenobiotics with the application of specific biosensors.

SUMMARY POINTS

- Risk assessment is made up of four steps: hazard identification, exposure assessment dose-response assessment and risk characterization.
- Knowledge of the environmental fate of xenobiotics in soils is necessary to assess their ecotoxicological risk.
- A wide range of biosensors can be used for ecotoxicological risk assessment of xenobiotics. The main classes include whole cell biosensors, chemosensors and supramolecular sensors.
- Biotechnology allows construction of biosensors specific for many different compounds, including important soil xenobiotics that also hold also relevance for human health.
- A correct application of biosensors for ecotoxicological risk assessment of xenobiotics in soil is important not only for assessing and promoting soil quality, but also for preventing xenobiotic-related human health issue.

ABBREVIATIONS

BTEX	:	benzene, toluene, ethylbenzene, xylenes
EC_{50}	:	Effective Concentration (for 50% of population)
GFP	:	green fluorescent protein
HPCD	:	hydroxypropyl-β-cyclodextrin
NOEC	:	No Observable Effect Concentration
PAHs	:	Polycyclic Aromatic Hydrocarbons
PCBs	:	PolyChloroBiphenyls

REFERENCES

Amin-Hanjani S, A Meikle, L Glover, J Prosser and K Killham. 1993. Plasmid and chromosomally encoded luminescence marker systems for detection of *Pseudomonas fluorescens* in soil. Mol Ecol 2: 47–54.

Andersen JB, C Sternberg, LK Poulsen, SP Bjorn, M Givskov and S Molin. 1998. New unstable variants of green fluorescent protein for studies of transient gene expression in bacteria. Appl Environ Microbiol 64: 2240–2246.

Applegate B, S Kehrmeyer and G Sayler. 1998. A chromosomally based *tod-luxCDABE* whole-cell reporter for benzene, toluene, ethybenzene, and xylene (BTEX) sensing. Appl Environ Microbiol 64: 2730–2735.

Bagni G, S Hernandez, M Mascini, E Sturchio, P Boccia and S Marconi. 2005. DNA biosensor for rapid detection of genotoxic compounds in soil samples. Sensors 5: 394–410.

Bahl MI, LH Hansen, TR Licht and SJ Sorensen. 2004. *In vivo* detection and quantification of tetracycline by use of a whole-cell biosensor in the rat intestine. Antimicrob Agents Ch 48: 1112–1117.

Bechor O, DR Smulski, TK Van Dyk, RA LaRossa and S Belkin. 2002. Recombinant microorganisms as environmental biosensors: pollutants detection by *Escherichia coli* bearing *fabA':: lux* fusions. J Biotechnol 94: 125–132.

Besselink HT, C Schipper, H Klamer, P Leonards, H Verhaar, E Felzel, AJ Murk, J Thain, K Hosoe, G Schoeters, J Legler and B Brouwer. 2004. Intra- and interlaboratory calibration of the DR CALUX bioassay for the analysis of dioxins and dioxin-like chemicals in sediments. Environ Toxicol Chem 23: 2781–9.

Björklund E, S Bøwadt, L Mathiasson and SB Hawthorne. 1999. Determining PCB sorption/desorption behavior on sediments using selective supercritical fluid extraction. 1. Desorption from historically contaminated samples. Environ Sci Technol 33: 2193–2203.

Boldt TS, J Sørensen, U Karlson, S Molin and C Ramos. 2004. Combined use of different Gfp reporters for monitoring single cell activities of a genetically modified PCB degrader in the rhizosphere of alfalfa. FEMS Microbiol Ecol 48: 139–148.

Boyd EM, K Killham and AA Meharg. 2001. Toxicity of mono-, di- and tri-chlorophenols to lux marked terrestrial bacteria, *Burkholderia* species *Rasc c2* and *Pseudomonas fluorescens*. Chemosphere 43: 157–166.

Bulich A and D Isenberg. 1981. Use of the luminescent bacterial system for the rapid assessment of aquatic toxicity. ISA Trans (United States) 20.

Burlage RS and C Kuo. 1994. Living biosensors for the management and manipulation of microbial consortia. Annu Rev Microbiol 48: 291–309.

Ceresa A, E Bakker, B Hattendorf, D Günther and E Pretsch. 2001. Potentiometric polymeric membrane electrodes for measurement of environmental samples at trace levels: new requirements for selectivities and measuring protocols, and comparison with ICPMS. Anal Chem 73: 343–351.

Cuypers C, T Grotenhuis, J Joziasse and W Rulken. 2000. Rapid persulfate oxidation predicts PAH bioavailability in soils and sediments. Environ Sci Technol 34: 2057–2063.

Daunert S, G Barrett, JS Feliciano, RS Shetty, S Shrestha and W Smith-Spencer. 2000. Genetically engineered whole-cell sensing systems: coupling biological recognition with reporter genes. Chem. Rev. 100: 2705–2738.

Desbrow C, E Routledge, G Brighty, J Sumpter and M Waldock. 1998. Identification of estrogenic chemicals in STW effluent. 1. Chemical fractionation and *in vitro* biological screening. Environ. Sci. Technol. 32: 1549–1558.

Elväng AM, K Westerberg, C Jernberg and JK Jansson. 2001. Use of green fluorescent protein and luciferase biomarkers to monitor survival and activity of *Arthrobacter chlorophenolicus* A6 cells during degradation of 4 chlorophenol in soil. Environ Microbiol 3: 32–42.

Füchslin HP, I Rüegg, JR Van Der Meer and T Egli. 2003. Effect of integration of a GFP reporter gene on fitness of Ralstonia eutropha during growth with 2, 4 dichlorophenoxyacetic acid. Environ Microbiol 5: 878–887.

Ganjali MR, P Norouzi, M Rezapour, F Faridbod and MR Pourjavid. 2006. Supramolecular based membrane sensors. Sensors 6: 1018–1086.

Garrison PM, K Tullis, JM Aarts, A Brouwer, JP Giesy and MS Denison. 1996. Species-specific recombinant cell lines as bioassay systems for the detection of 2,3,7,8-tetrachlorodibenzo-p-dioxin-like chemicals. Fundam Appl Toxicol 30: 194–203.

Hinwood AL, C Rodriguez, T Runnion, D Farrar, F Murray, A Horton, D Glass, V Sheppeard, JW Edwards and L Denison. 2007. Risk factors for increased BTEX exposure in four Australian cities. Chemosphere 66: 533–541.

Hamin-Hanjani S, A Meikle, LA Glover, JI Prosser and K Killham. 1993. Plasmid and chromosomally encoded luminescence marker system for detection of Pseudomonas fluorescens in soil. Mol Ecol 2: 47–54.

Inouye S and FI Tsuji. 1994. Aequorea green fluorescent protein: expression of the gene and fluorescence characteristics of the recombinant protein. FEBS Lett 341: 277–280.

Kelsey JW, BD Kottler and M Alexander. 1997. Selective chemical extractants to predict bioavailability of soil-aged organic chemicals. Environ Sci Technol 31: 214–217.

King JM, PM Digrazia, B Applegate, R Burlage, J Sanseverino, P Dunbar, F Larimer and GS Sayler. 1990. Rapid, sensitive bioluminescent reporter technology for naphthalene exposure and biodegradation. Science 249: 778–781.

Kurosawa S, JW Park, H Aizawa, SI Wakida, H Tao and K Ishihara. 2006. Quartz crystal microbalance immunosensors for environmental monitoring. Biosens Bioelectron 22: 473–481.

Layton A, M Muccini, M Ghosh and G Sayler. 1998. Construction of a bioluminescent reporter strain to detect polychlorinated biphenyls. Appl Environ Microbiol 64: 5023–5026.

Moriarty F. 1983. Ecotoxicology. Academic Press, London.

National Research Council. 1983. Risk assessment in the Federal Government: managing the process National Academy Press, Washington DC, USA.

Newman MC and MA Unger. 2003. Fundamentals of ecotoxicology. CRC Press, Florida, US.

Northcott GL and KC Jones. 2001. Partitioning, extractability, and formation of nonextractable PAH residues in soil. 2. Effects on compound dissolution behaviour. Environ Sci Technol 35: 1111–1117.

Palmer G, R McFadzean, K Killham, A Sinclair and GI Paton. 1998. Use of *lux*-based biosensors for rapid diagnosis of pollutants in arable soils. Chemosphere 36: 2683–2697.

Pancrazio J, J Whelan, D Borkholder, W Ma and D Stenger. 1999. Development and application of cell-based biosensors. Ann Biomed Eng 27: 697–711.

Prasher DC, VK Eckenrode, WW Ward, FG Prendergast and MJ Cormier. 1992. Primary structure of the *Aequorea victoria* green-fluorescent protein. Gene 111: 229–233.

Puglisi E, AJ Murk, HJ van den Berg and T Grotenhuis. 2007. Extraction and bioanalysis of the ecotoxicologically relevant fraction of contaminants in sediments. Environ Toxicol Chem 26: 2122–2128.

Puglisi E, G Fragoulis, AA Del Re, R Spaccini, A Piccolo, G Gigliotti, D Said-Pullicino and M Trevisan. 2008. Carbon deposition in soil rhizosphere following amendments with compost and its soluble fractions, as evaluated by combined soil-plant rhizobox and reporter gene systems. Chemosphere 73: 1292–1299.

Rabbow E, P Rettberg, C Baumstark-Khan and G Horneck. 2002. SOS-LUX- and LAC-FLUORO-TEST for the quantification of genotoxic and/or cytotoxic effects of heavy metal salts* 1. Anal Chim Acta 456: 31–39.

Rand GM and SR Petrocelli. 1985. Fundamentals of aquatic toxicology: methods and applications. FMC Corp., Princeton, NJ, USA.

Sanderson JT, J Aarts, A Brouwer, KL Froese, MS Denison and JP Giesy. 1996. Comparison of Ah receptor-mediated luciferase and ethoxyresorufin-O-deethylase induction in H4IIE cells: implications for their use as bioanalytical tools for the detection of polyhalogenated aromatic hydrocarbons. Toxicol Appl Pharm 137: 316–325.

Semple KT, A Morriss and G Paton. 2003. Bioavailability of hydrophobic organic contaminants in soils: fundamental concepts and techniques for analysis. Eur J Soil Sci 54: 809–818.

Shingler V and T Moore. 1994. Sensing of aromatic compounds by the DmpR transcriptional activator of phenol-catabolizing *Pseudomonas* sp. strain CF600. J Bacteriol 176: 1555–1560.

Sticher P, MC Jaspers, K Stemmler, H Harms, AJ Zehnder and JR van der Meer. 1997. Development and characterization of a whole-cell bioluminescent sensor for bioavailable middle-chain alkanes in contaminated groundwater samples. Appl Environ Microbiol 63: 4053–60.

Truhaut R. 1977. Ecotoxicology: objectives, principles and perspectives. Ecotox Environ Safe 1: 151–173.

Walker C, S Hopkin, R Sibly and D Peakall. 2001. Principles of ecotoxicology. Taylor and Francis, London, UK.

Weitz HJ, JM Ritchie, DA Bailey, AM Horsburgh, K Killham and LA Glover. 2001. Construction of a modified mini Tn5 lux*CDABE* transposon for the development of bacterial biosensors for ecotoxicity testing. FEMS Microbiol Lett 197: 159–165.

4

Biosensor-controlled Degradation of Organophosphate Insecticides in Water

Georges Istamboulie,[1,a] **Jean-Louis Marty,**[1,b]
Silvana Andreescu[2] **and Thierry Noguer**[1,*]

ABSTRACT

Due to their acute toxicity and high risk towards the population, some directives have been established to limit the presence of pesticides in water and food resources. It is thus mandatory to develop efficient decontamination techniques in combination with highly sensitive detection techniques allowing to accurately determine the level of contamination of waters. This chapter deals with the development of an enzymatic detoxification system based on bacterial phosphotriesterase for the degradation of organophosphate insecticides in waters. A detoxification column was designed based on the immobilization of phosphotriesterase on an activated agarose gel via covalent coupling. The column was tested using two widely used insecticides, chlorpyrifos

[1]Université de Perpignan Via Domitia, IMAGES, Perpignan, 66860 - France.
[a]E-mail: georges.istamboulie@univ-perp.fr
[b]E-mail: jlmarty@univ-perp.fr
[c]Email: noguer@univ-perp.fr
[2]Department of Chemistry, Clarkson University, Potsdam, NY 13699-5810 - USA.
E-mail: eandrees@clarkson.edu
*Corresponding author

List of abbreviations after the text.

and chlorfenvinfos, these compounds were selected as they are included in the list of priority substances established by the European Union Water Framework Directive. The efficiency of insecticide degradation was controlled using a highly sensitive biosensor based on a recombinant acetylcholinesterase, allowing the detection of organophosphate concentrations as low as 0.004 μg L^{-1}. It was shown that a column incorporating phosphotriesterase was suitable for the detoxification of solutions containing organophosphate insecticides, even at concentrations higher than authorized limits. The method was shown to be adapted to the decontamination of real samples of pesticides with concentrations up to 20 μg L^{-1}.

INTRODUCTION

Pesticides are widely used in agricultural crops, forests and wetlands as insecticides, fungicides, herbicides and nematocides. Many of them are considered to be particularly hazardous compounds and toxic because they inhibit fundamental metabolic pathways.

Due to their highly acute toxicity and risk towards the population, some directives have been established to limit the presence of pesticides in water and food resources. To regulate the quality of water for human consumption, the European Council directive 98/83/CE (Drinking Water Directive) has set a maximum admissible concentration of 0.1 μg.L^{-1} per pesticide and 0.5 μg.L^{-1} for the total amount of pesticides. However, this directive does not mention the admissible concentrations in waters intended for human consumption before treatment, recommending each European country to establish internal rules. In France, the decree no. 2001–1220 of 20 December 2001 has set a maximum admissible concentration of 2 μg.L^{-1} per pesticide and 5 μg.L^{-1} for total pesticides.

Organophosphates (OPs) are a class of synthetic pesticides developed during World War II, which are currently used as insecticides and nerve agents (Raushel 2002). With the restricted use of organochlorine insecticides, OPs have become the most widely used insecticides for agricultural, industrial, household and medical purposes. OPs poison insects and mammals by phosphorylation of the acetylcholinesterase (AChE) enzyme at nerve endings. Inactivation of this enzyme results in an accumulation of acetylcholine leading to an overstimulation of the effector organ (Aldridge 1950; Reigart and Roberts 1999).

Degradation of OPs is a mandatory step for the detoxification of waters for human consumption. Enzymes able to degrade OPs have been the focus of several research studies in the environmental decontamination/detoxification field (Sogorb and Vilanova 2002). OP-hydrolyzing enzymes have been found in bacteria, *Archaea* and *Eukarya*. The best-known and characterized enzyme is phosphotriesterase (PTE), first isolated from the soil bacterium *Pseudomonas diminuta*. However, its natural substrate and the main physiological function are still unknown (Merone et al. 2008). PTE has the ability of hydrolyzing most organophosphates, including chemical warfare agents like sarin or sonan (Raushel 2002; Ghanem and Raushel 2005). Due to these properties, PTE has been also described as a potential catalytic scavenger for the treatment of organophosphate poisoning (Masson et al. 1998).

The hazardous nature of OPs and their wide usage require development of highly sensitive detection techniques as well as efficient destruction methods of these compounds (Gill and Ballesteros 2000). Detection techniques are fundamental in order to accurately determine the level of contamination of waters by pesticides. Classical analytical methodologies are based on extraction, cleanup and analysis using gas chromatography (GC) or liquid chromatography (LC) coupled to sensitive and specific detectors (Ballesteros and Parrado 2004; Geerdink et al. 2002; Kuster et al. 2006; Lacorte et al. 1993). Although these methods are very sensitive, they are expensive, time consuming (involve extensive preparation steps) and are not adapted for *in situ* and real-time detection, often requiring highly trained personnel. In addition, these methods are not able to provide precise information concerning the toxicity of the sample.

AChE-based biosensors can be used as an alternative to conventional chromatographic techniques for the detection of OPs insecticides. A successful AChE biosensor for toxicity monitoring should offer comparable or even better analytical performances than the traditional chromatographic systems. Ideally, such sensors should be small, cheap, simple to handle and able to provide reliable information in real-time without or with a minimum sample preparation.

Another possible route for the detection of these insecticides is to use phosphotriesterase (PTE) biosensors (Mulchandani et al. 2001). Several PTE -based amperometric, potentiometric, or optical biosensors have been described (Mulchandani et al. 1999). Deo and coworkers have reported an amperometric biosensor for organophosphate pesticides based on a carbon nanotube (CNT)-modified electrode and PTE as a biocatalyst. In this sensor design, the detection is based on the electrochemical oxidation of *p*-nitrophenol, which is produced upon hydrolysis of paraoxon or methyl-parathion by PTE (Deo et al. 2005). However, it must be noted that this method can only be applied to the detection of a few insecticides including parathion, methyl-parathion and their oxidation products (paraoxon and paraoxon-methyl), due to the fact that other organophosphates do not generate *p*-nitrophenol upon hydrolysis.

PRINCIPLE OF OP INSECTICIDES DETOXIFICATION BY PTE

Catalytic Mechanism of PTE

The phosphotriesterase (PTE) (EC 3.1.8.1) is a member of the amidohydrolase superfamily (Holm and Sander 1997). Like other members of this superfamily PTE utilizes one or two divalent metal ions to activate a hydrolytic water molecule for a nucleophilic attack at tetrahedral phosphorus or trigonal carbon centres. The active site of the native enzyme contains two zinc ions per monomer. The enzyme retains catalytic activity when the native Zn^{2+} is replaced by Co^{2+}, Cd^{2+}, Ni^{2+}, or Mn^{2+}. The cobalt-substituted enzyme is the most active form (Omburo et al. 1992).

The natural substrate of PTE remains unknown. The catalytic mechanism for the hydrolysis of organophosphate triesters has been extensively studied using paraoxon as a reference substrate (Ghanem and Raushel 2005; Aubert et al. 2004). For other substrates, the catalytic efficiency and the rate-limiting step are dependent on the pK_a of the leaving group (Hong and Raushel 1996).

The kinetic constants (Km and Vm) of PTE for different organophosphate substrates are summarized in Table 4.1 (Istamboulie et al. 2009).

Table 4.1. Kinetic constants of PTE enzyme using eight different organophosphate insecticides (substrates). CPO = chlorpyrifos-oxon, CFV = chlorfenvinfos (in order to facilitate comparisons, the maximum velocity was expressed as paraoxon-ethyl-equivalent activity).

PTE	CPO	CPO-methyl	Paraoxon	Paraoxon-methyl	CFV	Dichlorvos	Omethoate, Fenamiphos
Km (mM)	0.04	0.06	0.25	1.2	0.017	0.31	Not substrate
Relative Vm (/paraoxon)	1.8	0.25	1	0.38	0.0002	0.003	

It has been demonstrated that PTE displays a higher affinity for chlorpyrifos-oxon (CPO) derivatives than for paraoxon compounds (lower Km), which are generally used in PTE catalysis studies. Both CPO and chlorfenvinfos (CFV) show a very good affinity for PTE, but even though CFV has a better affinity than CPO, its hydrolysis rate was 9,000-fold lower. The high affinity and the low degradation velocity of CFV indicate that this substrate is acting as a competitive inhibitor of PTE for the hydrolysis of other organophosphates (Istamboulie et al. 2009).

Biodetoxification of Water Contaminated by OP Using the PTE

Many non-selective techniques have been developed to clean-up OP-containing water, such as activated carbon adsorption, ozonation or biodegradation (Oturan 2000; Guivarch et al. 2003). Among these, Advanced Oxidation Processes (AOPs) have been successfully applied for degrading organic compounds present in polluted water (Spadaro et al. 1994).

Two different OPs were studied, chlorpyrifos and chlorfenvinfos, which are included in the list of priority substances in the field of water policy (decision 2455/2001/EC). Figure 4.1 presents the reactions catalyzed by PTE for the CPO and CFV. The main products of hydrolysis are O,O-diethyl phosphoric ester, 3,5,6-trichloro-2-pyridinol and 2- chloro-1-(2′,4′-dichlorophenyl)vinyl alcohol. It is important to mention that these detoxification products are not inhibitors of AChE and are not pointed out as harmful chemicals by the European regulation and WHO recommendations on water quality (Sogorb and Vilanova 2002).

It has been shown that the PTE-based detoxification column allows the degradation of OPs in a large range of concentrations. However, in order

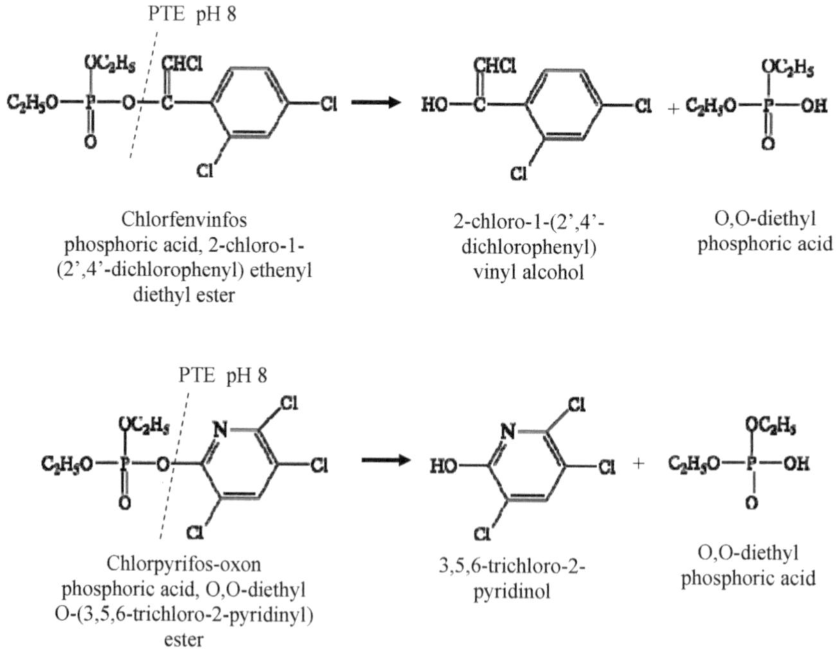

Figure 4.1. Molecular structures of CPO and CFV and their hydrolysis reactions catalyzed by phosphotriesterase (PTE).

to fit with the strict European and national norms, the real challenge is to find a method for *in situ* controlling the amount of pesticides in order to evaluate the efficiency of the treatment.

ACETYLCHOLINESTERASE-BASED BIOSENSORS

An alternative to the elaborated chromatographic methods is to determine pesticides using enzymatic assays, mostly based on AChE inhibition. Many detection kits have been successfully developed based on this principle. The most advanced systems have been designed as low cost disposable biosensors (Andreescu and Marty 2006; Istamboulie et al. 2007). Most advanced configurations involve the use of recombinant AChEs that were specifically engineered to enhance sensitivity and selectivity towards inhibitors (Istamboulie et al. 2007). The mechanism of inhibition of AChE by OP and carbamate compounds is well-known (Aldridge 1950). The inhibitor phosphorylates or carbamoylates the active site serine and the inhibition can be considered as irreversible in the first 30 min (Boublik et al. 2002).

$$K_d \qquad k_2$$
$$E + PX \leftrightarrow E^*PX \rightarrow EP + X$$

Where: E = enzyme, PX = carbamate or OP and X = leaving group.

This scheme can be simplified using the bimolecular constant $k_i = k_2/K_d$:

$$k_i$$
$$E + PX \rightarrow EP + X$$

Two types of cholinesterases (ChEs) are known and have been used for designing biosensors: AChE and butyrylcholinesterase (BuChE). BuChE has a similar molecular structure to AChE but is characterized by different substrate specificity: AChE preferentially hydrolyses acetyl esters such as acetylcholine, while BuChE hydrolyzes butyrylcholine. Another aspect that distinguishes AChE from BuChE is the AChE inhibition by excess of substrate. This property is related to substrate binding and the catalytic mechanism.

The most important step in the development of an enzyme biosensor is the stable attachment of the enzyme onto the surface of the working electrode. This process is governed by various interactions between the enzyme and the electrode material and strongly affects the performance of the biosensor in terms of sensitivity, stability, response time and reproducibility. The main methods used for depositing ChEs onto the working electrode surface are summarized in Table 4.2.

Apart from the natural substrates, ChEs are also able to hydrolyze esters of thiocholine such as acetylthiocholine, butyrylthiocholine,

Table 4.2. Principal methods used for ChEs immobilization in biosensors (Andreescu and Marty 2006); SPE: screen-printed electrodes; PVA: photocrosslinkable poly(vinyl alcohol); NTA: nitrilotriacetic acid; Con-A: concanavalin A.

Procedure	Electrode material/ immobilization agents	Observations
Adsorption	Graphite; SPE; Carbon nanotubes	Simple, Short response time; Poor operational and storage stability; Sensitive to changes in pH, temperature, ionic strength
Covalent binding	Pre-activated or functio-nalizable electrode surfaces: noble metals, graphite, SPE, Glass Glutaraldehyde, carbodiimide/NHS	Poor reproducibility, High amount of enzyme, Possible denaturation; Stable; Absence of diffusion barriers; Short response time
Self-assembled monolayers	–	Complex and difficult to reproduce; Possible electrode fouling
Physical entrapment	Noble metals; Graphite; SPE; Glassy carbon electrodes, Photopolymers (PVA-SbQ), Electro-polymerizable,polymers (pyrrole, aniline); Sol-gel	One-step procedure at ambient or low temperature; Simple; Many biocompatible polymers available; Suitable for a large variety of bio-receptors; Problems of reproducibility and control of pore size; Diffusion barriers
Affinity	Noble metals; SPE; Silica NTA-Ni, Carbohydrates	Need the presence of specific groups in the bio-receptor molecule (e.g. histidine, biotine, lectin...) Reusable surface; Low amount of enzyme; Controlled and orientated immobilization

propionylthiocholine, acetyl-β-methylthiocholine as well as *o*-nitrophenylacetate, indophenylacetate and α-naphtyl acetate. Many of these substrates have been used in different ChE biosensor configurations. AChE enzymes extracted from the *Drosophila melanogaster* and electric eel are commercially available and are the most widely used for biosensor fabrication. Most AChE biosensors have been used in amperometric detection mode by measuring the product of the enzymatic reaction at a constant potential and evaluating the inhibition degree after exposure to pesticides.

The first devices described were coupling a ChE with a choline oxidase, the detection being based on either the oxidation of H_2O_2 or the reduction of oxygen. This complicated system was further simplified using a synthetic substrate of AChE, acetylthiocholine, which produces an easily oxidizable compound, thiocholine, according to the reactions in Fig. 4.2.

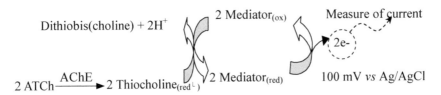

Figure 4.2. Principle of Amperometric detection of thiocholine: ATCh (Acetylthiocholine), AChE (Acetylcholinesterase).

The use of an appropriate mediator, like tetracyanoquinodiemethane (TCNQ), cobalt phtalocyanine (CoPC) or poly(3,4-ethylenedioxythiophene) (PEDOT) (Istamboulie et al. 2010b) allows decreasing the detection potential to values lower than 100 mV vs. Ag/AgCl, thus minimizing interfering oxidations. The mediator can be used in solution or incorporated in the working electrode material, the latest being the preferred and more convenient method. The most versatile method for manufacturing the electrode is the screen-printing method. This technology allows the production of screen-printed three-electrode system with a low cost and high reproducibility.

The detection principle of AChE biosensors is based on the inhibition of enzyme by OP insecticides, leading to reduced thiocholine production. Typically, amperometric measurements are performed in stirred PBS solution at pH values comprised between 7 and 8. After applying the appropriate potential for mediator oxidation, the current intensity is recorded in the presence of a saturating concentration of acetylthiocholine substrate. The time necessary to reach the plateau is 2–3 min. The measured signal corresponds to the difference of current intensity between the baseline and the plateau. The cell is washed with distilled water in between measurements. The pesticide detection is made in a three step procedure: first, the initial response of the electrode to acetylthiocholine (1 mM) is recorded two times, then the electrode is incubated in a solution containing a known concentration of insecticide, and finally the residual response of the electrode is recorded. The percentage of the inhibition is then correlated with the insecticide concentration. Based on this method, highly sensitive biosensors have been developed by our group using recombinant enzymes immobilized on magnetic micro-beads by nickel-histidine affinity. We have mainly focused our attention on two insecticides of interest: chlorpyrifos (CPO) and chlorfenvinfos (CFV). The developed sensors allowed the detection of pesticide concentrations as low as 1.3×10^{-11}M (0.004 µg.L^{-1}) (Istamboulie et al. 2007).

APPLICATION OF AChE-BASED SENSORS FOR CONTROLLING THE PTE-CATALYZED DEGRADATION OF OP INSECTICIDES

Our group has recently demonstrated the potential application of phosphotriesterase as a selective enzyme for degradation of organophosphate pesticides in polluted waters (Istamboulie et al. 2010a). A detoxification system was prepared using a 10 mL Omnifit column filled with 4 mL of Sepharose Gel 4B previously loaded with 500 IU of PTE. The column was connected via Teflon® tubes to a HPLC pump, and the flow rate was set to either 0.5 mL min^{-1} or 1 mL min^{-1}, equivalent to contact times of respectively 8 min and 4 min. Due to the high activity of PTE for CPO, the limiting factor lies in the ability of the column to efficiently degrade CFV as well as mixtures of CPO and CFV. The developed biosensors were used for the control of OP compounds in water before and after treatment using the column. The whole detoxification device coupling the PTE-based column and the biosensor allowing to monitor the column effluents is presented in Fig. 4.3 (Istamboulie et al. 2010a).

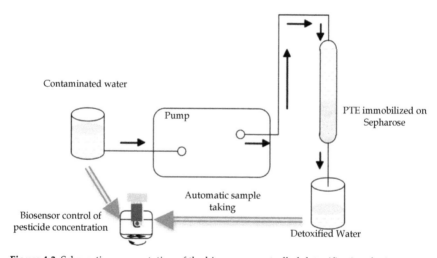

Figure 4.3. Schematic representation of the biosensor-controlled detoxification device.

Under optimum conditions, a column containing 500 IU of PTE was shown to immediately detoxify CPO solutions, whatever their concentration. In the case of CFV, the detoxification was much lower, due to the very slow hydrolysis of this pesticide by PTE (Table 4.3). However, CFV solutions with concentrations up to 15 µg L^{-1} could be efficiently detoxified, thus fulfilling the European regulations for human consumption water. Pesticides mixtures were also satisfactory degraded, even at concentrations higher

Table 4.3. Concentrations of CPO and CFV measured using a B394-AChE-based biosensor before and after treatment on a detoxification column loaded with 500 IU of PTE.

	Before treatment		After treatment - flow rate 1 mL min^{-1}		After treatment - flow rate 0.5 mL min^{-1}	
	Actual concentration µg L^{-1}	Inhibition %	Inhibition %	Calculated concentration µg L^{-1}	Inhibition %	Calculated concentration µg L^{-1}
CFV	0.1	43	0	< 0.005	0	< 0.005
	0.3	55	0	< 0.005	0	< 0.005
	0.5	60	0	< 0.005	0	< 0.005
	0.7	64	0	< 0.005	0	< 0.005
	1	67	0	< 0.005	0	< 0.005
	2	75	0	< 0.005	0	< 0.005
	5	84	5	< 0.005	0	< 0.005
	10	95	8	< 0.005	0	< 0.005
	12.5	100	43	0.1	0	< 0.005
	15	100	78	2	8	< 0.005
	20	100	90	10	35	< 0.1
CPO	334	100	0	< 0.004	0	< 0.004
CPO + CFV	334 (CPO) + 20 (CFV)	100	90	10	35	< 0.1 (CPO + CFV)

than authorized limits. Finally, the method was shown to be adapted to the decontamination of real samples previously spiked with pesticides concentrations up to 20 µg L^{-1}.

A very good stability was achieved for 2 wk of continuous use, even using real surface water spiked with the two OPs.

KEY FACTS OF OPS BIO-DETECTION AND BIODETOXIFICATION

- Phosphotriesterase (PTE) is an enzyme able to hydrolyze organophosphate triesters like organophosphate insecticides (OPs).
- The use of PTE immobilized in a column allows efficient and selective removal of OPs from water.
- The PTE column allows the decontamination of real samples of pesticides with concentrations up to 20 µg L^{-1}.
- An acetylcholinesterase (AChE) based sensor allows the determination of neurotoxic compounds including OPs.
- A biosensor based on recombinant AChE immobilized on magnetic micro-beads by nickel-histidine affinity allows the detection of very low concentrations of OPs, the limit of detection is as low as 1.3×10^{-11}M (0.004 µg.L^{-1}). The performance of the biosensor fits with the EU Water Framework Directive.

APPLICATION TO OTHER AREAS OF HEALTH AND DISEASE

The public health impacts of OPs are historic and of current and future importance. Their applications have allowed society to enjoy a wide range of foods in large quantities at low costs. At the same time these compounds are widely used to protect public health from epidemic diseases due to insect vectors. However, they have been associated with undesirable chronic health effects in adult populations and developmental effects in children.

Although some OPs compounds are highly toxic to humans, they generally break down rapidly in the environment and have been rarely found in groundwater. Despite the fact that final products of degradation are generally not toxic, the intermediary oxidized products are often much more toxic than the original OPs. The determination of all degradation products by traditional methods is quite difficult because of the uncertainty about which products to test. Alternatively, the monitoring of drinking water by AChE based biosensors allows the detection of global neurotoxic compounds in water (OPs, carbamate, heavy metals...). These tools are appropriate to protect human health from excess amounts of neurotoxic compounds, as well as diseases due to the accumulation of low amounts of these chemicals.

SUMMARY POINTS

- Biological detoxification columns (Omnifit, 10 mL) were designed by immobilizing 500 IU of PTE on Sepharose Gel 4B (4 mL), they were placed in a flow system.
- Two different OPs were studied, chlorpyrifos and chlorfenvinfos, which are included in the list of priority substances in the field of water policy (decision 2455/2001/EC).
- Biodetection experiments were performed using an amperometric biosensor based on a recombinant AChE immobilized on magnetic microbeads by nickel-histidine affinity. Screen-printed electrodes were used as transducing supports.
- Recombinant AChE based sensors allowed the detection of pesticide concentrations as low as 1.3×10^{-11}M (0.004 μg.L^{-1})
- The developed biosensors were used to monitor the efficiency of the biological detoxification, the method was shown to be adapted to the decontamination of real samples previously spiked with pesticides concentrations up to 20 μg L^{-1}, fulfilling the European regulations for human consumption water.
- The biological detoxification column showed a maximum efficiency during 2 wk in continuous use.

DEFINITIONS OF KEY TERMS

- Acetylcholinesterase (EC 3.1.1.7) is an important enzyme present in vertebrates and insects, it is mainly found at neuromuscular junctions and cholinergic nervous system, where its activity serves to terminate synaptic transmission which hydrolyze the neurotransmitter acetylcholine, producing choline and an acetate group; each molecule of AChE degrades about 25000 molecules of acetylcholine per second. AChE can be genetically modified in order to be more sensitive for some neurotoxin molecules.

- Phosphotriesterase (PTE) (EC 3.1.8.1), also known as organophosphorus hydrolase, aryldialkylphosphatase, and paraoxon hydrolase, is a member of the amidohydrolase superfamily. The natural substrate of this enzyme remains unknown, general substrates are aryldialkylphosphate and H_2O, producing upon hydrolysis a dialkyl phosphate and an aryl alcohol. This enzyme has attracted interest because of its potential use in the detoxification of chemical waste and warfare agents and its ability to degrade agricultural pesticides such as organophosphates. It acts specifically on synthetic organophosphate triesters and phosphorofluoridates.

- Organophosphates are esters of phosphoric acid, many of the most important biochemicals are organophosphates, including DNA and RNA as well as many cofactors that are essential for life. Synthetic organophosphates are also the basis of many insecticides, herbicides, and nerve gases. Organophosphate insecticides (OPs) have become the most widely used insecticides, they are normally used for agricultural, industrial, household and medical purposes. OPs poison insects and mammals by phosphorylation of the acetylcholinesterase enzyme (AChE) at nerve endings. Inactivation of this enzyme results in an accumulation of acetylcholine leading to an overstimulation of the effector organ.

- Screen-printed electrodes (SPEs) are fabricated using think film technology on planar substrates by consecutive depositions of conductive or insulating inks. In biosensor technology, electrochemical transducers are generally made of SPEs in a three-electrode system (working, auxiliary and reference electrodes). This simple and efficient method for producing electrodes has become highly attractive due to its versatility and cost-effectiveness.

- Inhibition-based enzyme sensors: conventional enzyme sensors are based on the catalytic conversion of a substrate in the corresponding product, which is generally determined by electrochemical or optical processes, allowing the direct determination of substrate concentration. In inhibition-based sensors, the target is a reversible or irreversible inhibitor of a specific enzyme, in this case the detection is based on a differential measurement in the presence of enzyme substrate with or without an inhibitor.

ABBREVIATIONS

AChE	:	Acetylcholinesterase
AOP	:	Advanced oxidation process
BuChE	:	Butyrylcholinesterase
CFV	:	Chlorfenvinfos
CNT	:	Carbon nanotube
CPO	:	Chlorpyrifos-oxon
ChE	:	Cholinesterase
CoPC	:	Cobalt phtalocyanine
E	:	Enzyme
GC	:	Gas chromatography
k_i	:	Inhibition constant
Km	:	Michaelis constant
LC	:	Liquid chromatography
OP	:	Organophosphate insecticide
PTE	:	Phosphotriesterase
PX	:	Carbamates or Organophosphates insecticides
TCNQ	:	Tetracyanoquinodiemethane
Vm	:	Maximum velocity

REFERENCES

Aldridge WN. 1950. Some properties of specific cholinesterase with particular reference to the mechanism of inhibition by diethyl p-nitrophenyl thiophosphate (E 605) and analogues. Biochemical J 46: 451–60.

Andreescu S and J-L Marty. 2006. Twenty years research in cholinesterase biosensors: From basic research to practical applications. Biomolecular Eng 23: 1–15.

Aubert SD, Y Li and FM Raushel. 2004. Mechanism for the hydrolysis of organophosphates by the bacterial phosphotriesterase. Biochemistry 43: 5707–5715.

Ballesteros E and MJ Parrado. 2004. Continuous solid-phase extraction and gas chromatographic determination of organophosphorus pesticides in natural and drinking waters. Journal of Chromatography A 1029: 267–73.

Boublik Y, P Saint-Aguet, A Lougarre, M Arnaud, F Villatte, S Estrada-Mondaca and D Fournier. 2002. Acetylcholinesterase engineering for detection of insecticide residues. Protein Eng 15: 43–50.

Deo RP, J Wang, I Block, A Mulchandani, KA Joshi, M Trojanowicz, F Scholz, W Chen and Y Lin. 2005. Determination of organophosphate pesticides at a carbon nanotube/ organophosphorus hydrolase electrochemical biosensor. Anal Chim Acta 530: 185–189.

Geerdink RB, WMA Niessen and UAT Brinkman. 2002. Trace-level determination of pesticides in water by means of liquid and gas chromatography. Journal of Chromatography A, 970: 65–93.

Ghanem E and FM Raushel. 2005. Detoxification of organophosphate nerve agents by bacterial phosphotriesterase. Toxicol Appl Pharmacol 207: 459–470.

Gill I and A Ballesteros. 2000. Degradation of organophosphorous nerve agents by enzyme-polymer nanocomposites: Efficient biocatalytic materials for personal protection and large-scale detoxification. Biotech and Bioeng 70: 400–10.

Guivarch E, N Oturan and MA Otura. 2003. Removal of organophosphorus pesticides from water by electrogenerated Fenton's reagent Environ Chem Lett 1: 165–168.

Holm L and C Sander. 1997. An evolutionary treasure: unification of a broad set of amidohydrolases related to urease. Proteins 28: 72–78.

Hong SB and FM Raushel. 1996. Metal—substrate interactions facilitate the catalytic activity of the bacterial phosphotriesterase. Biochemistry 35: 10904–10912.

Istamboulie G, S Andreescu, J-L Marty and T Noguer. 2007. Highly sensitive detection of organophosphorus insecticides using magnetic microbeads and genetically engineered acetylcholinesterase. Biosens Bioelectron 23: 506–12.

Istamboulie G, D Fournier, J-L Marty and T Noguer. 2009. Phosphotriesterase: A complementary tool for the selective detection of two organophosphate insecticides: Chlorpyrifos and chlorfenvinfos. Talanta 77: 1627–31.

Istamboulie G, R Durbiano, J-L Marty and T Noguer. 2010a. Biosensor-controlled degradation of chlorpyrifos and chlorfenvinfos using a phosphotriesterase-based detoxification column. Chemosphere 78: 1–6.

Istamboulie G, T Sikora, E Jubete, E Ochoteco, J-L Marty and T Noguer. 2010b. Screen-printed poly(3,4-ethylenedioxythiophene) (PEDOT): a new electrochemical mediator for acetylcholinesterase-based biosensors. Talanta 82: 957–961.

Kuster M, M López De Alda and D Barceló. 2006. Analysis of pesticides in water by liquid chromatography-tandem mass spectrometric techniques. Mass Spectrometry Reviews 25: 900–16.

Lacorte S, C Molina and D Barceló. 1993. Screening of organophosphorus pesticides in environmental matrices by various gas chromatographic techniques. Analytica Chimica Acta 281: 71–84.

Masson P, D Josse, O Lockridge, N Viguié, C Taupin and C Buhler. 1998. Enzymes hydrolyzing organophosphates as potential catalytic scavengers against organophosphate poisoning. J Physiol 92: 357–362.

Merone L, L Mandrich, M Rossi and G Manco. 2008. Enzymes with phosphotriesterase and lactonase activities in archaea, Curr Chem Biol 2: 237–248.

Mulchandani A, S Pan and W Chen. 1999. Fiber-optic enzyme biosensor for direct determination of organophosphate nerve agents. Biotechnol Prog 15: 130–134.

Mulchandani P, W Chen, A Mulchandani, J Wang and L Chen. 2001. Amperometric microbial biosensor for direct determination of organophosphate pesticides using recombinant microorganism with surface expressed organophosphorus hydrolase. Biosens Bioelectron 16: 433–437.

Omburo GA, JM Kuo, LS Mullins and FM Raushel. 1992. Characterization of the zinc binding site of bacterial phosphotriesterase. J Biol Chem 267: 13278–13283.

Oturan MA. 2000. An ecologically effective water treatment technique using electrochemically generated hydroxyl radicals for *in situ* destruction of organic pollutants: Application to the herbicide 2,4-D. J Appl Electrochem 30: 475–482.

Raushel FM. 2002. Bacterial detoxification of organophosphate nerve agents. Current Opinion in Microbiology 5: 288–95.

Reigart R and J Roberts. 1999. Recognition and Management of Pesticide Poisonings, US. Environmental Protection Agency, USA.

Sogorb MA and E Vilanova. 2002. Enzymes involved in the detoxification of organophosphorus, carbamate and pyrethroid insecticides through hydrolysis. Toxicol Lett 128: 215–228.

Spadaro JT, I Lorne and V Renganathan. 1994. Hydroxyl radical mediated degradation of azo dyes: evidence for benzene generation. Environ Sci Technol 28: 1389–1393.

5

Biosensors for Endocrine Disruptors: The Case of Bisphenol A and Catechol

Marianna Portaccio,[1,a,2] **Daniela Di Tuoro,**[1,b,2]
Umberto Bencivenga,[3] **Maria Lepore**[1,c,2] **and**
Damiano Gustavo Mita[1,d,2,3]

ABSTRACT

Experimental results obtained with novel biosensors constructed ad hoc for determination of endocrine disruptors in the environment are reported.

Endocrine disruptors are a class of chemicals, mainly of anthropic origin, with the ability of interfering with the normal function of the endocrine system and so inducing severe pathologies in living organisms and in their progeny. Thus, the detection and the determination of endocrine disruptors in ecosystems, food, beverages, and biological samples, as urine and blood are very necessary.

[1] Dipartimento di Medicina Sperimentale, Seconda Università di Napoli, Napoli, Italia.
[a] E-mail: marianna.portaccio@unina2.it
[b] E-mail: daniela.dituoro@libero.it
[c] E-mail: maria.lepore@unina2.it
[d] E-mail: mita@igb.cnr.it
[2] Consorzio Interuniversitario INBB, Sezione di Napoli, Napoli, Italia.
[3] Istituto of Genetica and Biofisica del CNR, Napoli, Italia.
E-mail: benciven@igb.cnr.it

List of abbreviations after the text.

Our interest in endocrine disruptors has focused mainly on Bisphenol A and Catechol and, we here report the characterization of some amperometric biosensors for their determination. Two different types of biosensors have been constructed: one is based on activated carbon rods and the other on carbon paste of different composition. The biorecognition element Laccase or Tyrosinase were separately immobilized. Devices based on modified carbon rods were used in flow injection analysis mode, while the others based on modified carbon paste were used in a batch electrochemical cell. For each biosensor type, different subtypes have been constructed. Sensitivity and detection limit for each of our biosensors have been experimentally determined and compared with the same parameters obtained by other authors with similar biosensors. Moreover we indicate the way by which the performance of our biosensors can be improved.

INTRODUCTION

Endocrine-disrupting chemicals (EDCs) are environmental contaminants, persistent and capable of interfering with endocrine system functions (Sumpter 1998). A major group of EDCs are called xenoestrogens since they exert estrogenic effects on the human body, including interference with reproductive, neurologic, and immunologic functions, and even carcinogenesis (Nilsson et al. 2001). Some of these xenocompounds have a structure similar to estrogens and hence exhibit an affinity for the human estrogen receptors (ERs). The strong similarity between xenoestrogens and estrogens is the main reason for the increase of some diseases, among which it is possible to identify infertility, functional alterations of sexual development and some types of cancer (Mueller 2004). The xenoestrogens list includes some phenolic compounds and, among these, Catechol and Bisphenol A (BPA). Bisphenol A (2,2-bis(4-hydroxyphenyl)propane), is a chemical substance widely employed as a monomer in the production of epoxy resins and polycarbonates, and as an antioxidant in PVC plastics. Epoxy resins are used as coating of inner surfaces of cans containing food and beverages. Polycarbonates are used in the manufacture of plastic containers for food, such as infant feeding bottles and tableware. PVC is also used in a variety of products which come in contact with food, such as cling film used for food packaging. The migration of BPA from epoxy-coated surfaces, polycarbonate plastics and PVC products into food simulants or food has been reported (Lopez-Cervantes and Paseiro-Losada 2003; Nerin et al. 2003; Goodson et al. 2004). In addition, it has been reported that BPA exhibits estrogenic activity in *in vitro* assays at concentrations of 10–25 nM (2–5 ng/mL), competing with estradiol for binding to estrogen receptors (Krishnan et al. 1993). Moreover, using oral administration, studies *in vitro* on mice (vom Saal et al. 1998) have shown that a BPA dose of 2 ng/g of body weight affected the size of reproductive organs of male offspring fed by this substance during pregnancy, and a dose of BPA 20 ng/g of body weight significantly decreased efficiency of sperm production by 20% relative to control males.

Thus, the detection and the determination of phenol and phenol derivatives in ecosystems, food, beverages, and biological samples, as urine and blood, are imperative necessities. For the measurement of the concentration of phenolic compounds many methods, such as colourimetry, gas chromatography, liquid chromatography, capillary electrophoresis and spectrophotometric analysis, have been developed. However, these techniques require tedious procedures for sample preparation and well-trained technicians. This limits the application of these methods in routine measurements and in the screening on-site of these pollutants. For these reasons, the electrochemical detection of phenolic compounds by means of biosensors has received much attention because of their simplicity, high sensitivity and quick detection (Li et al. 2005; Tembe et al. 2006; Zhang et al. 2009). Amperometric biosensors, based on oxidoreductase enzymes, tyrosinase or laccase in particular, have been proved to be sensitive and convenient tools for this purpose (Tillyer and Gobin 1991; Wang et al. 1994; Liu et al. 2000; Wang et al. 2000a; Li et al. 2005). Oxidoreductases catalyse two reactions: an ortho-hydroxylation of phenols to Catechol and a further oxidation of Catechols to ortho-quinones, both in the presence of molecular oxygen, according to the following scheme:

$$\text{Phenol} + O_2 \xrightarrow{\text{oxidoreductase}} \text{Catechol}$$

$$\text{Catechol} + O_2 \xrightarrow{\text{oxidoreductase}} o\text{-quinone}$$

The strong oxidizing power of oxygen makes the overall reaction irreversible. In biosensors the electrochemical reduction of the enzymatic reaction product, the o-quinone, is used as a detection reaction by which an electrical signal proportional to the phenol concentration is obtained.

$$o\text{-Quinone} + H^+ + 2e^- \longrightarrow \text{Catechol}$$

It should be mentioned that the Catechol produced in the electrochemical reaction is also taking place in the enzymatic reaction, so that an "enhancing effect" is observed.

In Fig. 5.1 a scheme of functioning of a tyrosinase-based electrode is reported.

In this chapter we present a survey of some our results obtained with different electrodes types utilizing immobilized laccase or tyrosinase for the determination of Catechol and Bisphenol A. For the Catechol we used laccase immobilized on graphite rods or tyrosinase entrapped in

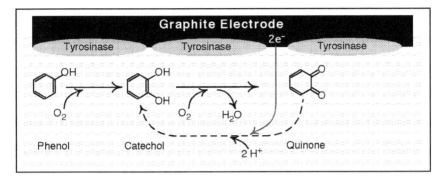

Figure 5.1. Schematic representation of the oxidoreductase reaction at the interface tyrosinase based biosensor and phenol polluted aqueous solution. Origin of the electrical response.

carbon paste with different concentration of thionine, as mediator. For BPA, tyrosinase was entrapped in a carbon paste mixture in the presence or absence of thionine. The electrochemical response of our electrodes is compared, when possible, with the results obtained by other authors. The way in which the performance of our biosensors can be improved is also indicated.

APPARATUS AND METHODS

Regardless of the used enzymes and of the target EDCs, two types of electrodes were employed: the first type, denoted by A, was a carbon rod properly activated to bind the enzyme; the second type, denoted by B, was a carbon paste in which the bioelement was entrapped, in the presence or absence of thionine, which was used as mediator for the enzymatic reaction.

The type A electrode operated in FIA mode, while the type B electrode operated in a batch.

The FIA Electrochemical Cell

The electrochemical cell was a three electrode cell where the enzyme modified graphite electrode, a graphite rod (4 mm in diameter) purchased from Agar Scientific (Agar Scientific Limited, 66a, Cambridge Road Stansted, Essex CM24 8DA, England), acted as a working electrode and the platinum electrode (type M241Pt) as a counter electrode. All measurements were carried out versus an Ag/AgCl reference electrode (type REF321), kept at –100mV versus the working electrode. The platinum and the Ag/AgCl electrodes were purchased from Radiometer Analytical (Radiometer-Analytical. SAS,Villeurbanne CEDEX, Lyon, France). The potential difference was ensured by means of a low current potentiostat/galvanostat model 2059 from Amel (Amel, Milan, Italy) interfaced to a PC through a board (PCI-6221) purchased from National Instruments Corporation (National Instruments, Austin, TX, USA). Continuous flows of the washing buffer solution (0.1M sodium acetate, pH 5.0; T=25°C) or of the mixture containing the Catechol to be determined were injected by means of a peristaltic pump through the electroanalytic cell under the control of an electrovalve from RS Components (RS Components s.p.a., Cinisello Balsamo, Milan, Italy). The injected volume was 200 μL and the electrical response, constituting the output signal from the biosensor, was acquired using the Labview software package, purchased from the National Instrument Corporation (National Instruments, Austin, TX, USA). The software accounted for the value of the background current, which was continuously subtracted from the subsequent value of the measurement. The electrical current produced by the oxidation of the substrate by the immobilized enzyme is proportional to the reaction rate, which, in turn, is function of the substrate concentration.

Three electrode types, indicated by A1, A2 and A3 in dependence of the employed enzyme immobilization method, have been used. The enzyme immobilization phase for all electrode types was preceded by a cleaning operation of the electrode surface using gamma alumina powder, after which the electrode was washed and sonicated in a 5% (v/v) ethanol aqueous solution. Enzyme immobilization on the electrode A1 was carried out by absorption, while it was performed through covalent bond on electrodes A2 and A3. In the latter cases, Hexamethylendiamine (HMDA) was used as a spacer to bind the enzyme to the functional carboxylic groups induced on the graphite electrode by treatment under a potential difference (A2 electrode) or with nitric acid (A3 electrode). The procedure to immobilize the laccase on the activated graphite rods are illustrated in Fig. 5.2. All details on the assemblage of the apparatus and on the type of electrical response of the biosensors are reported in Portaccio (Portaccio et al. 2006).

The Batch Electrochemical Cell

All experiments were conducted in a three electrode electrochemical cell with a volume of 10 mL (0.1M phosphate, pH 6.5, containing 0.1M KCl) with the enzyme modified carbon paste electrode (electrode type B) as the working electrode, the Ag/AgCl electrode as the reference electrode and the platinum wire as an auxiliary electrode. The working electrode was operated at −150 mV and the transient currents were allowed to decay to a steady-state value. A magnetic stirrer and a stirring bar provided the convective transport in the electrolytic cell. Four sub type of B electrodes were used: B1, B2, B3 and B4. The first three electrodes had the same composition: carbon

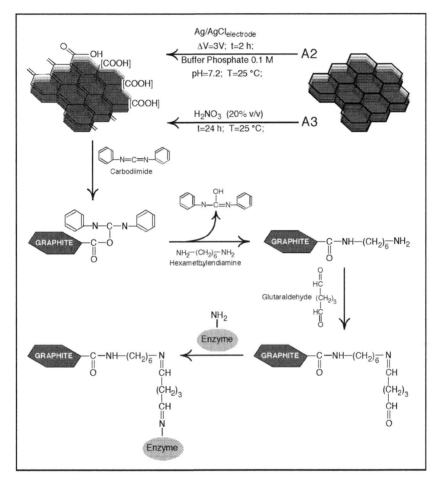

Figure 5.2. Functionalization, activation and enzyme immobilization on carbon rods employed as working electrode (Electrode A2 and A3) in an electrochemical cell operating under FIA mode.

powder, mineral oil and tyrosinase. The differences were in the percentage of tyrosinase, whereas the ratio between the carbon powder and the mineral oil was constant and equal to 1.25. This means that the B1 electrode type was constructed by mixing 2.5% of enzyme, 54.1% of carbon powder and 43.4% of mineral oil; the B2 electrode type was constructed by mixing 5.0% of enzyme, 53% of carbon powder and 42% of mineral oil; the B3 electrode type was constructed by mixing 10.0% of enzyme, 50% of carbon powder and 40% of mineral oil. The resulting pastes were packed into the well of the working electrode constituted by a Teflon tube (3 mm in diameter) with an electrical contact provided by a copper wire.

Concerning the preparation of the B4 electrode type the carbon paste was prepared by hand mixing graphite powder, thionine (when it was the case), mineral oil and tyrosinase in an appropriate weight ratio. The resulting pastes were packed into the well of the working electrode constituted, as usual, by the Teflon tube (3 mm in diameter) with an electrical contact provided by a copper wire. The Teflon tube, this time, was packed with a double layer of carbon paste: an inner layer containing only graphite powder mixed to mineral oil, and an outer layer composed of carbon paste, tyrosinase and thionine (when indicated). The inner layer (about 2/3 of total volume) was prepared by mixing 60% of graphite powder and 40% of mineral oil, for a total weight of 4.8 mg. The outer layer (about 1/3 of the total volume) with a total weight of 2.5 mg, was prepared by a carbon paste obtained from two steps. During the first step, 600 mg of graphite powder are mixed for 10 min with 1mL of thionine (1mM or 10 mM), and the thionine modified powder was put in an oven at 60°C for about 15 hr. In the second step, the thionine/graphite powder was mixed with mineral oil and tyrosinase in the ratio (w/w) of 50:40:10, respectively. For each electrode we used about 0.25 mg of tyrosinase which correspond to about 500 U. To prepare thionine free electrode we followed only the second step, mixing graphite powder, mineral oil and enzyme in the ratio used before. All details on the assemblage of the apparatus and on the type of electrical response of the biosensors are reported in Mita (Mita et al. 2007) and in Portaccio (Portaccio et al. 2010).

RESULTS AND DISCUSSION

FIA Electrochemical Cell Results

As previously reported the FIA electrochemical cell was used to determine with the electrodes of A type the Catechol concentration in aqueous solutions. Once established that the optimum value of the peak current for our biosensors occurs at pH 5.0, in order to obtain calibration curves for each biosensor type experiments were conducted at different Catechol concentration at pH 5.0 and at T= 25°C. The results are displayed in Figs. 5.3A, B and C, where the calibration linear range of each biosensor type is reported as value of the peak current as a function of the Catechol concentration. Figure 5.3A refers to biosensor A1, Fig. 5.3B to biosensor A2, and Fig. 5.3C to biosensor A3, respectively. In each of these figures, the inset represents the electrical response obtained in correspondence of the whole explored Catechol concentration range. The results in Figs. 5.3A, B and C clearly show that: (1) the electrical response for each of the three biosensor types resembles the Michaelis–Menten behaviour; (2) there are remarkable differences in the extension of the calibration curves and in the

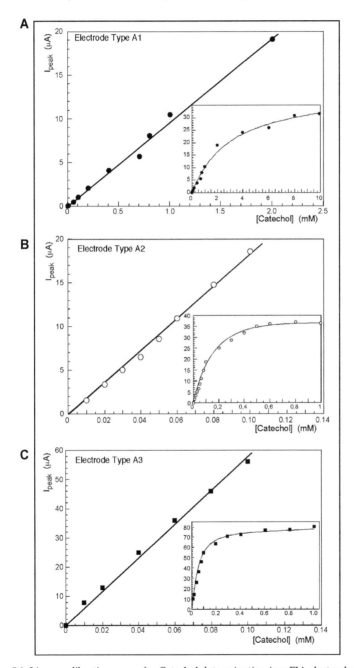

Figure 5.3. Linear calibration range for Catechol determination in a FIA electrochemical cell using biosensors type A1 **(A)**, type A2 **(B)** and type A3 **(C)**. The insets are the reproduction of the biosensors electrical responses for a large Catechol concentration.

sensitivities. The sensitivities, as well known, are the slopes of the calibration curves. In particular we obtained a sensitivity equal to 9.7 µA/mM, 196 µA/mM and 490 µA/mM for A1, A2 and A3 biosensor, respectively. The values for A2 and A3 biosensor types are higher than those obtained by other authors. In fact, Haghighi et al. 2003 obtained for sensitivity a value equal to 68.6 µA/mM, while Freire et al. (2001) and Gomes and Rebelo (2003) reported still lower values equal to 16.1 µA/mM and 0.2 µA/mM, respectively. Concerning the linear range extension, an interval up to 2 mM has been obtained for A1 biosensor while for A2 and A3 biosensors resulted in a range of up to 0.1. Our data show that sensitivies are higher when the laccase is covalently bound to the electrode in comparison to the value obtained with the adsorbed laccase. The opposite is true when the extension of the linear range is considered. When the apparent electrical K_m values, obtained from the Michaelis–Menten curves, are reported as a function of the sensitivities, the results reported in Fig. 5.4 emerge. The data in Fig. 5.4 clearly show that small values of sensitivity correspond to high values of the electrical constant K_m, i.e. to small affinity values. On the contrary, high values of sensitivity correspond to high values of the electrical affinity, i.e. to small values of K_m. It is therefore evident that, in order to design laccase-based biosensors for the determination of phenolic compounds (such as Catechol), one must adapt the immobilization method depending on the required performance: high sensitivity or high extension of the calibration linear range.

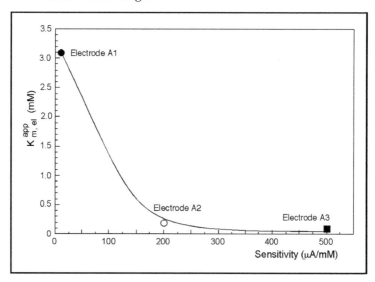

Figure 5.4. Values of the electrochemical affinity constants, calculated from the Michaelis–Menten electrical behaviour of each biosensor (insets in Fig. 5.3), as a function of respective sensitivity.

Batch Electrochemical Cell Results

As previously reported the batch electrochemical cell was used with the electrodes of B type to separately determine the Catechol or Bisphenol A concentrations in aqueous solutions. In particular, with the B1, B2 and B3 electrodes we have studied the electrochemical response to BPA in the concentration range from 3 to 20 μM. The results of this investigation are displayed in Fig. 5.5A, where the electrochemical currents are reported as a function of BPA concentration for three different enzyme percentages. Data in Fig. 5.5A show that: (i) each electrode type displays their own linear behaviour; (ii) at each BPA concentration the electrochemical signals are not linear function of enzyme concentration, but increase faster than enzyme concentration (Fig. 5.5B). This behaviour indicates that under the conditions used to prepare the electrodes, no protein-protein interactions occurred. Similar results have been obtained by other authors (Erdem et al. 2000) who observed a non-linear increase in electrochemical signal when different percentages of horseradish peroxidase were entrapped in a carbon paste electrode. The same non-linear behaviour is observed when the slopes of the straight lines in Fig. 5.5A (i.e. the electrode sensitivities: 10.02 nA μM^{-1} for 2.5% tyrosinase, 22.82 nA μM^{-1} for 5% tyrosinase and 64.48 nA μM^{-1} for 10% tyrosinase) are reported (Fig. 5.5C) as a function of enzyme concentration in the carbon paste. The results above appear to indicate that it is possible to modulate both the electrochemical response of a carbon paste biosensor to BPA and its relative sensitivity on the basis of the amount of immobilized tyrosinase, and consequently on the basis of the carbon paste composition.

The next step in the study involved the modulation of the electrical tyrosinase carbon paste response by different thionine concentration. To this aim the B4 electrode type was used in the electrochemical batch cell it represented. Thionine is an artificial organic dye derivative of phenothiazine, and it is used as a mediator since its formal potential is between 0.08 and −0.25 V, near the redox potential of many biomolecules (Shobha Jeykumari et al. 2007). Cosnier (Cosnier et al. 2001) reported the mediated electrochemical detection of Catechol using thionine covalently bound to a poly(dicarbazole) backbone. Thionine has also been used as polythionine by means of electropolymerization of the monomer. Dempsey (Dempsey et al. 2004) developed a biosensor using thionine electropolymerized on a glassy carbon electrode in the presence of tyrosinase for the detection of synthetic estrogens (BPA) and phenolic compounds.

To study the effects induced by the presence of thionine three different working electrodes have been built, varying the thionine concentrations (0, 1 mM and 10 mM) and leaving constant the amount of graphite powder, mineral oil and tyrosinase. Fig. 5.6A shows the results obtained using the

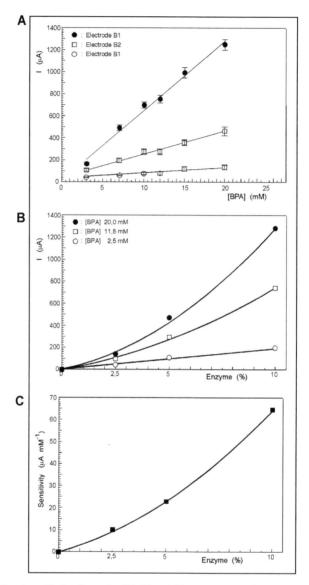

Figure 5.5. Results with the electrodes B1, B2 and B3 operating in a batch electrochemical cell for determination of BPA.

(A):   Linear calibration range for each electrode type: Currents as a function of BPA concentration.

(B):   Currents measured at fixed BPA concentration by means of the three electrodes characterized by the different amount of entrapped enzyme: 2.5% (electrode B1), 5% (electrode B2) and 10% (electrode B3).

(C):   Sensitivity of each biosensor as a function of the amount of entrapped enzyme: 2.5% (electrode B1), 5% (electrode B2) and 10% (electrode B3).

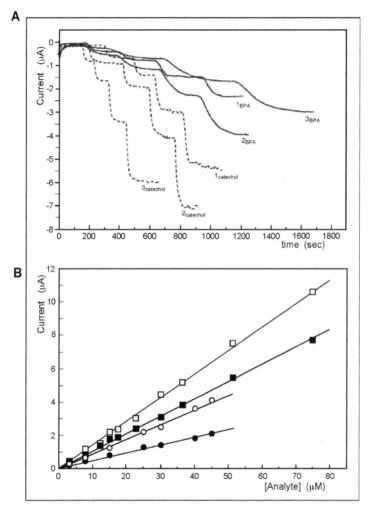

Figure 5.6. Measures with the B4 electrode in the case of Catechol and Bisphenol A.
(A): Repeatability and linearity with the analyte concentration. Current responses to catechol (dotted lines) and to BPA (continuous lines) at thionine free electrode (curves 1), at 1mM thionine electrode (curves 2) and at10mM thionine electrode (curves 3).
(B): Linear calibration curves for Catechol (■,□) and BPA (●,○). The full symbols refer to biosensors constructed with thionine in the carbon paste; the empty symbols refer to biosensors constructed without thionine in the carbon past.

three electrodes subjected to a work potential of −200 mV when Catechol was used as analyte. Once the background current was measured, four injections of standard amounts of Catechol were added: the first two equal to get a final concentration of 7.5 µM, for each injection, to assess the repeatability of the biosensor response. The third and the fourth injection were to reach a final concentration of 30 µM and 52.5 µM, respectively, in order to assess the proportionality of the electrical signal to the analyte concentration. Repeatability and proportionality were clearly shown for each of the three electrodes. As expected, equal injections give signals equal within 3% and proportional electrode prepared with 1mM thionine produces higher current values than those observed with the thionine-free electrode or those obtained with the electrode prepared with 10mM thionine. The average response time was about 100 s. Analogous results (still in Fig. 5.6A) have been obtained when BPA was used as analyte in the same concentration used before for Catechol. In this case the average response time was about 200 s. Comparison among curves in Fig. 5.6A indicates that: (1) the electrical currents in the presence of Catechol are higher than those measured with BPA, Catechol being a better substrate of tyrosinase enzyme as reported in literature (Andreescu and Sadik 2004; Mita et al. 2007); (2) the electrodes modified with thionine give higher responses, demonstrating the electrocatalytic properties of thionine towards the reaction product of BPA with tyrosinase.

Having ascertained that thionine increases the biosensors response and that the highest increase is obtained at the concentration of 1 mM, we prepared two electrodes one with 1mM thionine and the other thionine-free in order to obtain linear calibration curves for both analytes. In Fig. 5.6B the responses of the two biosensors are displayed as a function of the analyte concentration. In the case of Catechol for both electrodes the detection limit is 0.15 µM, calculated according to the standard definition $S/N = 3$, where S is the current signal and N is the noise. The linear range is from 0.15 to 75 µM, while the sensitivity is 139.6 nA/ µM or 104.4 nA/ µM for 1 mM thionine (□) or thionine-free (■) electrode, respectively. It is worth mentioning that the sensitivity values obtained for our biosensors are higher than those reported in the literature for Catechol: 114 nA/ µM (Li et al. 2006), 70.2 nA/µM (Rajesh et al. 2004) or 26.23 nA/ µM (Wang and Dong 2000).

In Fig. 5.6B we have also reported (as a function of the analyte concentration) the responses of the two biosensors when BPA was used as analyte. For both the electrodes the detection limit is 0.15 µM and the linear range is from 0.15 to 45 µM, while the sensitivity is 85.4 nA/ µM or 51.1 nA/ µM for 1 mM thionine (○) and thionine-free (●) electrode, respectively. Also in the case of BPA, it is worth mentioning that the sensitivity values displayed by our biosensors are higher than those reported in the literature for BPA by Andreescu (Andreescu and Sadik 2004) (S = 20.91 nA/µM) or by Dempsey (Dempsey et al. 2004) (S = 0.4 nA/µM), who also used thionine as an electrochemical mediator or also those reported in the literature for phenol compounds: 17.1 nA/µM (Rajesh et al. 2004) or 2.42 nA/µM (Wang et al. 2000).

The intraelectrode repeatability never exceeded the 2% (n = 5) while the interelectrode reproducibility was around 7%. Concerning the stability, the carbon paste electrode was stable for about 1 mon of working. The electrodes were discarded when their response to a Catechol solution of 5µM decreased 7% relative to the initial one.

Before concluding we describe an experiment in which the functioning of our B4 thionine-based biosensor is validated during a bioremediation process of an aqueous solution polluted by BPA. 10 mL of 1 mM BPA aqueous solution were allowed to react in a bioreactor with a membrane on which laccase was immobilized. The immobilization method and the modus operandi of bioreactor are illustrated in Mita et al. 2009. At zero time 200µL were withdrawn from the reaction vessel for subsequent analysis. 100 µL of those were diluted in 10 mL of buffer solution and put in contact with the biosensor, while 25 µL were processed in the HPLC. Every 10 min the same operation was repeated until 100 min. In Fig. 5.7 the average results of five experiments are reported. Data in the figure show: (i) the time decrease of the current intensity measured by the biosensor during the bioremediation process (continuous line and right scale); (ii) the time decrease of BPA concentration (left scale) as measured by HPLC (o) and as obtained from the respective calibration curve reported in Fig. 5.6B. It is interesting to note how: (i) a reduction of 29% in the current value corresponds to a reduction of 25% in the concentration value; and (ii) the BPA concentration values

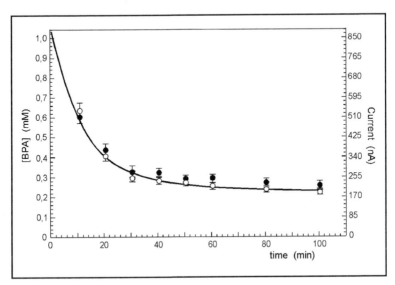

Figure 5.7. Time decrease of BPA concentration during experiments of bioremediation with laccase immobilized. Measure of BPA concentration were simultaneously done by Biosensor B4 (right scale and ●) and by HPCL (left scale and ○).

measured by HPLC are very similar to those obtained from the calibration curve. The conclusion is that the functioning of our thionine-tyrosinase based biosensor appears to be excellent.

CONCLUSIONS

Determination of Catechol and Bisphenol A concentrations in aqueous systems has been carried out with seven types of different biosensors constructed ad hoc. As carriers for enzyme immobilization functionalized carbon rods or activated carbon paste of difference composition have been used. Laccase or tyrosinase, in the presence or absence of thionine, were used as biological recognition elements. For all the biosensors types linear calibration curves were obtained. The electrical parameters of our biosensors, i.e. sensitivity, extension of the linear range and detection limit, resulted comparable to or, in some cases, better than those found in the literature.

To end the B4 biosensor has been successful applied in following an experiment of bioremediation of water polluted by BPA. No significant differences were found when the measure of the decrease of BPA during the time was performed with biosensors or by the means of an HPLC.

APPLICATIONS TO OTHER AREAS OF HEALTH AND DISEASE

Other specific areas of interest are: Endocrinology, General and Comparative Endocrinology, Toxicology, Ecotoxicology, General Pathology, Food Science, Packaging, Immunology, Haematology. The reasons for this interest are mainly because Bisphenol A and Catechol are xenoestrogens of phenolic origin widely present in the environment and in living organisms. For example BPA and other endocrine disruptors have been found in biological samples, blood and urine, and also in people apparently healthy. Endocrine disruptors reach the humans mainly through diet and also at low concentrations induce severe pathologies ranging from reproduction to cancer, from neurological and cardiac disorders to infertility. The molecular mechanisms by which these chemicals act on living organisms are based on their ability of interfering with the endocrine system and for this reason they are named endocrine disruptors chemicals (EDCs). Just to give an example, we recently published (Signorile et al. 2010) the occurrence of endometriosis in mice exposed to Bisphenol A during their prenatal life and lactation.

On the basis of these considerations it is important to determine the EDCs in the environment and in biological fluids, such as urine and blood, of living organisms. Classical analytical methods (HPLC, GC-MS, and HPLC-

MS/MS) are expensive, require skilled operators and are time consuming for samples treatment. On the contrary, the employment of biosensors is becoming a competitive measurement technique either in the environment or in living organisms. Few papers have been published on the construction of biosensors for Catechol and Bisphenol A.

KEY FACTS OF BIOSENSORS AND ENDOCRINE DISRUPTORS

- The release of some chemical pollutants such as pesticides, flame retardants, alkylphenols, polychlorinated biphenyls, phthalates and metals into the environment has increased in recent years.
- Some of these substances are called "endocrine disruptors" (EDCs) due to their ability to interfere with hormonal activity.
- EDCs are harmful even at very low doses. In both animals and humans, these compounds lead to increased incidence of endocrine-related cancers, increased risk of cardiovascular diseases, reduced fertility and changes in developmental processes.
- Foetuses and children are living organism that are mainly at risk.
- Endocrine disruptors reach living organisms through the air, soil, water and food.
- The detection and the determination of phenol and phenol derivatives in ecosystems, food, beverages, and biological samples, as urine and blood, are of great importance.
- The EDCs detection is currently done using classic analytical techniques such as gas chromatography (GC) and HPLC. Mass spectrometry is also considered as a useful technique. However, all these technologies are expensive and time-consuming.
- Electrochemical biosensors, on the contrary, represent a rapid and less expensive method and have the advantage of high sensitivity, potential for miniaturization, and the possibility of *in situ* analysis.

DEFINITIONS

- *Endocrine Disruptors Chemicals*: An endocrine disruptor is a synthetic chemical product, when absorbed into the body, mimics or blocks the action of hormones, and disrupts the body's normal functions. This disruption takes place altering normal hormone levels, halting or stimulating the production of hormones, or changing the way by which hormones travel through the body, thus affecting the functions controlled by these hormones. Chemicals known as human endocrine disruptors include diethylstilbesterol (the drug DES), dioxin, PCBs,

DDT, and some other pesticides. Many chemicals, particularly pesticides and plasticizers, are suspected endocrine disruptors on the basis of studies on animal models.

- *Bisphenol A*: Commonly abbreviated as BPA, is an organic compound with two phenol functional groups. It is used to make polycarbonate plastic and epoxy resins, along with other applications. The estrogenic activity of BPA is known since the mid 1930s. Concerns about the use of Bisphenol A in consumer products were regularly reported since 2008 after several governments issued reports questioning its safety. A 2010 report from the United States Food and Drug Administration (FDA) raised further concerns regarding exposure of foetuses, infants and young children. In September 2010, Canada became the first country to declare BPA as a toxic substance. In the European Union and in Canada the employment of BPA in polycarbonate baby bottles has been recently banned.

- *Catechol*: Known as pyrocatechol or 1,2-dihydroxybenzene, is an organic compound with the molecular formula $C_6H_4(OH)_2$. It is the *ortho* isomer of three isomeric benzenediols. This colourless compound occurs naturally in trace amounts. About 20 million kg are produced annually, mainly as a precursor for the production of pesticides, flavours, and fragrances.

- *Laccase*: Laccases are copper-containing oxidase enzymes that are found in many plants, fungi, and microorganisms. The copper is bound in several sites: Type 1, Type 2, and/or Type 3. The ensemble of types 2 and 3 copper is called a trinuclear cluster Laccase and acts on phenols and similar molecules, performing a one-electron oxidation, which remain poorly defined. It has been proposed that laccases play a role in the formation of lignin by promoting the oxidative coupling of lignols, a family of naturally occurring phenols. Laccases can be polymeric, and the enzyme active form can be a dimer or trimer. Other laccases, such as ones produced by the fungus *Pleurotus ostreatus*, play a role in the degradation of lignin, and can therefore be included in the broad category of ligninases.

- *Tyrosinase*: Known also as monophenol monooxygenase, tyrosinase is an enzyme that catalyzes the oxidation of phenolic compounds such as tyrosine. Tyrosinase is a copper-containing enzyme present in plant and animal tissues and catalyzes the production of melanin and other pigments from tyrosine by oxidation. The typical reaction is similar to that occurring during the blackening of a peeled or sliced potatoes exposed to air. In humans, the tyrosinase enzyme is encoded by the *TYR* gene.

- *Flow Injection Analysis*: This is an approach to chemical analysis that is accomplished by injecting a plug of sample into a flowing carrier stream. A sample (analyte) is injected into a carrier solution which mixes through radial and convection diffusion with a reagent for a period of time (depending on the flow rate and the coil length and diameter) before the sample passes through a detector to waste. A peristaltic pump is the commonly used pump in FIA instruments. FIA can be used for both medical and industrial analyses.

SUMMARY POINTS

- Compared with the large number of proposed biosensors for glucose determination, the production of biosensors for endocrine disruptors is limited.
- The reasons for this probably are based on the little knowledge and awareness on the documented harmful effects induced by endocrine disruptors, also at low concentrations, in living organisms.
- To fill this gap in this chapter the construction of two types of biosensors, able to separately determine low concentrations of Bisphenol A and Catechol, is presented.
- Several biosensors, different for their modus operandi, have been prepared.
- One biosensor type operates in FIA mode, the other type in batch.
- Laccase and Tyrosinase have been used as biorecognition elements.
- Carbon rods or carbon paste formulations have been used for enzyme immobilization.
- Sensitivity and detection limit have been estimated for each biosensor type and compared, when possible, with analogous biosensors.

ABBREVIATIONS

BPA	:	Bisphenol A
DDT	:	Dichlorodiphenyltrichloroethane
DES	:	Diethylstilbestrol
EDCs	:	Endocrine Disruptors Chemicals
ERs	:	Restrogen Receptors
FDA	:	Food and Drug Administration
FIA	:	Flow Injection Analysis
HMDA	:	Hexamethylenediamine
PCBs	:	Polychlorinated biphenyls
PVC	:	Polyvinyl Chloride

REFERENCES

Andreescu S and OA Sadik. 2004. Correlation of Analyte Structures with Biosensor Responses Using the Detection of Phenolic Estrogens as a Model Anal Chem 76: 552–560.

Cosnier S, S Szunerits, RS Marks, JP Lellouche and K Perie. 2001. Mediated electrochemical detection of catechol by tyrosinase-based poly(dicarbazole) electrodes. J Biochem Biophys Meth 50: 65–77.

Dempsey E, D Diamond and A Collier. 2004. Development of a biosensor for endocrine disrupting compounds based on tyrosinase entrapped within a poly(thionine) film. Biosens Bioelectron 20: 367–77.

Erdem A, A Pabbuccuoglu, B Meric, H Kerman and M Ozsoz. 2000. Electrochemical Biosensor Based on Horseradish Peroxidase for the Determination of Oxidizable Drugs. Turk. J. Med. Sci. 30: 349–354.

Freire RS, N Duran and LT Kubota. 2001. Effects of fungal laccase immobilization procedures for the development of a biosensor for phenol compounds. Talanta 50: 681–685.

Gomes SA and MJ Rebelo. 2003. A new laccase biosensor for polyphenols determination. Sensors 3: 166–175.

Goodson A, H Robin, W Summerfield and I Cooper. 2004. Migration of Bisphenol A from can coatings--effects of damage, storage conditions and heating. Food Addit Contam 21: 1015–26.

Haghighi B, L Gorton, T Ruzgas and LJ Jonsson. 2003. Characterization of graphite electrodes modified with laccase from *Trametes versicolor* and their use for bioelectrochemical monitoring of phenolic compounds in flow injection analysis. Anal Chim Acta 487: 3–14.

Krishnan AV, P Stathis, SF Permuth, L Tokes and D Feldman. 1993. Bisphenol-A: an estrogenic substance is released from polycarbonate flasks during autoclaving. Endocrinol 132: 2279–86.

Li N, MH Xue, H Yao and JJ Zhu. 2005. Reagentless biosensor for phenolic compounds based on tyrosinase entrapped within gelatine film. Anal Bioanal Chem 383: 1127–32.

Li YF, ZM Liu, YL Liu, YH Yang, GL Shen and RQ Yu. 2006. A mediator-free phenol biosensor based on immobilizing tyrosinase to ZnO nanoparticles. Anal Biochem 349: 33–40.

Liu Z, B Liu, J Kong and J Deng. 2000. Probing trace phenols based on mediator-free alumina sol-gel-derived tyrosinase biosensor. Anal Chem 72: 4707–12.

Lopez-Cervantes J and P Paseiro-Losada. 2003. Determination of Bisphenol A in, and its migration from, PVC stretch film used for food packaging. Food Addit Contam 20: 596–606.

Mita DG, A Attanasio, F Arduini, N Diano, V Grano, U Bencivenga, S Rossi, A Amine and D Moscone. 2007. Enzymatic determination of BPA by means of tyrosinase immobilized on different carbon carriers. Biosens Bioelectron 23: 60–5.

Mita DG, N Diano, V Grano, M Portaccio, S Rossi, U Bencivenga, I Manco, C Nicolucci, M Bianco, T Grimaldi, L Mita, T Georgieva and T Godjevargova. 2009. The process of thermodialysis in bioremediation of waters polluted by endocrine disruptors. J Mol Catal B: Enzym 58: 199–2007.

Mueller SO. 2004. Xenoestrogens: mechanisms of action and detection methods. Anal Bioanal Chem 378: 582–7.

Nerin C, C Fernandez, C Domeno and J Salafranca. 2003. Determination of potential migrants in polycarbonate containers used for microwave ovens by high-performance liquid chromatography with ultraviolet and fluorescence detection. J Agric Food Chem 51: 5647–53.

Nilsson S, S Makela, E Treuter, M Tujague, J Thomsen, G Andersson, E Enmark, K Pettersson, M Warner and JA Gustafsson. 2001. Mechanisms of estrogen action. Physiol Rev 81: 1535–65.

Portaccio M, S Di Martino, P Maiuri, D Durante, P De Luca, M Lepore, U Bencivenga, S Rossi, A De Maio and DG Mita. 2006. Biosensors for phenolic compounds: the Catechol as substrate model. J Mole Catal B: Enzym 41: 97–102.

Portaccio M, D Di Tuoro, F Arduini, M Lepore, DG Mita, N Diano, L Mita and D Moscone. 2010. A thionine-modified carbon paste amperometric biosensor for Catechol and Bisphenol A determination. Biosens Bioelectron 25: 2003–8.

Rajesh A, W Takashima and K Kaneto. 2004. Amperometric tyrosinase based biosensor using an electropolymerized PTS-doped polypyrrole film as an entrapment support. React Funct Polym 59: 163–169.

Shobha Jeykumari DR, SA Ramaprabhu and S Sriman Narayanan. 2007. Thionine functionalized multiwalled carbon nanotube modified electrode for the determination of hydrogen peroxide. Carbon 45: 1340–1353.

Signorile PG, EP Spugnini, L Mita, P Mellone, A D'Avino, M Bianco, N Diano, L Caputo, F Rea, R Viceconte, M Portaccio, E Viggiano, G Citro, R Pierantoni, V Sica, B Vincenzi, DG Mita, F Baldi and A Baldi. 2010. Pre-natal exposure of mice to Bisphenol A elicits an endometriosis-like phenotype in female offspring. Gen Comp Endocrinol 168: 318–25.

Sumpter JP. 1998. Xenoendorine disrupters—environmental impacts. Toxicol Lett 102–103: 337–42.

Tembe S, M Karve, S Inamdar, S Haram, J Melo and SF D'Souza. 2006. Development of electrochemical biosensor based on tyrosinase immobilized in composite biopolymeric film. Anal Biochem 349: 72–77.

Tillyer CR and PT Gobin. 1991. The development of a Catechol enzyme electrode and its possible use for the diagnosis and monitoring of neural crest tumours. Biosens Bioelectron 6: 569–73.

vom Saal FS, PS Cooke, DL Buchanan, P Palanza, KA Thayer, SC Nagel, S Parmigiani and WV Welshons. 1998. A physiologically based approach to the study of Bisphenol A and other estrogenic chemicals on the size of reproductive organs, daily sperm production, and behavior. Toxicol Ind Health 14: 239–60.

Wang B and S Dong. 2000. Organic-phase enzyme electrode for phenolic determination based on a functionalized sol–gel composite. J Electroanal Chem 487: 45–50.

Wang B, J Zhang and S Dong. 2000a. Silica sol-gel composite film as an encapsulation matrix for the construction of an amperometric tyrosinase-based biosensor. Biosens Bioelectron 15: 397–402.

Wang J, L Fang and D Lopez. 1994. Amperometric biosensor for phenols based on a tyrosinase-graphite-epoxy biocomposite. Analyst 119: 455–8.

Zhang J, J Lei, Y Liu, J Zhao and H Ju. 2009. Highly sensitive amperometric biosensors for phenols based on polyaniline-ionic liquid-carbon nanofiber composite. Biosens Bioelectron 24: 1858–63.

6

Biosensors for Detection of Heavy Metals

Pavlina Sobrova,[1,a] Libuse Trnkova,[2]
Vojtech Adam,[1,b] Jaromir Hubalek[3] and
Rene Kizek[1,c,*]

ABSTRACT

Current development of miniature analytical instruments for detecting heavy metals in various samples is very dynamic. The most interesting representatives of these instruments are biosensors, which combine the physico-chemical transducer with biological components. Biosensors operate on different principles and their preparation can lead to the acquisition of tools with high sensitivity and selectivity against individual heavy metal ions or their mixtures. In this chapter, various types of biosensors divided according to their biological components for the detection of heavy metals ions are discussed. Generally, biosensors for heavy metals detection can be divided into two groups, based on their biological part: i) protein biosensors and ii) biosensors based on nucleic acids. Biosensors using nucleic acids represent a new and

[1]Mendel University in Brno, Department of Chemistry and Biochemistry, Zemedelska 1, 61300 Brno, Czech Republic.
[a]E-mail: pavlina.sobrova@seznam.cz
[b]E-mail: vojtech.adam@mendelu.cz
[c]E-mail: kizek@sci.muni.cz
[2]Masaryk University, Brno, Department of Chemistry; E-mail: libuse@chemi.muni.cz
[3]University of Technology in Brno, Department of Microelectronics;
E-mail: hubalek@feec.vutbr.cz
*Corresponding author

List of abbreviations after the text.

quickly developing branch. Protein biosensors can be divided into enzymatic and affinity ones, while enzymatic biosensors are based on activation or inhibition of enzymes activity, the affinity biosensors use specific antibodies. Affinity biosensors using immunodetection belong to the other method for metal ion determination, which—in comparing with traditional detection methods - brings certain advantages, such as high sensitivity and selectivity. They are theoretically useful for other metal complexes, where the antibody can be prepared. Biosensors belong to the most powerful bioanalytical tools and represent the future in development of instruments which can be used in medicine and pharmaceutical research, environment and BioDefence techniques and in monitoring of food quality and safety.

INTRODUCTION

Heavy metals belong to the group of substances, which on one hand include chemical elements beneficial and essential for organism function but on the other hand elements also with an unascertained biological function which predominantly cause toxic effects in many organisms. Numerous research facilities aim their attention at direct effects of heavy metals in various species. The toxicity of heavy metals is related to the bonding possibility with different types of biomolecules, which alter their prevalent biological function and, thus, damage life-sustaining biochemical processes. As a result, eventually death of the organism may be observed.

Homeostasis, transcription and a lot of other processes are impaired by the presence of toxic metal ions. Plants are able to survive under metal-ions toxic conditions thanks to the number of defence mechanisms including synthesis of phytochelatins (Supalkova et al. 2007). Bacteria, invertebrates and vertebrates protect themselves against metal ions via synthesis of low molecular mass proteins rich in cysteine called metallothioneins (Fig. 6.1). The concentration of metal ions in the environment must be monitored since they are able to accumulate in organisms through the food chain. Therefore, finding of rapid, accurate and robust methods, protocols and instruments for their detection represents a real problem. The development of new instruments, able to carry out *in situ* and on-time analysis with sensitivity close to laboratory equipment, together with their miniaturization is of great interest. Electrochemistry—due to high sensitivity—offers numerous methods for metal ions detection. Possibility to miniaturize the whole detection system, which is then easily portable and able to carry out *in situ* analysis, is the second reason to consider electrochemistry. Last but not least, the combination of a physico-chemical transducer as an electrode with biological substance is the third reason to choose electrochemistry. In this way, a biosensor can be designed (Fig. 6.2), which next to the other above mentioned advantages of electrochemistry, brings to the analytical system

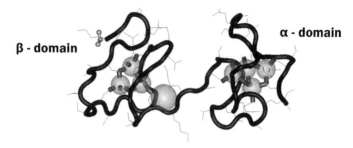

Figure 6.1. Structure of metallothionein. The N-terminal part of the protein is marked as α-domain, which has three binding sites for divalent ions. β-Domain (C-terminal part) has the ability to bind four divalent ions of heavy metals. Due to the property of MT being metal-inducible and, also, due to their high affinity to metal ions, homeostasis of heavy metal levels is probably their most important biological function.

Color image of this figure appears in the color plate section at the end of the book.

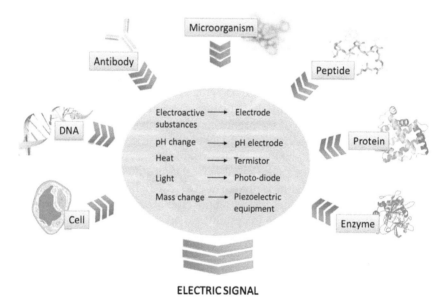

Figure 6.2. Principle of biosensor function. Bioreceptor (e.g. enzyme, peptide, antibody) is properly bond to physico-chemical transducer (electrodes, thermistor, piezoelectric device), which is able to detect the change in heat, light and mass and others.

the heightened selectivity factor (Kizek et al. 2003). The scheme of design of the electrochemical biosensor is shown in Fig. 6.3. Other possibilities of heavy metals biosensing are represented by optical materials in optode form and surface analysis or detection of temperature, pH or weight changes.

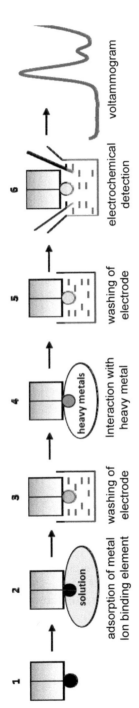

Figure 6.3. Scheme of adsorptive transfer stripping technique used for detection of heavy metals; **(1)** renewing of the hanging mercury drop electrode (HMDE) surface; **(2)** adsorbing of metal ion binding element (e.g. MT, phytochelatins) in a drop solution onto the HMDE surface; **(3)** washing electrode in sodium chloride (0.5 M, pH 6.4) at open circuit; **(4)** interaction of heavy metal (cadmium and/or zinc) in a drop solution with the protein modified HMDE surface at open circuit; **(5)** washing electrode in sodium chloride (0.5 M, pH 6.4); **(6)** measurement of MT by differential pulse voltammetry in 0.5 M sodium chloride, pH 6.4.

BIOSENSORS FOR HEAVY METALS DETERMINATION

Generally, biosensors for heavy metals detection can be divided into two groups based on their biological part as i) protein biosensors and ii) biosensors based on nucleic acids. Biosensors using nucleic acids are a new and rapidly developing branch. The very first publication on this issue focused on lead (II) ions (Liu and Lu 2003). Protein biosensors can be divided into enzymatic and affinity ones, which include antibodies-based biosensors. A short overview of each group and its application for heavy metals is given in the following sections.

Enzymatic Biosensors

Many families of enzymes such as oxidases, dehydrogenases, phosphatases, kinases and ureases have been used for heavy metals detection. The detection is based on activation or inhibition of enzyme activity. Detected heavy metal ion activates the enzyme in case the ion is a part of the enzyme structure, or inhibits it when the ion is able to bond to the active centre of employed enzyme and, thus, inactivates it (Verma and Singh 2005). The biosensor for zinc (II) ions detection based on phosphatases alkali activation has been developed because this ion is a component of the enzyme active centre. The whole system was implemented in micro-injected fluid analysis coupled with calorimetric sensor. Enzyme was covalently immobilized and the authors declare that they were able to determine zinc (II) ions within the range from micromolar to milimolar concentration. The response time was 3 min, which is very important for measurements of larger set of samples. Furthermore, the authors achieved the notable stability of the biosensor, which is very important for analysis of samples without losing the sensitivity and selectivity for a period of longer than 2 mon (Satoh 1991). Biosensors for heavy metal ions determination based on inhibition of enzyme activity use more than one enzyme compared to those based on activation. Oxidases and dehydrogenases belong to the most commonly used enzymes in these types of biosensors. These enzymes are immobilized due to reticulate gelatinous film or affinity interaction with a special type of membrane. For mercury (II) ions determination, L-glycerol phosphate oxidase coupled with Clarc electrode was used; the detection limit 20 µM was achieved. The biosensor was able to regenerate using ethylenediaminetetraacetic acid (EDTA) and dithiothreitol (Gayet et al. 1993). The same authors used pyruvate monoamino oxidase to compare it with the previous enzymes and achieved three times lower detection limit (50 nM mercury (II) ions). It was also possible to regenerate the biosensor by the same chemicals (Gayet et al. 1993). Furthermore, the biosensor consisting of enzymatic system including L-lactate dehydrogenase and L-lactate oxidase as the substance non-sensitive for heavy metals—was developed and proved to be a good choice for various metal ions detection. The detection limits were as follows: 1.0 µM $HgCl_2$, 0.1 µM $AgNO_3$, 10 µM $CdCl_2$, 10 µM $ZnCl_2$, 50 µM Pb $(CH_3COO)_2$ and 250 µM $CuSO_4$ (Gayet et al. 1993; Verma and Singh 2005). This system was also applied for detection of Hg(II), Ag(I), Pb(II), Cu(II) and Zn(II) ions, where Fennouh et al. obtained twice the lower limit detection as compared with the previous study (Fennouh et al. 1998). It was possible to regenerate the biosensor in mixture of EDTA, KCN and dithiothreitol. In addition, inhibitive influence of chromium (III) ions to L-lactate dehydrogenase, hexokinase and pyruvate kinase was utilized for chromium (III) sensitive biosensors. Moreover, various interferences including other metal ions were used in this study and the results were evaluated using artificial neural networks (Cowell et al. 1995). The specific inhibition of peroxidase by mercury (II) ions was observed after enzyme immobilization in chitosan. The attained concentration interval was from 0.02 to 1000 µM Hg(II) (Shekhovtsova et al. 1997). Other large groups of enzymatic biosensors are based on urease. Optical biosensor based on urease immobilized on glass pores was developed for mercury (II) ions determination. However, the concentration interval, in which the biosensor operates, was only from one to 10 µM (Andres and Narayanaswamy 1995). One of the one-shot approaches using urease was established on combination with ammonia sensitive optode and ammonium ions sensitive optode. Limits of detection were as follows: Ag (I) 0.18 µM, Hg(II) 0.35 µM, and Cu(II) 3.94 µM. The study also showed that the above mentioned metals exhibited synergic effects to inhibition, which was confirmed by the highest inhibition of metal ions mixture in comparison with single metals (Preininger and Wolfbeis 1996). Ion-sensitive field-effect transistors (ISFETs) in combination with urease were used besides optodes. Based on the suggested system the detection limits for Ag(I), Hg(II) and Cu(II) were down to units of µM. Furthermore, the authors succeeded in finding a way to modify the specific biosensor to be sensitive to Hg(II) only. The suggested method employs NaI as a masking agent for Ag(I) and EDTA for Cu(II) (Volotovsky et al. 1997). Urease inhibition by mercury (II) ions was also studied using the potentiometric biosensor (Krawczyk et al. 2000). Interaction between urease and nickel (II) ions is another promising way to modify the biosensor for heavy metals ion detection (Verma and Singh 2006). Recently, it was shown that urease system and glutamate dehydrogenase may be employed for mercury (II), copper (II), cadmium (II) and zinc (II) ions detection. Amperometric determination was used for the suggested method (Rodriguez et al. 2004).

Affinity Biosensors

The wide spectrum of metal-binding proteins from naturally occurring to artificial ones, prepared by using protein engineering, which is mostly specific for one metal ion, is used for non-enzymatic or affinity based biosensors employed for heavy metal ions determination. Heavy metal binding proteins as SmtA metallothionein, MerR regulation protein, MerP periplasmatic protein and phytochelatin AC20 are mostly used for biosensors construction for different heavy metal ions, e.g. mercury (II), copper (II), cadmium (II), zinc (II) and lead (II) in wide concentration range from fM to mM. These biosensors have good sensitivity and selectivity, and also acceptable stability time (approximately 2 wk) besides wide concentration intervals (Castillo et al. 2004). Biosensors based on synthetic phytochelatin and electrochemical capacity detection were successfully used for Hg(II), Cd(II), Pb(II), Cu(II), and Zn(II) ions determination in concentration from 100 fM to 10 mM. Biological component was regenerated in EDTA with 15 d stability (Bontidean et al. 2003). Moreover, phytochelatin 2 was used for cadmium (II) and lead(II) ions determination (Adam et al. 2005). The other approach used in biosensing of heavy metal ions is fusion of SmtA metallothionen from nostoc (families of cyanobacteria) with glutathione-S-transferase. Such modified metallothionein demonstrated wide selectivity to heavy metals (Zn(II), Cd(II), Cu(II) and Hg(II)) with high sensitivity up to fM. Glutathione-S-transferase-SmtA electrode was based on electric capacity determination and it was regenerated with EDTA and stored for 16 d (Corbisier et al. 1999). Rabbit metallothionein was successfully employed as a biological agent of biosensor for cadmium (II) and zinc (II) ions (Adam et al. 2007b), palladium (II) ions (Adam et al. 2007a), silver (I) ions (Krizkova et al. 2010), and cisplatin (Huska et al. 2009) determination. Sensitive biosensor for *in situ* detection of Cu(II) based on fluorescently labelled human carbon anhydrase II and optode, was developed by Zeng et al. Detection limit of this sensor was 0.1 pM. The disadvantage of this biosensor was its short stability, which was only 12 hr (Zeng et al. 2003). Changela et al. reported remarkable stability of biosensors using CueR protein, *E. coli* transcription activator, on copper (II) ions determination. These biosensors were able to detect the analyte at concentration of 10^{-21} molar, which is on the level of molecules (Changela et al. 2003). Mouse metallothionein was employed as a fluorescent agent of biosensor for cadmium (II) ions determination (Varriale et al. 2007). Moreover, mutant of green-fluorescent protein (GFP) called BFPms1 was prepared using protein engineering, which was employed for metal ion determination based on fluorescent characteristics change. Zn(II) and Cu(II) were bonded with this protein, when the zinc(II) ions bond caused the conformation change resulting in increasing fluorescent intensity while the copper(II) ions bond scavenged the fluorescent of given mutant protein (Barondeau et al. 2002). Immunodetection belongs to other method for metal ion determination, which compared to traditional detection methods has some advantages such as high sensitivity and selectivity. Such biosensors are theoretically useful for other metal complexes, where it is possible to prepare the antibody (Verma and Singh 2005). Monoclonal antibodies were prepared for EDTA complex with cadmium (II), mercury (II), copper (II), nickel (II), lead (II), cobalt (II), and silver (I) ions. Prepared antibodies had the highest affinity to Cd(II) with detection limit 100 µM (Blake et al. 1998). Other authors prepared monoclonal antibodies for this complex determination with three times lower detection limit (Khosraviani et al. 1998). In addition, the antibodies for cobalt (II) and uranium (VI) ions determination using different monoclonal antibodies specific for complexes of DTPA-Co(II) and 2,9-dicarboxyl-1,10-phenanthroline-U(VI) were prepared (Blake et al. 2001).

APPLICATION TO OTHER AREAS OF HEALTH AND DISEASE

Biosensors belong to the most powerful bioanalytical tools and represent the future in development of instruments, which may be used in medicine, pharmaceutical research, environmental protection, BioDefence area and for monitoring of food quality and safety (Fig. 6.4). The increasing interest of researchers in biosensing is well documented in Fig. 6.5. From the above mentioned areas, biosensors play an important role in clinical diagnostics. Platinum complexes were accidently discovered to suppress cell division and became one of the most successful antitumour drugs used in chemotherapy of various malignancies. The cytotoxic lesion of the platinating agents is thought to be the platinum intrastrand crosslink that forms on the DNA, although treatment activates a number of signal transduction pathways. Treatment with these agents is characterized by resistance, both acquired and intrinsic. The relevant biosensor could be a new tool to study the interaction between platinum based cytostatics and DNA, which would cast the light on cytotoxicity of these drugs.

Next to medicine, environmental protection is also of great interest. One of the biggest problems associated with heavy metals in the environment is their potential for bioaccumulation. Heavy metals can thus be accumulated through the food chain with the top represented by predators. The food chain network in the sea, which is one of the most damaged environments, is shown in Fig. 6.6. Coastal fish (such as the smooth toadfish) and seabirds (such as the Atlantic Puffin) are often monitored for the presence of such contaminants. Biosensors help to monitor food safety and quality and to detect environmental pollution.

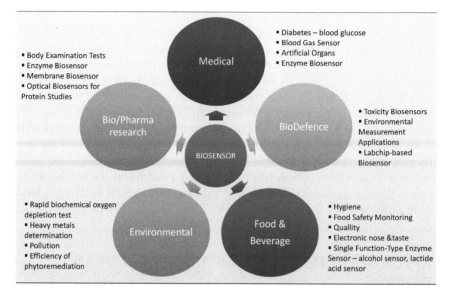

Figure 6.4. Biosensors and their application in different fields of use. This scheme shows five of the most important areas of interest. Medical applications show the importance in utilization of biosensor for monitoring diabetic patients' glucose levels, blood gas level and verification of artificial organs function. The pharmaceutical research industry is driving the need for new rapid assay biosensors to speed the progress of drug discovery. War or terrorism demands new rapid detection biosensors against bio warfare agents for military and civil defence applications. Biosensors also help in monitoring food quality and safety and detect environmental pollution.

KEY FACTS

- Heavy metal pollution can arise from many sources but usually comes from the mining and heavy industry.
- Smoking tobacco is the most important single source of cadmium exposure in the general population.
- Coastal fish and seabirds are sensitive to heavy metal bioaccumulation and thus these species are often monitored.
- Platinum complexes (cisplatin, oxaliplatin) are used in chemotherapy, and show good activity against certain tumours.
- The blood glucose biosensor is a common example of a commercial biosensor.

DEFINITIONS

- *Heavy metals*: are naturally occurring elements. Some of them (zinc, copper, iron) are essential to maintain the metabolism of an organism.

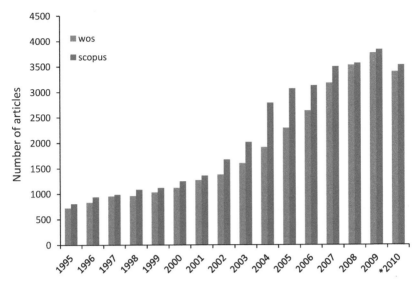

Figure 6.5. Scientific interest in biosensors. The number of full length articles, reviews, meeting abstracts and proceedings papers having "biosensor*" in titles, abstracts and keywords per year according to Web of Science and SCOPUS (5. 1. 2011).

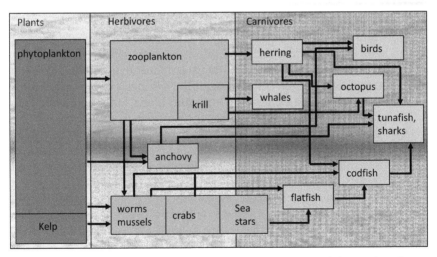

Figure 6.6. Food chain network. Entrance of heavy metals to the sea-food chain and its advance from plants to herbivores and subsequently to carnivores.

However, others such as cadmium, lead, and mercury are very toxic. At higher concentrations both groups of heavy metals (toxic and essential) can cause poisoning.

- *Metallothionein*: belongs to group of intracellular, low molecular mass and cysteine-rich proteins with molecular weight from 6 to 10 kDa. MTs consist of two binding domains (α and β) assembled from cysteine clusters. Cysteine sulphydryl moieties participate in covalent bindings with heavy metals.
- *Enzyme*: are proteins that catalyze chemical reaction.
- *Sensor*: is a device that measures a physical quantity and converts it into a signal which can be read by an observer or by an instrument.
- *Biosensor*: An analytical device comprising a biological recognition element (e.g. enzyme, receptor, DNA, antibody, or microorganism) in intimate contact with an electrochemical, optical, thermal, or acoustic signal transducer that together permit analyses of chemical properties or quantities. It shows potential development in some areas, including environmental monitoring.
- *Monoclonal antibody*: an antibody produced by a single clone of cells (specifically, a single clone of hybridoma cells) and therefore a single pure homogeneous type of antibody. Monoclonal antibodies can be produced in large amounts in the laboratory and are a cornerstone of immunology.

SUMMARY POINTS

- From the physiological point of view, heavy metals are elements with both essential and toxic effects on an organism.
- Plants are able to survive under metal ions toxic conditions thanks to the defence mechanisms including synthesis of phytochelatins.
- Metallothionein-cysteine rich protein binds metal ions and, thus, protecting an organism against their influence.
- Electrochemistry represents a suitable technique for metal ions detection due to its high sensitivity and the possibility to miniaturize the whole detection system, which is thus easily portable and able to carry out *in situ* analysis.
- Combination of physico-chemical transducer as an electrode with a biological substance is suitable way for biosensor design.
- Biosensors for heavy metals are divided into two groups based on their biological part: i) protein biosensors and ii) biosensors based on nucleic acid.
- Protein biosensors are further divided into enzymatic and affinity (antibodies based) biosensors.
- Enzymatic biosensors are based on activation or inhibition of enzymes activity.
- Affinity biosensors are based on detecting of changes in protein structure due to binding of heavy metal ion.

ACKNOWLEDGEMENT

This work was supported by grants NANIMEL GA CR 102/08/1546, and NANOSEMED GA AV KAN208130801.

ABBREVIATIONS

BFPms1	:	Green fluorescent protein mutant
CueR	:	Copper-responsive Metalloregulatory Protein
DNA	:	Deoxyribonucleic Acid
EDTA	:	Ethylenediaminetetraacetic Acid
GFP	:	Green fluorescent protein
ISFET	:	Ion-sensitive field-effect transistor
MerR	:	Metalloregulatory protein
SmtA	:	Metallothionein protein from *Escherichia Coli*

REFERENCES

Adam V, J Zehnalek, J Petrlova, D Potesil, B Sures, L Trnkova, F Jelen, J Vitecek and R Kizek. 2005. Phytochelatin modified electrode surface as a sensitive heavy-metal ion biosensor. Sensors 5: 70–84.

Adam V, P Hanustiak, S Krizkova, M Beklova, J Zehnalek, L Trnkova, A Horna, B Sures and R Kizek. 2007a. Palladium biosensor. Electroanalysis 19: 1909–1914.

Adam V, S Krizkova, O Zitka, L Trnkova, J Petrlova, M Beklova and R Kizek. 2007b. Determination of apo-metallothionein using adsorptive transfer stripping technique in connection with differential pulse voltammetry. Electroanalysis 19: 339–347.

Andres RT and R Narayanaswamy. 1995. Effect of the coupling reagent on the metal inhibition of immobilized urease in an optical biosensor. Analyst 120: 1549–1554.

Barondeau DP, CJ Kassmann, JA Tainer and ED Getzoff. 2002. Structural chemistry of a green fluorescent protein Zn biosensor. J Am Chem Soc 124: 3522–3524.

Blake DA, RC Blake, M Khosraviani and AR Pavlov 1998. Immunoassays for metal ions. Anal Chim Acta 376: 13–19.

Blake DA, RM Jones, RC Blake, AR Pavlov, IA Darwish and HN Yu. 2001. Antibody-based sensors for heavy metal ions. Biosens Bioelectron 16: 799–809.

Bontidean I, J Ahlqvist, A Mulchandani, W Chen, W Bae, RK Mehra, A Mortari and E Csoregi. 2003. Novel synthetic phytochelatin-based capacitive biosensor for heavy metal ion detection. Biosens Bioelectron 18: 547–553.

Castillo J, S Gaspar, S Leth, M Niculescu, A Mortari, I Bontidean, V Soukharev, SA Dorneanu, AD Ryabov and E Csoregi. 2004. Biosensors for life quality—Design, development and applications. Sens Actuator B-Chem 102: 179–194.

Changela A, K Chen, Y Xue, J Holschen, CE Outten, TV O'Halloran and A Mondragon. 2003. Molecular basis of metal-ion selectivity and zeptomolar sensitivity by CueR. Science 301: 1383–1387.

Corbisier P, D van der Lelie, B Borremans, A Provoost, V de Lorenzo, NL Brown, JR Lloyd, JL Hobman, E Csoregi, G Johansson and B Mattiasson. 1999. Whole cell- and protein-based biosensors for the detection of bioavailable heavy metals in environmental samples. Anal Chim Acta 387: 235–244.

Cowell DC, AA Dowman, T Ashcroft and I Caffoor. 1995. The detection and identification of metal and organic pollutants in potable water using enzyme assays suitable for sensor development. Biosens Bioelectron 10: 509–516.

Fennouh S, V Casimiri, A Geloso-Meyer and C Burstein. 1998. Kinetic study of heavy metal salt effects on the activity of L-lactate dehydrogenase in solution of immobilized on an oxygen electrode. Biosens Bioelectron 13: 903–909.

Gayet JC, A Haouz, A Gelosomeyer and C Burstein. 1993. Detection of heavy metal salts with biosensors built with an oxygen electrode coupled to various immobilized oxidases and dehydrogenases. Biosens Bioelectron 8: 177–183.

Huska D, I Fabrik, J Baloun, V Adam, M Masarik, J Hubalek, A Vasku, L Trnkova, A Horna, L Zeman and R Kizek. 2009. Study of Interactions between Metallothionein and Cisplatin by using Differential Pulse Voltammetry Brdicka's reaction and Quartz Crystal Microbalance. Sensors 9: 1355–1369.

Khosraviani M, AR Pavlov, GC Flowers and DA Blake. 1998. Detection of heavy metals by immunoassay: Optimization and validation of a rapid, portable assay for ionic cadmium. Environ Sci Technol 32: 137–142.

Kizek R, J Vacek, L Trnkova, B Klejdus and V Kuban. 2003. Electrochemical biosensors in agricultural and environmental analysis. Chem Listy 97: 1003–1006.

Krawczyk TKV, T Moszczynska and M Trojanowicz. 2000. Inhibitive determination of mercury and other metal ions by potentiometric urea biosensor. Biosens Bioelectron 15: 681–691.

Krizkova S, D Huska, M Beklova, J Hubalek, V Adam, L Trnkova and R Kizek. 2010. Protein-based electrochemical biosensor for detection of silver (I) ions. Environ Toxicol Chem 29: 492–496.

Liu JW and Y Lu 2003. A colorimetric lead biosensor using DNAzyme-directed assembly of gold nanoparticles. J Am Chem Soc 125: 6642–6643.

Preininger C and OS Wolfbeis. 1996. Disposable cuvette test with integrated sensor layer for enzymatic determination of heavy metals. Biosens Bioelectron 11: 981–990.

Rodriguez BB, JA Bolbot and IE Tothill. 2004. Urease-glutamic dehydrogenase biosensor for screening heavy metals in water and soil samples. Anal Bioanal Chem 380: 284–292.

Satoh I. 1991. An apoenzyme thermistor microanalysis for zinc (II) ions with use of an immobilized alkaline phosphatase reactor in a flow system. Biosens Bioelectron 6: 375–379.

Shekhovtsova TN, SV Muginova and NA Bagirova. 1997. Determination of organomercury compounds using immobilized peroxidase. Anal Chim Acta 344: 145–151.

Supalkova V, J Petrek, J Baloun, V Adam, K Bartusek, L Trnkova, M Beklova, V Diopan, L Havel and R Kizek. 2007. Multi-instrumental investigation of affecting of early somatic embryos of spruce by cadmium(II) and lead(II) ions. Sensors 7: 743–759.

Varriale A, M Staiano, M Rossi and S D'Auria. 2007. High-affinity binding of cadmium ions by mouse metallothionein prompting the design of a reversed-displacement protein-based fluorescence biosensor for cadmium detection. Anal Chem 79: 5760–5762.

Verma N and M Singh. 2005. Biosensors for heavy metals. Biometals 18: 121–129.

Verma N and A Singh. 2006. A Bacillus sphaericus based biosensor for monitoring nickel ions in industrial effluents and foods. J Autom Methods Manag Chem4.

Volotovsky V, YJ Nam and N Kim. 1997. Urease-based biosensor for mercuric ions determination. Sens Actuator B-Chem 42: 233–237.

Zeng HH, RB Thompson, BP Maliwal, GR Fones, JW Moffett and CA Fierke. 2003. Real-time determination of picomolar free Cu (II) in seawater using a fluorescence based fiber optic biosensor. Anal Chem 75: 6807–6812.

7

Whole Cell Biosensors: Applications to Environmental Health

S. Buchinger[1,a] and G. Reifferscheid[1,b]

ABSTRACT

Whole cell biosensors allow the detection of a biological effect rather than the detection of a specific molecular structure which can be done e.g. by biosensors that are based on antibodies. Whole cell biosensors are of great interest because they allow a complementary approach to chemical analysis. With living cells it is possible to address either general toxicity which can be monitored by the measurement of physiological parameters or specific toxic effects by making use of biosensors that are based on reporter gene assays. This chapter focuses on the latter type of whole cell biosensors and exemplifies the basic concepts on biosensors for the detection of genotoxicity and endocrine disruption. Both toxicological endpoints are of concern for human health because of their long term effects that might contribute to the initiation and progression of cancer, developmental disorders or, in case of the endocrine disruptors, impact on reproductive success. The combination of a gene promoter that is involved in the biological pathway of interest as a sensor element with a reporter gene for the quantification of the

[1]German Federal Institute of Hydrology (BfG), Division of Qualitative Hydrology, Am Mainzer Tor 1, 56068 Koblenz, 56068, Germany.
[a]E-mail: Buchinger@bafg.de
[b]E-mail: Reifferscheid@bafg.de

List of abbreviations after the text.

promoter activity has a high degree of freedom. Both, the sensor element and the reporting element can be combined freely according to the specific (toxic) effect of interest and the measurement technology of demand. Consequently, a broad variety of different whole cell biosensors was developed in the past decade—not all of them could be included in the present chapter.

INTRODUCTION

As per the definition of Lowe (Lowe 2007) a biosensor is "an analytical device, which converts the concentration of the target substance into an electrical signal through a combination of biological recognition system associated with a physico-chemical transducer". Usually, a biomolecule like an antibody that specifically binds to an antigen—the analyte—or an enzyme that is, e.g. inhibited by a specific compound is used for the "biological recognition". In these cases the biosensor is specific for target structures, i.e. single substances or compound classes like pesticides or heavy metals as described elsewhere in the current book. In this respect biosensors have to compete with classical analytical methods like coupled gas chromatography/mass spectrometry.

If intact, living cells are used in a biosensor setup as the biological entity (whole cell biosensors) instead of isolated biomolecules, a completely different analytical approach is possible that is most valuable in environmental science and environmental health. By monitoring the activity of a living cell that is exposed to an environmental sample it is possible to detect biological (toxic) **effects** that might be caused by contaminants in the sample. Therefore, whole cell biosensors generate complementary information to chemical analysis and can help to bridge the gap between chemistry and (eco)toxicology. In addition, whole cell biosensors measure the toxicity of a sample in an integral manner because the living cells respond to all contaminants in a mixture that cause same effects even if some constituents are not yet identified by chemical analysis.

Among whole cell biosensors, the most promising ones to date are those based on microbia (Su et al. 2011). Eukaryotic cells can be used as well in a biosensor setup; however, the number of applications is limited. In general, whole cell biosensors can be divided into two groups that are based on different principles (Fig. 7.1). The first group of biosensors detects general physiological changes of the cells with cell death as the most extreme answer. With the second group of biosensors it is possible to detect specific toxic effects like genotoxicity or endocrine disruption which have a higher relevance for human health compared to the former group of sensors. Whole cell biosensors for the detection of specific effects are based on the principle of reporter gene assays—the fusion of a sensing element

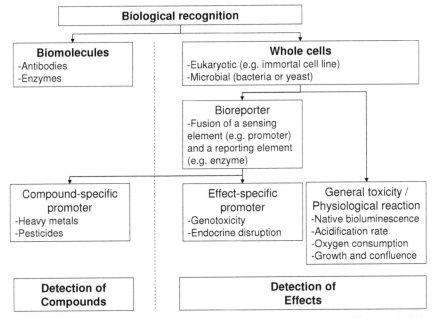

Figure 7.1. Schematic overview of the possibilities for biological recognition. The use of whole cells as the biological part of a biosensor allows the detection of biological effects rather than the detection of compounds. Biological effects can be detected in terms of general toxicity by the monitoring of physiological reactions due to an exposition or in terms of specific toxic effects like genotoxicity or endocrine disruption by making use of tailored bioreporters. Bioreporters are transformed cells that contain DNA-elements consisting out of a fusion of a sensing element with a reporting element.

which is most frequently a gene promoter and a reporting element. The reporting element—e.g. an enzyme coding gene—is expressed under the control of the sensing element. In general its expression is proportional to the activation of the biological process in which the sensing element is involved. Some promoters are activated by specific compounds or compound classes such as heavy metals. If such a promoter is fused to a reporting element a compound-specific bioreporter is generated in contrast to an effect-specific biosensor (Fig. 7.1).

In this chapter the basic concepts for effect-specific whole cell biosensors will be explained and examples for applications will be given for the endpoints genotoxicity and estrogenic potential (endocrine disruption).

WHOLE CELL BIOSENSORS FOR THE DETECTION OF GENOTOXICITY

Genotoxicity is linked to the terms mutagenicity and cancerogenicity. Environmental pollutants like polycyclic aromatic hydrocarbons (PAHs), nitroarenes or heterocyclic aromatic amines are genotoxic. Some of these compounds are persistent and bio-accumulative in addition. Therefore, the toxicological endpoint genotoxicity is highly relevant for environmental and human health.

The genotoxic potential of pure compounds and environmental samples can be evaluated by various *in vitro* bioassays based on both, eukaryotic and prokaryotic cells. A large number of different methods use a SOS-dependent reporter gene assay in which a LexA-responsive promoter like *recA*, *sulA* or *umuDC* is fused to a reporter gene. Alternatively, SOS-independent promoters like *alkA* which is induced by DNA-alkylation can be used as well. The nature of the reporter gene is variable allowing the use of various detection techniques for the quantification of the reporter gene expression. Very often colorimetric assays are used in which the reporter is an enzyme that catalyzes the reaction of a colorless substrate to a colored product. Examples for such reporters are the alkaline phosphatase (*phoA*) or the β-galactosidase (*lacZ*) (Quillardet et al. 1982). With respect to a biosensor application a colorimetric detection is comparatively complex. Therefore, most of the biosensors that are described in literature are based on bioluminescence, fluorescence or direct amperometric detection (Table 7.1). The integration of cells with the already mentioned reporter genes *lacZ* or *phoA* in a biosensor context is possible by the use of a substrate the reaction product of which can be oxidized. The basic biological and chemical processes of a bacterial whole cell biosensor for the detection of genotoxic effects by an amperometric quantification of the reporter gene activity is shown schematically in Fig. 7.2. After the exposition to the model genotoxicant 6-nitrochrysense (6-NC) the bacterial SOS-response is initiated resulting in the expression of the SOS-genes and the reporter gene *lacZ* because of its fusion to a SOS-responsive promoter. The electrochemical signal detection of the *lacZ* expression is possible via *para*-aminophenyl-β-D-galactopyranoside (pAPG) which enters the cell without disruption of the cellular structure (Biran et al. 2000). The glycosidic bond of the pAPG is cleaved by the reporter enzyme β-galactosidase under the formation of galactose and para-aminophenol (pAP). The generated pAP is converted electrochemically by a two electron oxidation to *para*-quinonimine. Matsui et al. 2006 demonstrated electrochemical signal detection after SOS-induction with 2-(2-furyl)-3-(5-nitro-2-furyl)acrylamide (AF-2), mitomycin C (MMC) and 2-aminoanthracene (2-AA) by scanning electrochemical microscopy of collagen embedded *Salmonella typhimurium* TA1535 *pSK1002* (Matsui et al. 2006). According to the authors this approach resulted in considerable lower detection limits for the analyzed compounds compared to the standard *in vitro* assay that is performed in microtiter plates. However, direct electrochemical signal detection instead of the scanning electrochemical microscopy is preferable since the use of a simple set of electrodes would

Table 7.1. List of sensor elements and reporter elements which are used in biosensor applications for the detection of genotoxicity.

	Sensor elements			Reporter elements				
	Promoter	Function of native target gene	Gene	Protein function	Substrate	Transducer	Signal	Literature
SOS-dependent	recN	double stranded DNA break repair	phoA	alkaline phosphatase	pAPP	electrode	amperometric	Ben-Yoav et al. (2009)
	recA	RecA recombinase			pAPG			Matsui et al. (2006)
	umuDC	DNA polymerase V (error prone DNA repair)	lacZ	β-galactosidase	galacton	photo detector	light/(chemo) luminescence	
	sulA	inhibition of cell division	luc	luciferase	luciferin		light/(bio) luminescence	Fine et al. 2006*
	cda	colicin D (inhibition of bacterial protein synthesis)	luxAB	luciferase	RCHO			Ptitsyn et al. (1997) Polyak et al. (2001) Baumstark-Khan et al. (2007)
SOS-independent	alkA	N3-methyladenine DNA glycosylase II (repair of DNA alkylation)	luxCDABE	luciferase + substrate regeneration	–			For review see: Woutersen et al. (2010)
	nrdA	ribonucleoside diphosphate reductase	gfp	green fluorescence protein	–		light/ fluorescence	Norman et al. (2005)
			uidA	glucuronidase	FD-GlcU			Dreier et al. (2002)

In principle every sensor element (i.e. promoter) can be combined with every reporter element in order to tailor a fusion construct according to demand and application; literature is given for the various reporter elements. For a more detailed overview see (Biran et al. 2009). pAPP: para-aminophenolphosphate; pAPG: para-aminophenol-β-D-galactopyranoside; galacton: 1,2 dioxetane substrate containing a glycosidic bond to galactose; RCHO: long-chain aliphatic aldehyde; FD-GlcU: fluorescein di-β-d-glucuronide; *in combination with an estrogene responsive element for the detection of estogenic activity.

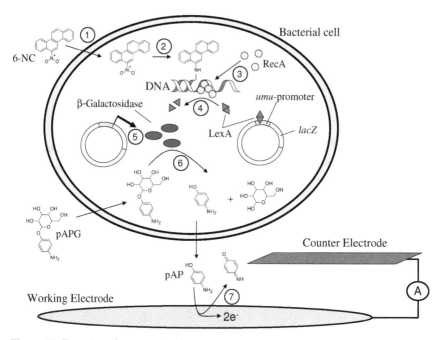

Figure 7.2. Detection of genotoxicity by an amperometric whole cell biosensor (scheme). **(1)** The genotoxic compound 6-nitrochrysene (6-NC) enters the bacterial cell. **(2)** The 6-NC is metabolized within the bacterial cell and forms bulky DNA-adducts which results in sections of single stranded DNA because of, e.g. stalled replication forks. **(3)** The protein RecA polymerizes at the single stranded DNA. **(4)** The RecA-filament induces the cleavage of the transcriptional repressor LexA. **(5)** The reporter plasmid contains a fusion of the LexA-dependent *umu*-promoter and *lacZ* that encodes the enzyme β-galactosidase. After the cleavage of LexA the promoter is released and *lacZ* is expressed. **(6)** *para*-aminophenol-β-D-galactopyranoside (pAPG) enters the cell and its glycosidic bond is cleaved by the expressed β-galactosidase. **(7)** The reaction product *para*-aminophenol (pAP) diffuses out of the bacterial cell and is oxidized at the surface of the working electrode to *para*-quinonimine. The resulting current between the working electrode and the counter electrode is proportional to the induction of the bacterial SOS-response and thus a measure for the genotoxicity of the sample.

greatly reduce the complexity, size and costs of such a biosensor. As shown in Fig. 7.3 the electrochemical detection can be done with simple, commercial screen printed electrodes (SPE) that consist out of a working electrode at which the reaction of interest takes place, a counter electrode for a closed electrical circuit and an electrode with a stable half cell potential, e.g. Ag/ AgCl, that serves as a reference to adjust the potential of the working electrode. After the exposition of the bacteria to the genotoxic compound 2-amino-3-methylimidazo[4,5-f]quinoline (IQ)—a pyrolysis product of proteins that is formed by frying of fish or meat and is detectable in rivers —the electrochemical signal detection is started by the addition of pAPG that is cleaved to pAP by the reporter enzyme β-galactosidase. No pAP

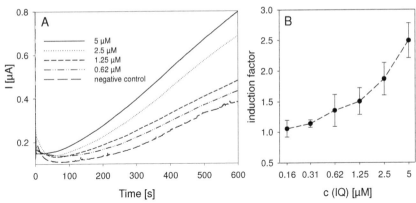

Figure 7.3. Chronoamperometric measurement of *Salmonella typhimurium* TA1535 *pSK1002* exposed to 2-amino-3-methylimidazo[4,5-f]quinoline (IQ). **(A)** The substrate pAPG was added at time point zero to exposed bacteria and the electric current between the working- and the counter electrode was monitored over 600 s using a screen printed electrode connected to a potentiostat. The potential of the working electrode was applied to 400 mV against Ag/AgCl (reference electrode). **(B)** Dose response relationship for IQ derived from the chronoamperometric measurement. The induction factor is the quotient of the slope (dI/dt) over the respective slope of the negative control. Data points show the mean ± SD of eight independent measurements (unpublished data).

can exist at the surface of the working electrode if a suitable potential is applied, because every pAP molecule is immediately oxidized. Therefore, the concentration of the pAP at the working electrode is zero and raises with increasing distance to the electrode until the concentration of the bulk solution is reached. The current that is generated by the oxidation of the pAP is proportional to the diffusive flux of the pAP towards the electrode and thus proportional to the concentration gradient of pAP. At the beginning of the measurement the concentration of the pAP in the bulk solution rises and consequently the concentration gradient that drives the diffusion. The formation of pAP is of pseudo-zero-order and depends only from the concentration of the reporter enzyme β-galactosidase if the pAPG concentration is high enough to guarantee substrate saturation of the enzyme. Because of this the concentration of the pAP increases over time proportional to the concentration of the reporter enzyme the expression of which is controlled by the strength of the SOS-response, i.e. the strength of the genotoxic stress. The pAP concentration in strongly induced bacteria increases faster compared to weak inductions or uninduced bacteria and therefore the slope dI/dt is a measure for the genotoxicity of a sample. This behavior is demonstrated in Fig. 7.3A. It is evident, that the slope of the electrical current over time (chronoamperometric measurement) increases with the concentration of the genotoxic compound IQ. Usually, the results of reporter gene assays are expressed as induction factors. For the chronoamperometric measurement the induction factor is the quotient of the slope dI/dt of induced bacteria over the respective slope of the negative control (uninduced bacteria). Figure 7.3B shows a dose response relationship in terms of induction factor versus IQ concentrations that was derived from biosensor measurements. The construction of an amperometric biosensor based on miniaturized µ-electrodes with reaction volumes in the nl-range was reported by (Ben-Yoav et al. 2009).

One of the key elements that determine the sensitivity of biosensors that are based on reporter gene assays is the ratio of the reporter gene expression in induced and uninduced cells. In general every promoter has its basal activity, even in an inactive state. In case of the LexA-dependent promoters of the SOS-response this is due to the reversible and imperfect binding of the transcriptional repressor LexA to its DNA-binding motive called SOS-box. The various SOS-promoters differ in number, sequence and position of this SOS-box. The binding affinity of LexA to the promoter and thus the repression of the promoter increases with the number of binding motives and their homology to the consensus sequence of the SOS-box. On the other hand, the activity of a promoter that is to be used as the sensing element in a biosensor should be high after induction. The TATAAT-box at position –10 base pairs upstream of the initiation of transcription and the TTGACA-box at position –35 are DNA-sequences which drive the activity of a bacterial promoter. If these motives show a high homology to the consensus sequence the promoter is strong, i.e. it has a high affinity to activating transcription factors and the RNA-polymerase. The interplay of SOS-box and the activating motives in SOS-responsive promoters was analyzed in detail by Norman et al. (Norman et al. 2005). In this study *gfp* was used as the reporter which allows the use of non-invasive fluorescence detection for the quantification of the reporter expression. The SOS-promoters *recA*, *sulA*, *umuDC* and *cda* were comparatively evaluated by means of basal promoter activity and inducibility. It was found that the *cda* promoter shows the highest response to the known carcinogen N-methyl-N'-nitro-N-nitrosoguanidine followed by *recA*, *sulA* and *umuCD* in descending order. The *recA* promoter shows high expression rates due to near consensus sequences at –35 and –10 but a high basal activity because of only one SOS-box with low homology to the consensus. In contrast the *umuCD* promoter is well repressed because of two near consensus SOS-boxes but its inducibility is comparatively low. According to the study the *cda* promoter combines both, a high expression rate after induction and a low basal activity due to a strong binding of the LexA-repressor.

As an alternative to the amperometric signal detection described above, several reporter proteins for optical signal detection can be used with the advantage that these methods are non-invasive and work without substrate addition. The green fluorescent protein and the bacterial luciferase (*luxCDABE*-operon) are most commonly used (Table 7.1). Biosensors that are based on the bacterial bioluminescence are of special interest as they are often proposed for online biosensors. The first reports about biosensors for the monitoring of genotoxicity that are based on bioluminescent reporter gene assays in *E. coli* were published by Ptitsyn et al. 1997 and Vollmer et al. 1997. In a further study van der Lelie et al. 1997 used the strain *Salmonella typhimurium* which is associated with the detection of genotoxic effects since the early 70s of the last century because of the initial work of Bruce Ames on the Salmonella/mammalian-microsome mutagenicity test (Ames et al. 1975). A number of further strains have been generated in the past decade; some milestones of these developments are listed below.

- Elasri and Miller 1998 used *Pseudomonas aeruginosa* as a host for the reporter gene construct *recA::luxCDABE* because it was expected to be more robust compared to *E. coli*. Furthermore, *P. aeruginosa*—an opportunistic human pathogen that causes cystic fibrosis—can survive in thick biofilms of alginate that is often used for the immobilization of living cells in whole cell biosensors.
- Polyak et al. 2001 used an optical fiber connected to a photodetector for the measurement of the bioluminescence that was emitted by immobilized sensor cells. The bacteria were immobilized in alginate layers by the alternating immersion of the fiber tip in a mixture of a sodium alginate solution with the bacterial cell suspension and a sterile calcium chloride solution. The functionality of the sensor fiber was tested with the model genotoxicant MMC.
- Baumstark-Khan et al. 2007 developed a combined reporter system using bioluminescence and fluorescence. They cloned the complete *lux*-operon from the marine photobacterium *P. leiognathi* under the control of a SOS promoter. In addition, the gene *gfp* was fused to a constitutive promoter resulting in a parallel detection method for general cytotoxicity. By this approach false negative test results for genotoxicity due to the masking effect of acute toxicity can be detected.
- Song et al. 2009 achieved the control for acute cytotoxicity by the construction of a *recA::lux* fusion that was integrated in the chromosome of *Acinetobacter baylyi* (ADP1). The background bioluminescence of this strain is high enough to monitor cell viability. Furthermore, it is reported that ADP1 is more robust compared to *E. coli* in terms of maintenance and storage.
- Yagur-Kroll and Belkin 2010 split the *luxCDABE* genes of *Photorhabdus luminescens* into the genes encoding the bacterial luciferase *luxAB* and *luxCDE* that encode for enzymes that are involved in the regeneration of the substrate. The highest sensitivity resulted if the genes *luxCDE*

were constitutively expressed whereas the genes *luxAB* were under the control of an inducible promoter. This finding indicates that the substrate of the bacterial luciferase is limited if all genes of the lux-operon are expressed simultaneously.

All whole cell biosensors for the detection of genotoxins that are based on bacteria share one drawback, namely the different metabolic activation of xenobiotics compared to vertebrates. The metabolization of xenobiotics occurs in vertebrates mainly in the liver. This tissue contains high concentrations of cytochrome-P450-dependent monooxygenases that catalyze the oxidation of organic compounds by molecular oxygen. By this oxidation step lipophilic compounds are functionalized, e.g. by the formation of hydroxyl residues that serve for a subsequent conjugation to small hydrophilic endogenous metabolites. By this reaction sequence the hydrophilicity of a compound is increased in order to facilitate its renal excretion. However, this metabolization goes along with a potential bio-activation and thus toxification of the xenobiotic. One of the most prominent examples is the metabolic activation of the PAH benzo[a]pyrene to the ultimative cancerogen benzo[a]pyrene-7,8-dihydrodiol-9,10-epoxide by the cytochrome-P450-dependent monooxygenases CYP1A1 and CYP1B1. In the standard bioassays this is usually done by the addition of a so called S9-fraction which is prepared from the liver of rodents after exposure to CYP-inducing chemicals (Ames et al. 1975). The S9-fraction is composed of a complex mixture of enzymes involved in the metabolism of xenobiotics, in particular the microsomal bound cytochrome-P450-dependent monooxygenases. Neglecting the bio-activation of xenobiotics with respect to the detection of genotoxic effects can cause false negative test results. The aspect of metabolization can be implemented into a biosensor by two different strategies, namely the addition of S9-enzymes and needed cofactors or second, the heterologous expression of the responsible enzymes in the reporter cells. Both approaches have advantages and disadvantages. In case of the addition of S9-enzymes to the biosensor these enzymes must be stabilized and/or immobilized like the living cells which can be achieved by freeze drying procedures. However, in this case the biosensor must be designed for single usage; a setup for an online measurement would not be possible. The heterologous expression of metabolizing enzymes in the reporter cells is an attractive alternative which is already demonstrated by various studies (Yamazaki et al. 1992). In this case online measurements would be possible because all enzymes and cofactors that are needed for metabolic activation are provided by the bacterial cell. But, the bio-activation of xenobiotics is a multi-step process in which many different enzymes are involved. Up to date this enzymatic network can be rebuild only in part by heterologous expression in bacteria.

WHOLE CELL BIOSENSORS FOR THE DETECTION OF XENOESTROGENS

It has been known for nearly two decades that endocrine disrupting chemicals (EDCs) that act as (xeno)hormones in the environment may have adverse effects on the reproduction of affected organisms and sensitive populations. Data from controlled field experiments support the hypothesis that EDCs in the aquatic environment have the potential to impact the reproductive health and persistence of various fish species (Kidd et al. 2007). Although controversial, ongoing research underlines the relevance of the subject for potential human health effects (Solomon and Schettler 2000).

Different types of hormonal effects can be distinguished. It has become increasingly apparent that estrogen receptor-mediated effects triggered by environmental agents such as ethinylestradiol (EE2), isomers of nonylphenole (NP), bisphenol-A (BPA) and diethylphthalate (DEP) are of high relevance. Chemical analysis unequivocally demonstrates that emissions from sewage treatment plants contribute significantly to environmental pollution by substances that bind specifically to the estrogen receptor. The relevant receptors are transcriptional regulatory proteins and members of the nuclear receptor superfamily (Evans 1988). The isolation of the respective human genes allows the development of reporter gene assays in which ligand-bound nuclear hormone receptors bind specific DNA response elements. One of the most prominent *in vitro* test systems for estrogenic activity is the yeast estrogen screen (YES) using *Saccharomyces cerevisiae* which can be integrated in a biosensor setup in a modified version (McDonnell et al. 1991). The basic principle of the YES is shown in Fig. 7.4. An estrogen receptor (ER) agonist binds to human ER that is constitutively expressed in the yeast cell. The ligand binding induces a conformational change of the ER which leads to the formation of ER-homodimers. These homodimers are imported in the nucleus of the yeast cell and bind to a promoter that contains at least one estrogen responding element (ERE). In contrast to the above mentioned LexA the ER is a transcriptional activator that induces the expression of the reporter gene after promoter binding. By this, the transcription of downstream genes like luciferase (Fine et al. 2006), β-galactosidase (McDonnell et al. 1991) or phytase (*phyK*) (Hahn et al. 2006) that are part of a recombinant genetic construct is initiated.

Other biosensors for the detection of EDCs are based on the determination of the ligand-receptor interaction or the receptor dimerization which can be measured by several techniques like surface plasmon resonance, quartz crystal microbalance or immuno-analytical methods. These approaches belong to the group of structure-specific biosensors and cannot distinguish between agonistic and antagonistic effects—a differentiation which is possible with whole cell biosensors. A further advantage of a biological

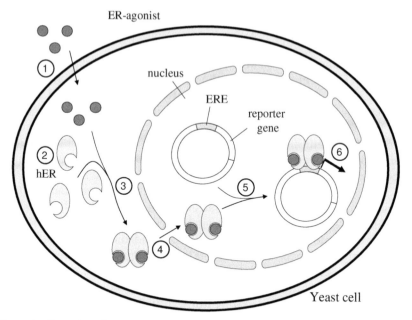

Figure 7.4. Detection of estrogenic potentials by genetically engineered yeast (scheme). (1) The ER-agonist (xenoestrogen) diffuses into the cell. (2) The human estrogen receptor (hER) is heterologously expressed (e.g. via a plasmid which contains the respective gene under the control of a constitutive promoter (not shown)). (3) The ER-agonist binds to the estrogen receptor and induces a dimerization of the hER. (4) Due to receptor binding and dimerization a nuclear localization signal is exposed and the dimer is imported in the cell nucleus. (5) The receptor dimer binds to and activates a promoter that contains an estrogen responding element. The reporter construct might be integrated in the yeast genome or can be part of a plasmid as shown here. (6) The expression of the reporter gene is proportional to the activation of the human estrogen receptor by the (xeno)estrogen.

cell sensor over immuno-analytical and chemical analysis is the possibility to measure the activities of chemicals in mixtures which can result in antagonistic, additive or synergistic effects.

Hahn et al. 2006 constructed a recombinant yeast strain (*Arxula adeninivorans*) that was engineered to co-express the human estrogen receptor alpha and a *Klebsiella*-derived phytase (*phyK*) reporter gene (A-YES). The native substrate of the phytase is phytic acid (myo-inositol hexakisphosphate) but it accepts as well *para*-nitrophenol phosphate and *para*-aminophenol phosphate allowing a colorimetric or amperometric detection, respectively. In case of the *para*-aminophenol phosphate the phytase catalyzes the cleavage of the phosphate whereby *para*-aminophenol (pAP) is generated. The amperometric detection of pAP is described above. This method has two advantages compared to the standard yeast estrogen screen. First, the yeast *A. adeninivorans* is a robust strain that is osmo- and thermotolerant. According to Hahn et al. it can be cultured in the presence of even 17.5% NaCl. The second advantage is the secretion of the reporter enzyme phytase by the yeast cell. This allows a non-invasive detection of the reporter gene activity which is a pre-requisite for the usage in a whole cell biosensor. A detection limit of 10 ng/l for 17β-estradiol-like activity was achieved, which is comparable to the classic YES with *S. cerevisiae*.

Fine et al. 2006 reported the construction of a fiber-optic biosensor that utilize luminescent yeast cells entrapped in hydrogel matrices based on calcium alginate or polyvinyl alcohol (PVA). They used genetically modified cells of *S. cerevisiae* that carry a plasmid for the expression of the human estrogen receptor alpha and a plasmid for the expression of the reporter gene *luc*—the luciferase gene of the American firefly—under the control of a human estrogen responsive element (hERE). In this report the shelf life of the immobilized cells was characterized. Figure 7.5 shows the results of this experiment by Fine et al. 2006. The modified yeast cells were immobilized in alginate beads by dropping a mixture of 2% (w/v) sodium alginate and yeast cell suspension in 0.5 M calcium chloride. Alginic acid is a copolymer

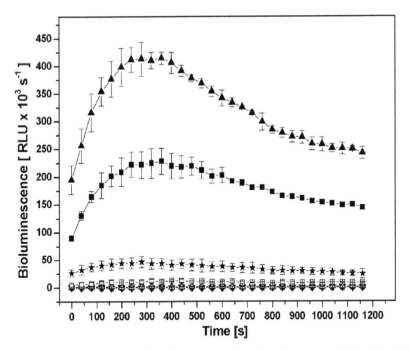

Figure 7.5. Luminometric detection of estrogenic effects by alginate embedded cells. With permission from Elsevier Limited. Luminescence kinetics curves for β-E2 induced alginate beads after 4 wk of storage in slow freeze conditions against (/) their control equivalent samples (0.1% EtOH) at –80°C (▲/△), –20°C (▼/▽), and 4°C (★/☆), and in fast freeze conditions (liquid nitrogen treatment) at –80°C (■/□) and –20°C (●/○) (Fine et al. 2006). [RLU: relative light units. Data points show mean ± SD of three independent sets each in triplicate.]

consisting out of homopolymeric blocks of α-L-gulcuronate and β-D-mannuronate linked by glycosidic bonds. The regions of homopolymeric blocks build higher structures that are comparable to β-sheets in proteins. These structures are fixed by the complexation with divalent cations (Ca^{2+}) which leads to the gelation of the alginate. The prepared beads were frozen in 50% (v/v) glycerol solution either by a slow freezing process or a fast freezing process with liquid nitrogen and stored for 4 wk under various conditions (–80°C, –20°C and 4°C (no fast freezing)). After thawing the beads were exposed to 10 nM of 17β-estradiol (β-E2) and the kinetic of the bioluminescence was monitored after addition of the substrate luciferine. From Fig. 7.5 it is obvious that storage temperature and freezing conditions have a high impact on the functionality of the system. The best performance resulted after the storage at –80°C. Interestingly, the cells after fast freezing with liquid nitrogen were less induced compared to the slow cooling in the freezer. The storage at 4°C and especially –20°C leads to a virtually complete loss of function. This example illustrates the outstanding importance of immobilization and long term storage of cells in an active state for whole cell biosensors.

SUMMARY POINTS

- Whole cell biosensors contain living cells as the biological part of the biosensor. Single cell organisms like bacteria or yeast but as well cell cultures from a multicellular organism like immortal cell lines can be used. Such biosensors detect (toxic) effects rather than compounds—the question addressed is not "what does the sample contain" but "how toxic is the sample". In this respect they display the sum effect of a sample even in the case of unknown composition, because all present compounds will impact the living cell in the sensor simultaneously.
- Specific toxicologic effects like genotoxicity or endocrine disruption can be detected by "bioreporters" that are composed by the fusion of a sensing element (mostly a gene promoter) with a reporting element (e.g. an enzyme-coding gene or a gene encoding a fluorescent protein). The nature of the transducer is variable and depends on the reporting element and the application, e.g. a photo detector for the measurement of bioluminescence in an online biosensor setup.
- Genotoxicity is to be differentiated from mutagenicity and cancerogenicity, but these effects are strongly linked together. Therefore, genotoxicity is a toxicological endpoint of high relevance. Genotoxic effects can be probed by whole cell biosensors by making use of genetically engineered microorganisms. Most frequently fusions of various SOS-responsive promoters and reporter genes,

the activity of which can be measured either amperometrically or via bioluminescence with a photo detector are used. Biosensors that are based on *luxCDABE* can be used in online systems because the quantification of the reporter enzyme can be done non-invasive and without addition of a substrate. One challenge that has to be further addressed in the future is the implementation of the metabolic activation of xenobiotics in microbial biosensors.

- Endocrine disrupting chemicals interfere with the endogenous hormonal regulations and might cause long-term chronic effects in humans. Although the impact of contaminants that act as endocrine disruptors on human health is under discussion there is a great deal of concern about such chemicals in the environment. One important mode of action of endocrine disruptors is their binding to and activation of the estrogen receptor that belongs to the family of nuclear hormone receptors and acts as a transcription factor. The receptor binding can be quantified with various biosensor techniques that are based, e.g. on plasmon resonance. However, such biosensors do not allow the detection of the biological consequence of the receptor binding, namely the transactivation of a target gene which can be addressed by whole cell biosensors. Again, biosensors that use bioluminescence for the quantification of the reporter are frequently used, but the use of the enzyme phytase that is excreted after expression in *A. adeninivorans* is a good alternative that allows amperometric signal detection.

- The immobilization and stabilization of the living cells in an active state is a challenge of outstanding importance for the construction of whole cell biosensors that are thought to work outside a laboratory environment. Most frequently encapsulation in hydro-gels like alginate or polyvinyl alcohol is used for this purpose. Despite all the progress that was achieved in the past decades further research is necessary. In sum, the development of whole cell biosensors is an exciting and interdisciplinary field that covers topics in microbiology, cell biology, molecular biology, biochemistry, physics and engineering.

APPLICATIONS TO AREAS OF HEALTH AND DISEASE

It is worth pointing out that the relevance of *in vitro* assays—and a whole cell biosensor is nothing else than an automated *in vitro* assay with a coupled signal detection—for *in vivo* effects like the impact of a contaminant on human health is to be discussed in general.

Whole cell biosensors which either detect genotoxic stress or effects of endocrine disruptors or biosensors that use human cell lines have the highest relevance for human health and disease. Genotoxic compounds cause DNA-lesions that are usually reversible because of the induction of various repair mechanisms like base- or nucleotide-excision repair. But the primary DNA-damage increases the mutation frequency of a stressed cell resulting in a changed DNA-sequence. In contrast to the primary DNA-damage mutations are irreversible and associated with the initiation and promotion of cancer. Thus, the terms genotoxicity, mutagenicity and cancerogenicity are distinct but strongly linked together. Up to now there is no biosensor available that detects mutations but as already mentioned it is possible to detect primary DNA-damage with specific whole cell biosensors.

Exposition of laboratory animals and wildlife to endocrine disrupting chemicals lead to numerous adverse phenotypes like feminization of males, birth defects, decreased sperm density and reproductive failure. Whereas the impact of endocrine disruptors on wildlife populations is evident the risk for human health is still under debate. However, some studies report adverse effects of endocrine disruptors on humans. One example is the observation of semen abnormalities in males after exposure to the estrogen receptor agonist diethylstilbestrol.

KEY FACTS OF EFFECT SPECIFIC WHOLE CELL BIOSENSORS

- Whole cell biosensors that detect specific toxicological effects are based on the concept of a **reporter gene assay** which is used in order to quantify the activity of a gene promoter. If a promoter is chosen that is involved in the regulation of an (adverse) cellular pathway, it is possible to determine the activation of this pathway that is linked to a biological effect.
- One of the first whole cell biosensors that used **bioluminescent bacteria** was developed by (Lee et al. 1992) for the detection of toxic compounds. Five years later the first systems for the detection of genotoxic effects by the fusion of SOS-promoters to the *luxCDABE*-operon were developed.
- **Cancer** causes worldwide more than 10% of all deaths with life style and environmental pollution as the main causes. Carcinogenesis is a multi step process that is initiated by DNA-mutations. In this respect genotoxic environmental contaminants are of high relevance for human health.
- Many genotoxic xenobiotics cannot damage the DNA directly, but they are bio-activated in the liver. This process is termed **metabolic activation**. Functional groups that are introduced in (lipophilic)

compounds in order to increase their water-solubility might attack DNA-bases.

- The effect of **endocrine disruptors** on human health is still under discussion. However, some trends in human health effects that are potentially related to endocrine function are reported, such as the sperm count that decreases in the US by –3%/ml per year and even –5.3%/ml per year in Europe (Solomon and Schettler 2000).

DEFINITIONS

- *Bioluminescene:* is the emission of light by organisms. It is based on chemiluminescence. Chemiluminescence occurs if a product of a reaction is formed in an excited state and the energy dissipates not as heat but as light. Some eukaryotic organisms like fireflies produce an enzyme called luciferase that catalyzes the oxidation of its substrate luciferin—a reaction that causes chemiluminescence. Bioluminescent bacteria express the *lux*-operon (*luxCDABE*).

- *Endocrine disruptors*: "are exogenous compounds that have the potential to interfere with hormonal regulations and the normal endocrine system and consequently [may] cause health effects in animals and humans" (Bonefeld-Jorgensen et al. 2007).

- *Estrogen-receptor (ER)*: refers to a group of intracellular receptors out of the family of nuclear hormone receptors which is activated by its native agonist 17β-estradiol (estrogen) but as well by synthetic steroid hormones (e.g. 17α-ethinylestradiol that is used in contraceptive pills) or xenoestrogens like nonylphenol-isomers or various phtalates that are used as plasticizers.

- *Genotoxicity*: describes the capacity of a compound to cause DNA-damage. A so called primary DNA-damage is a change in the chemical structure of the DNA like DNA-alkylation, oxidation, bulky adducts formation, crosslinks or strand-breaks. These lesions can be repaired by various mechanisms (e.g. bacterial SOS-response). Secondary DNA-damage is a change in the DNA-sequence (mutation) possibly resulting in miscoding of proteins. Mutations cannot be repaired by cellular mechanisms.

- *gfp*: is a gene coding the green fluorescence protein (GFP). GFP was initially described in the jellyfish *Aequorea victoria* and emits green light (509 nm) when it is excited by blue light (395 nm). Today, several variations of the GFP exist with altered protein stability or enhanced fluorescence.

- *lacZ*: is a gene coding for the enzyme β-galactosidase that is part of the bacterial *lac*-operon. Its native function is the cleavage of the glycosidic bond in the disaccharide lactose, but it also accepts artificial substrates like *para*-aminophenol-β-D-galactopyranoside (pAPG).

- *luxCDABE*: is an operon which is composed out of the genes *luxAB* encoding the bacterial luciferase which catalyzes the co-oxidation of long-chain aliphatic aldehydes and flavin mononucleotide in its reduced state ($FMNH_2$). The genes *luxCDE* are involved in the regeneration of the required substrate.
- *SOS-response*: is induced in bacteria by genotoxic stress. It is activated by the occurrence of single stranded DNA which results from DNA-adducts or crosslinks. The single stranded DNA is detected by the RecA-protein which forms a protein filament along the DNA lesion. This protein filament cleaves the transcription factor LexA that represses the expression of genes which are involved in the DNA repair like *uvrA, uvrB* (nucleotide excision repair) or *umuC, umuD* (error prone DNA repair) as the most prominent ones. In total more that 30 genes are involved in the SOS-response; *recA* and *lexA* the genes coding for the proteins that drive the SOS-response are target genes of the LexA-repressor themselves resulting in feedback loops for the control of the SOS-response (Sassanfar and Roberts 1990).

ABBREVIATIONS

2-AA	:	2-aminoanthracene
AF-2:	:	2-(2-furyl)-3-(5-nitro-2-furyl)acrylamide
BPA	:	bisphenol-A
CYP	:	cytochrome-P450-dependent monooxygenase
DEP	:	diethylphthalate
EDC	:	endocrine disrupting chemical
β-E2	:	17β-estradiol
EE2	:	α-ethinylestradiol
ERE	:	estrogen responsive element
$FMNH_2$	:	reduced flavin mononucleotide
GFP	:	green fluorescence protein
hER	:	human estrogen receptor
IQ	:	2-amino-3-methylimidazo[4,5-f]quinoline
MMC	:	mitomycin C
6-NC	:	6-nitrochrysense
NP	:	nonylphenol
PAH	:	polycyclic aromatic hydrocarbon
pAPG	:	*para*-aminophenol-β-D-galactopyranoside
PVA	:	polyvinyl alcohol
SPE	:	screen printed electrode
YES	:	yeast estrogen screen

REFERENCES

Ames BN, J McCann and E Yamasaki. 1975. Methods for detecting carcinogens and mutagens with the Salmonella/mammalian-microsome mutagenicity test. Mutat Res 31(6): 347–64.
Baumstark-Khan C, E Rabbow, P Rettberg and G Horneck. 2007. The combined bacterial Lux-Fluoro test for the detection and quantification of genotoxic and cytotoxic agents in surface water: results from the Technical Workshop on Genotoxicity Biosensing. Aquat Toxicol 85(3): 209–18.
Ben-Yoav H, A Biran, R Pedahzur, S Belkin, S Buchinger, G Reifferscheid and Y Shacham-Diamand. 2009. Whole cell electrochemical biosensor for water genotoxicity bio-detection. Electrochimica Acta 54(25): 6113–6118.
Biran A, R Pedahzur, S Buchinger, G Reifferscheid and S Belkin. Genetically Engineered Bacteria for Genotoxicity Assessment. pp 161–186. In: D Barceló and PD Hansen. [eds.] 2009. Biosensors for Environmental Monitoring of Aquatic Systems. Springer-Verlag Berlin Heidelberg, Germany.
Biran I, R Babai, K Levcov, J Rishpon and EZ Ron. 2000. Online and in situ monitoring of environmental pollutants: electrochemical biosensing of cadmium. Environ Microbiol 2(3): 285–90.
Bonefeld-Jorgensen EC, M Long, MV Hofmeister and AM Vinggaard. 2007. Endocrine-disrupting potential of bisphenol A, bisphenol A dimethacrylate, 4-n-nonylphenol, and 4-n-octylphenol *in vitro*: new data and a brief review. Environ Health Perspect 115 Suppl 1: 69–76.
Dreier J, EB Breitmaier, E Gocke, CM Apfel and MG Page. 2002. Direct influence of S9 liver homogenate on fluorescence signals: impact on practical applications in a bacterial genotoxicity assay. Mutat Res 513(1-2): 169–82.
Elasri MO and RV Miller. 1998. A Pseudomonas aeruginosa biosensor responds to exposure to ultraviolet radiation. Appl Microbiol Biotechnol 50(4): 455–8.
Evans RM. 1988. The steroid and thyroid hormone receptor superfamily. Science 240(4854): 889–95.
Fine T, P Leskinen, T Isobe, H Shiraishi, M Morita, RS Marks and M Virta. 2006. Luminescent yeast cells entrapped in hydrogels for estrogenic endocrine disrupting chemical biodetection. Biosens Bioelectron 21(12): 2263–9.
Hahn T, K Tag, K Riedel, S Uhlig, K Baronian, G Gellissen and G Kunze. 2006. A novel estrogen sensor based on recombinant Arxula adeninivorans cells. Biosens Bioelectron 21(11): 2078–85.
Kidd KA, PJ Blanchfield, KH Mills, VP Palace, RE Evans, JM Lazorchak and RW Flick. 2007. Collapse of a fish population after exposure to a synthetic estrogen. Proc Natl Acad Sci USA 104(21): 8897–901.
Lee S, K Sode, K Nakanishi, JL Marty, E Tamiya and I Karube. 1992. A novel microbial sensor using luminous bacteria. Biosens Bioelectron 7(4): 273–7.
Lowe CR, Introduction to biosensor and biochip technologies. pp 7–22 In: RS Marks, DC Cullen, I Karube, CR Lowe and HH Weetall. [eds.] 2007. Handbook of Biosensors and Biochips. Wiley, Chichester, UK.
Matsui N, T Kaya, K Nagamine, T Yasukawa, H Shiku and T Matsue. 2006. Electrochemical mutagen screening using microbial chip. Biosens Bioelectron. 21(7): 1202–9. Epub 2005 Jun 20.
McDonnell DP, Z Nawaz, C Densmore, NL Weigel, TA Pham, JH Clark and BW O'Malley. 1991. High level expression of biologically active estrogen receptor in Saccharomyces cerevisiae. J Steroid Biochem Mol Biol 39(3): 291–297.
Norman A, L Hestbjerg Hansen and SJ Sorensen. 2005. Construction of a ColD cda promoter-based SOS-green fluorescent protein whole-cell biosensor with higher sensitivity toward

genotoxic compounds than constructs based on recA, umuDC, or sulA promoters. Appl Environ Microbiol 71(5): 2338–46.

Polyak B, E Bassis, A Novodvorets, S Belkin and RS Marks. 2001. Bioluminescent whole cell optical fiber sensor to genotoxicants: system optimization. Sensors and Actuators B: Chemical 74(1-3): 18–26.

Ptitsyn LR, G Horneck, O Komova, S Kozubek, EA Krasavin, M Bonev and P Rettberg. 1997. A biosensor for environmental genotoxin screening based on an SOS lux assay in recombinant *Escherichia coli* cells. Appl Environ Microbiol 63(11): 4377–84.

Quillardet P, O Huisman, R D'Ari and M Hofnung. 1982. SOS chromotest, a direct assay of induction of an SOS function in *Escherichia coli* K-12 to measure genotoxicity. Proc Natl Acad Sci USA 79(19): 5971–5.

Sassanfar M and JW Roberts. 1990. Nature of the SOS-inducing signal in Escherichia coli. The involvement of DNA replication. J Mol Biol 212(1): 79–96.

Solomon GM and T Schettler. 2000. Environment and health: 6. Endocrine disruption and potential human health implications. CMAJ 163(11): 1471–6.

Song YZ, GH Li, SF Thornton, IP Thompson, SA Banwart, DN Lerner and WE Huang. 2009. Optimization of Bacterial Whole Cell Bioreporters for Toxicity Assay of Environmental Samples. Environmental Science & Technology 43(20): 7931–7938.

Su L, W Jia, C Hou and Y Lei. 2011. Microbial biosensors: A review. Biosensors and Bioelectronics 26(5): 1788–1799.

van der Lelie D, L Regniers, B Borremans, A Provoost and L Verschaeve. 1997. The VITOTOX test, an SOS bioluminescence Salmonella typhimurium test to measure genotoxicity kinetics. Mutat Res 389(2-3): 279–90.

Vollmer AC, S Belkin, DR Smulski, TK Van Dyk and RA LaRossa. 1997. Detection of DNA damage by use of *Escherichia coli* carrying recA'::lux, uvrA'::lux, or alkA'::lux reporter plasmids. Appl Environ Microbiol 63(7): 2566–71.

Woutersen M S Belkin, B Brouwer, AP van Wezel and MB Heringa. 2010. Are luminescent bacteria suitable for online detection and monitoring of toxic compounds in drinking water and its sources? Anal Bioanal Chem.

Yagur-Kroll S and S Belkin. 2010. Upgrading bioluminescent bacterial bioreporter performance by splitting the lux operon. Anal Bioanal Chem.

Yamazaki H, Y Oda, Y Funae, S Imaoka, Y Inui, FP Guengerich and T Shimada. 1992. Participation of rat liver cytochrome P450 2E1 in the activation of N-nitrosodimethylamine and N-nitrosodiethylamine to products genotoxic in an acetyltransferase-overexpressing Salmonella typhimurium strain (NM2009). Carcinogenesis 13(6): 979–85.

Bacterial Whole Cell Bioreporters in Environmental Health

Yizhi Song,[1,a,2] Dayi Zhang,[2,c] Karen Polizzi,[3] Eko Ge Zhang,[4] Guanghe Li[1,b] and Wei E. Huang[2,d,*]

ABSTRACT

Environmental contaminants and pathogens pose a serious risk to human health. Bacterial whole cell biosensors (BWBs) have several unique advantages in the detection of toxicity and bioavailability. BWBs can also be rapid, sensitive, semi-quantitative, cost-effective and easy to use. BWBs can be configured for cytotoxicity or genotoxicity assays, but also to quantitatively detect particular compounds of interest through transcriptional, translational, or posttranslational means. Collectively,

[1] School of Environment, Tsinghua University, Beijing 100084, P.R. China.
[a] E-mail: song.ezhi@gmail.com
[b] Email: ligh@mail.tsinghua.edu.cn
[2] Kroto Research Institute, Department of Civil and Structural Engineering, North Campus, University of Sheffield, Broad Lane, Sheffield S3 7HQ, UK.
[c] E-mail: d.zhang@shef.ac.uk
[d] Email: w.huang@shef.ac.uk
[3] Division of Molecular Biosciences & Centre for Synthetic Biology and Innovation, Imperial College London, London SW7 2AZ, UK; E-mail: k.polizzi@imperial.ac.uk
[4] Fetal and Maternal Medicine, Imperial College NHS Trust, London SW7 2AZ, UK; Email: GeEko.Zhang@Imperial.nhs.uk
*Corresponding author

List of abbreviations after the text.

seven different types of BWBs have been developed to detect a broad range of ions, metals, drugs, toxins, hormones, specific compounds and pathogens. Increasingly BWBs are also being applied to drug screening, pathogen detection and biomarker monitoring associated with human health. Since BWBs use live cells the efficiency of the sensor will be determined by the cellular physiological state which will affect signal pathway including chemical transport, chemical-protein, DNA-protein interactions and reporter gene activation. It results in BWBs suffering from problems of robustness, reliability and reproducibility. Recent progress include the use of synthetic biology to optimize sensitivity, specificity, and signal to noise ratio, the use of nanoparticles to functionalize BWBs and confer novel properties, and the multiplexed detection of contaminants and pathogens using BWBs arrays. Recent advances in synthetic biology and nanotechnology, will endow BWBs with novel functions and significantly improve properties, bringing their deployment into the field for online and *in situ* monitoring closer to reality.

INTRODUCTION

Human beings are facing severe health risks as human activity releases significant amount of waste and hazardous material into the environment. Water-soil pathogens transported through catchments and water supply, and other pathogens transmitted via human-human and human-animal contacts pose serious problems to human health. A broad range of contaminants has been designated by the European Union (EU) and U.S. Environmental Protection Agency (EPA) as priority pollutants because of their carcinogenic and mutagenic effects on humans and animals. The assessment of human health risks associated with environmental contamination and diseases requires a rapid, cost-effective, reliable and *in situ* method to detect contaminants and pathogens. Armed with recent advances in synthetic biology and nanotechnology, bacterial whole-cell biosensors (BWBs) could be developed to address these challenges.

A "biosensor" is a system that detects the presence of a substrate using a biological component which then provides a signal that can be quantified (Gu et al. 2004). BWBs employ live cells (usually genetically engineered bacteria) as detection elements and they have two unique advantages: providing information on toxicity and bioavailability, which directly link environmental contamination to human health risk. Most environmental samples are mixtures of complex contaminants. The additive, antagonistic, and synergistic effects caused by complex physical or chemical interactions would make the risk assessment unpredictable if only chemical analysis (e.g. GC-MS or HPLC) were used to estimate the toxicity and bioavailability of contaminated samples. It has been shown that even when the concentration of each compound in a mixture was below the individual toxicity effect there could be an additive effect as a mixture that was detrimental to fishes (Schwarzenbach et al. 2006). Chemical analysis of contaminated soil and water samples usually requires sample pretreatment and extraction, which makes the inert and active portions of contaminants indistinguishable from each other, compromising the risk assessment which is concerned with the active portion (bioavailability). In contrast, BWBs assay detects contaminants' active or bioavailable portions, and it directly links contamination to biological effects and makes the toxicity and bioavailability assessment more relevant to human health risk (Song et al. 2009; Sorensen et al. 2006). Other advantages of BWBs include 1) it is much lower cost than animal tests and eliminates the need of animals sacrificed for the toxicity assessment; 2) it can be quantitative (DeFraia et al. 2008), very sensitive (down to fM level), rapid, and easy to use (Massai et al. 2011; Virta et al. 1995); and 3) it requires minimal sample pretreatment and can be used for *in situ* or potentially online detection of pathogens and contaminants.

The use of whole cells as biosensors for toxicity detection was first developed in the laboratory nearly 30 years ago (Bulich and Isenberq 1981; Quillardet et al. 1982). BWBs can also detect pathogens by sensing small molecules related to pathogens (Massai et al. 2011). Since the live cells of BWBs were used to sense chemicals, all elements involved in signal pathway, including chemical transport, chemical-protein, DNA-protein interactions and reporter gene activation, can be affected by the cellular physiological state. It is therefore a challenge to construct BWBs with high reliability, robustness and reproducibility. Recent advances in molecular cell biology, synthetic biology and nanotechnology, could improve BWBs sensing performance in the laboratory and ultimately allow their use in real world situations. In this chapter, we will review the recent trends of BWBs in environmental health. We will focus on synthetic biology for optimization of biosensors, bionanotechnology for functionalization of BWBs and whole cell biosensor arrays for high throughput biosensing.

PRINCIPLES OF BACTERIAL WHOLE-CELL BIOSENSORS

The biological detection element can be made of nucleic acids or proteins, and a number of biosensing systems can be constructed which exploit different aspects of molecular biology to produce the signal (van der Meer and Belkin 2010). Figure 8.1 illustrates seven types of BWBs, which usually consist of regulatory gene(s), regulated or constitutive promoter(s) and reporter gene(s).

Cytotoxic BWBs detect general toxicity that causes the detrimental effects on cellular and enzymatic activities (Fig. 8.1A). The reporter gene of cytotoxic BWBs is usually fused to a constitutive promoter (P_{const}) and

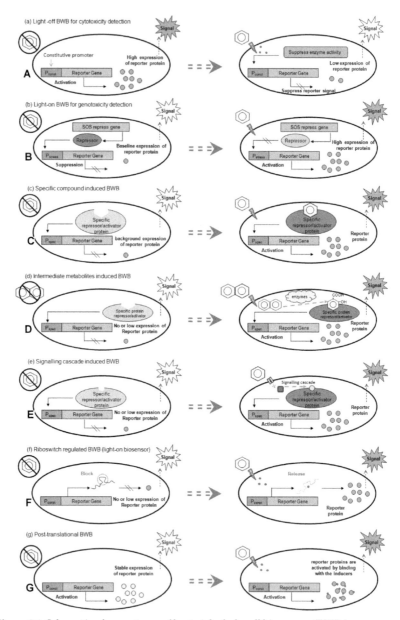

Figure 8.1. Schematic of seven types of bacterial whole cell biosensors (BWBs). *Color image of this figure appears in the color plate section at the end of the book.*

toxicity is measured as a decrease in reporter gene expression (a 'light-off' system). Genotoxic BWBs detect toxicity that leads to DNA damage (Fig. 8.1B). Genotoxicity is measured as an increase in the expression of a reporter gene (a 'light-off' system) that is fused with the "SOS" responding promoter, e.g. *recA* promoter (Davidov et al. 2000; Song et al. 2009).

Transcription-based biosensors activate transcription of mRNA in response to the presence of the signal and consist of an activatable promoter fused to a reporter gene. Transcription can be activated either through direct binding of a specific compound or its intermediate metabolite to a regulatory protein (Fig. 8.1C and 8.1D); or through a signalling cascade activated in response to a receptor binding event at the cell membrane (Fig. 8.1E). In the presence of a specific compound, a regulatory protein undergoes a conformational change via chemical-protein interactions and activates a regulated promoter (P_{spec}) to express a reporter gene (Fig. 8.1C). Since the regulatory protein usually responds to specific compounds, it enables the BWBs to selectively detect the targeted compounds (Huang et al. 2005; Huang et al. 2006; Huang et al. 2008). A specific compound can also be indirectly detected by constructing a gene regulation system which is activated by the intermediate metabolites instead of the original compound itself (Fig. 8.1D). For example, naphthalene can be detected by sensing salicylate which is an intermediate metabolite of the naphthalene degradation pathway (King et al. 1990). For signalling cascade induced BWBs (Fig. 8.1e), a compound interacts with sensory proteins (e.g. membrane proteins), which subsequently transmit signals (e.g. phosphorylation) to trigger a transcription regulator (Kiel et al. 2010). However, since these type BWBs rely on both transcription and translation to occur before the signal is manifested; transcription-based biosensors show a somewhat delayed response and are the slowest of BWBs.

With translation-based biosensors (Fig. 8.1F), the mRNA is constantly produced within the cell, but its translation is dependent on either the presence or absence of the signal in question. Most translation-based biosensors rely on reversible mRNA secondary structure (riboswitches) to control translation. These biosensors can be designed to be turned on or off in the presence of the signal depending on where the mRNA secondary structure is formed upon binding of the signal molecule. Translation-based biosensors have a faster response time than transcription-based biosensors because the mRNA is already present in the cell. Riboswitches (part of mRNA) can be designed to directly bind the target compound which then either triggers or suppresses gene expression (Fig. 8.1F). For example a riboswitch has been designed to activate protein translation in response to atrazine (Sinha et al. 2010).

BWBs can be constructed based on changes in protein behaviour or spectral properties in response to the signal (Palmer et al. 2011). These can also be called "post-translational" biosensors (Fig. 8.1G). Post-translational biosensors manifest the fastest response because they are always present within the cells and do not rely on the relatively slow processes of transcription and translation. Usually the response time is of the order of time it takes for a small molecule to bind to an enzyme. In the simplest configuration, the molecule of interest can interact directly with the protein signal, causing a change in spectral properties (Fig. 8.1G). Biosensors of this type have been created for sensing small molecules such as hydrogen peroxide, divalent metal ions, and pH changes (Palmer et al. 2011). The second type of post-translational biosensor relies on Förster (or Fluorescence) Resonance Energy Transfer (FRET) (Fig. 8.1G). FRET relies on the physical proximity of the two protein fluorophores. Ligand binding causes a conformational change in the binding domain, altering the spatial orientation of the fluorescent proteins, leading to a detectable change in spectral properties (Palmer et al. 2011). Since the proportion of biosensors with ligand bound depends on the concentration of the ligand within the cell, the spectral changes correlate to changes in ligand concentration. FRET biosensors have been developed for many metabolites including divalent metal ions, amino acids, and sugars (VanEngelenburg and Palmer 2008).

The output signal in BWBs is produced through expression of a reporter gene. A variety of reporter genes have been used including β-galactosidase gene *lacZ*, luciferase gene (*luxAB* and *luc*) and fluorescent-protein genes (GFP, YFP, CFP and RFP). Whilst fluorescent protein genes are ideal to visualize BWBs in single cells, bioluminescence genes such as *luxAB* and *luc* are more suitable to quantitative detection. Several recent reviews have discussed the reporter genes in details, (Daunert et al. 2000; van der Meer and Belkin 2010).

SYNTHETIC BIOLOGY FOR OPTIMIZATION OF BIOSENSORS

Synthetic biology is an emerging methodology for the development of new biological organisms. Through an approach that often utilizes computational modelling in the design of new pathways, coupled with the use of well-characterized, standardized "parts" (genes) which are used to construct devices (groups of parts) and systems of desired function, synthetic biology aims to create new systems using an engineering design cycle (Fig. 8.2). The desired behaviour is specified, parts are chosen and their behaviour modelled in silico, systems are constructed, and tested to see if the design

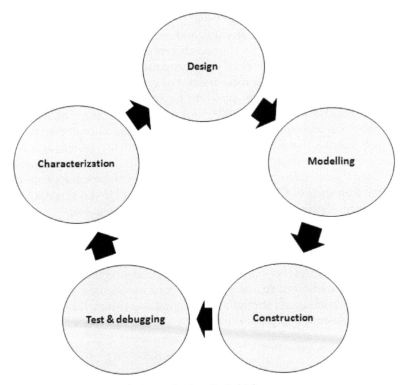

Figure 8.2. The engineering design cycle of synthetic biology.

was successful, followed by iteration to improve the design (Fig. 8.2) (Kitney et al. 2007). Synthetic biology can be applied to the design of new whole cell biosensors with improved properties such as signal-to-noise ratio, sensitivity, specificity, or speed of signal generation (van der Meer and Belkin 2010).

Chassis for BWBs

The chassis is the bacterial host of the genetic elements used in biosensing, which is the "hardware" of the BWBs. The choice of chassis needs to consider the following: 1) is the bacterial host easy to manipulate genetically? 2) will the inserted genes be correctly expressed and gene regulation in the host be reliable and reproducible? 3) will the host survive and perform in the specific detection environment (e.g. extremely high or low pH, salt, temperature and toxicity)? 4) does the host have a suitable system to access the target compounds (e.g. chemotaxis sensing, chemicals transport into the cells)? 5) is the host robust enough to be stored and functional for a period of time? 6) what is the risk if the host is released to environment?

Escherichia coli is the most popular chassis since the regulations of its genes have been intensively studied, and gene manipulation methods are well-established. However, for the detection of environmental contaminants and toxicity, other environmental microbes could be a good chassis in terms of ecological relevance and robustness. For example, BWBs based on the versatile chemo-heterotroph *Acinetobacter* sp. ADP1 (Huang et al. 2005; Huang et al. 2006; Huang et al. 2008; Song et al. 2009; Zhang et al. 2011) have been shown to perform well in the detection of several specific compounds. Figure 8.3 shows that ADP1 is able to actively emulsify, search, attach and sense oil in contaminated water, while this ability is absent in *E. coli*. As ADP1 was isolated from soil, ADP1 based BWBs can be robust enough to be stored on shelf (4°C) for a long period of time (>45 days) without any significantly change in the biosensing performance (Fig. 8.4).

Plasmids carrying a DNA cassette containing a promoter and a reporter gene are most common in BWBs construction and application. Plasmids are easy to operate in terms of genetic modification, transformation and expression in various host cells. The modularized structure also allows the ease of construction of fusion gene and gene expression. The high copy number of plasmids can strengthen the signal due to high level expression. However, plasmid based biosensors have several drawbacks: 1) plasmids

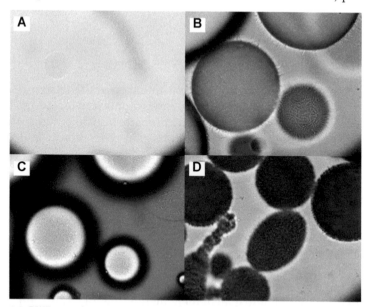

Figure 8.3. BWBs searching and sensing oil **(A)** *E. coli* DH5α mixed with mineral oil. **(B)** and **(C)** Oil sensing BWBs (*Acinetobacter* ADPWH_alk) mixed with mineral oil with different optical focuses. **(D)** ADPWH_alk mixed with crude oil. *E. coli* randomly distributed between water and oil, showing no affinity to oil droplets, while *Acinetobacter* actively searched and bound to oil droplets of mineral and crude oils.

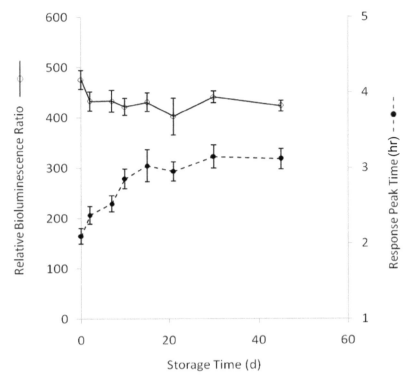

Figure 8.4. Storage and performance test of a salicylate responsive BWBs ADPWH_lux. The BWBs responsive ratio and activation time remained unchanged during a period of 45 dy storage at 4°C.

require selective pressure to be maintained, usually by adding antibiotics. 2) multiple copies of the promoter-binding-region on plasmids compete with the binding site on the chromosome, causing less expression of the inserted gene cassette (Applegate et al. 1998). 3) it has been demonstrated that the plasmid-based gene expression inevitably suffers from instability due to plasmid segregational and structural instability and allele segregation (Tyo et al. 2009). 4) the risk of leaking plasmids into the environment is higher than chromosome-based gene cassette because plasmids contain the whole elements for the replication and can be readily taken by other organisms in the environment while chromosomal DNA uptake requires natural competent and existing homologous fragments, which tends to be species-specific and hence less likely. Chromosomally-based BWBs have been used in detecting metals, antibiotics, pesticides, herbicides or a group of genotoxins. Table 8.1 lists some of the studies in which a fusion gene cassette was inserted into a chromosome and successfully expressed by the host cells.

Table 8.1. List of some biosensor constructed on chromosome of the host strain.

target	promoter	reporter gene	host cell	Chromosomal integration methods	reference
BTEX	P$_{tod}$	luxCDABE	*Pseudomonas putida*	mini-Tn5 transposon	(Applegate et al. 1998)
2,4-D	tfdRP$_{DII}$	luxCDABE	*Ralstonia eutropha*	mini-Tn5 transposon	(Hay et al. 2000)
tetracycline	P$_{tet}$	luxCDABE, lacZYA, gfp	*Escherichia coli*	mini-Tn5 transposon	(Hansen and Sorensen 2000b)
mercury	P$_{mer}$/P$_{merT}$	luxCDABE, lacZYA, gfp/egfp	*Escherichia coli, Pseudomonas putida*	mini-Tn5 transposon	(Hansen and Sorensen 2000a)
styrene	P$_{sty}$	lacZ	*Pseudomonas sp.*	mini-Tn5 transposon	(Alonso et al. 2003)
iron	P$_{fepA}$	luxCDABE	*Pseudomonas putida*	mini-Tn5 transposon	(Mioni et al. 2003)
PCB	P$_{m}$	Gfp	*P. fluorescens.*	mini-Tn5 transposon	(Liu et al. 2007)
salicylate	P$_{sal}$	luxCDABE, gfp	*Acinetobacter baylyi*	homogulos recombination	(Huang et al. 2005)
genotoxins	P$_{recA}$	luxCDABE	*E.col, Salmonella typhimurium/ Acinetobacter baylyi*	bacteriophage transduction/ homogulos recombination	(Davidov et al. 2000; Song et al. 2009)

Note:
2,4-D: 2,4-Dichlorophenol
BTEX: Benzene, toluene, ethylbenzene, and xylene
PCB: polychlorinated biphenyls

Synthetic Biology Design for BWBs

Synthetic biology employs an engineering approach to design, model, construct, test and characterize BWBs (Fig. 8.2). A few interesting cases of BWBs are shown below:

BWBs search and sense atrazine (Sinha et al. 2010)

An evolved riboswitch that is responsive to the herbicide atrazine was used to create a translation-based whole cell biosensor that could sense the presence of atrazine, move to the site of the contamination, and then degrade the toxin. The riboswitch was selected from a randomized library and placed in control of the translation of the bacterial chemotaxis protein *CheZ*. Therefore, cells were only able to activate chemotaxis in response to atrazine and would migrate to areas of high atrazine concentration. The same cells were given the ability to express an enzyme which converts atrazine to a lower toxicity compound. Thus, the whole cell biosensor was programmed to "seek and destroy" the contaminant (Sinha et al. 2010).

Imperial iGEM, 2010 (Parasight) (Post-translational biosensor, synthetic biology)

A post-translational biosensor for the detection of parasites in water was designed as part of the International Genetically Engineered Machines (iGEM) synthetic biology competition. In this example, a tetrameric enzyme was constitutively produced, but fused to another protein which would not allow it to form its active oligomeric state. The whole cell biosensor consists of three modules. The first to detect the release of a parasite invasion protease is composed of a surface tethered ligand which when cleaved by the protease activates cell signalling. The second is a two-component signalling pathway which produces a second protease. The second protease then processes the tetrameric enzyme, removing the piece that blocks oligomerization and activating the biosensor. Since one protease molecule can process many tetrameric reporters, the signal response is much faster than that of a transcriptional or translational biosensor, and is predicted to act on the order of minutes (iGEM 2010).

Cadmium Toggle switch, (Wu et al. 2009) (Synthetic Biology and Transcriptional sensor)

A transcription-based biosensor for cadmium was optimized using a synthetic biology approach. Rather than just fusing a cadmium responsive promoter to a reporter gene, it was found that using a toggle switch decreased the background signal and increased the sensitivity of the biosensor to low levels of cadmium. By designing in extra gene regulation, the properties of the biosensor were improved, illustrating the power of synthetic biology in optimizing whole cell sensors (Wu et al. 2009).

BIONANOTECHNOLOGY FOR ENDOWING NEW FUNCTIONS TO BWBS

Besides genetic modification, nanoparticles (NPs) derived from chemistry and physical sciences provide an additional new toolbox to modify BWBs. Bionanotechnology or nanobiotechnology, which couples nanoparticles (NPs ~1–100 nm) with biological system, is an emerging cutting-edge research area which promises to create novel devices with unprecedented features. Nano-scale NPs are small enough to exhibit new and interesting properties (e.g. quantum entanglement, unique photonic, electronic and catalytic properties) that bulk materials lack (Whitesides 2005). NPs functionalized with various molecules enable NPs to access the cell membranes or enter into the cytoplasm, where NPs specifically recognize and interact with targeting biomolecules (Katz and Willner 2004). NPs share similar sizes with many biomolecules such as enzymes and DNA in the range of 2–50 nm, this structurally compatible property makes NPs to straight away interfere with the DNA-protein and protein-protein interactions. The combination of nanotechnology and microbiology endows bacterial cells with new features which are predicted to have many useful applications in medicine, biofuel production and environmental monitoring and bioremediation. For example, *Acinetobacter* BWBs functionalized with magnetic nanoparticles (MNPs) can be controlled by a magnetic field, which enables in detecting specific compounds in complex samples such as sediments or soils (Zhang et al. 2011). The MNPs coupled BWBs were viable and functional as well as those untreated cells and they were used to sense salicylate, toluene/xylene and alkanes (Fig. 8.5). The coupling efficiency of MNPs to cells was greater than 99%. MNPs functionalized BWBs can be remotely controlled by a magnetic field and be redispersed when the magnet is removed. The MNPs coupled BWBs can be regarded as millions of remotely controllable mini-biosensing devices that can probe the chemical or biological information in wastewater, groundwater or complex environments where scientific instruments are unable to readily access. These BWBs can be easily recollected for further analysis after exposure to a contaminated environment (Zhang et al. 2011).

NPs modification, like genetic engineering, could be used to change bacterial properties. NPs could become a new kind of building blocks to cells. Convergence of genetic engineering and nanotechnology would create

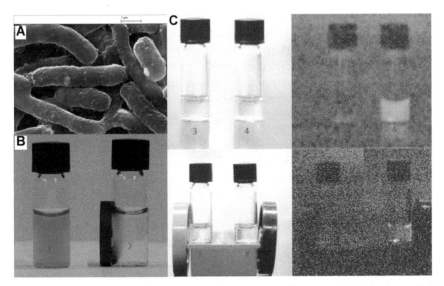

Figure 8.5. Viability and functionality of BWBs (*Acinetobacter* ADPWH_lux) coupled with magnetic nanoparticles (MNPs). **(A)** Scanning Electron Microscope (SEM) image of Acinetobacter ADPWH_lux bioreporters coupled with MNPs. **(B)** Most MNPs coupled cells originally suspended in water (1) were attracted to the side of a permanent magnet (2) after 5 min. **(C)** MNPs coupled cells were present with 0 µM (3) and 100 µM salicylate induction (4). Salicylate induced MNPs coupled cells were attracted to the magnet side (left images were under ordinary light and right in the dark).

Color image of this figure appears in the color plate section at the end of the book.

a powerful toolbox that extends DNA-based biotechnology to a further way modifying existing biological systems or even creating bio-hybrid and semi-biotic devices to carry out novel tasks. Although bionanotechnology has already been applied to microbiology and environmental science for water treatment and microbial detection, it yet reach its full potential. The NPs could also provide biocatalytic functions to cells. For example, magnetite (Fe$_3$O$_4$) NPs were previously thought to be inert but recent studies indicate they have peroxidase-like activity (Wei and Wang 2008). Bacteria functionalized with NPs would be a powerful tool, opening a new frontier for biosensors.

MULTIPLE SENSING USING BWBs ARRAYS

In most cases, contaminants in the environment are present as mixtures but not a single compound. To obtain a global picture of the impact of environmental pollution on human health, it is useful to detect multi-contaminants, toxicity and pathogens in-parallel. To address this challenge, BWBs arrays are proposed which are made up of a large number of well-characterized BWBs onto a solid surface, and simultaneously detect multi-contaminant and pathogen mixtures in environmental soil or water samples (Elad et al. 2008; van der Meer and Belkin 2010). Since each BWBs in the array will have a different genetic configuration, the first challenge is to make every BWBs responsive under the same conditions. Second, the BWBs array should avoid cross-sensing. This requires each BWBs to be well-defined and characterized. In an attempt to prove the concept, we employed five different BWBs to make a simple array and applied it to test a mixture which contained 100 µM sodium salicylate, 10 µM sodium benzoate and 100 nM MMC. The BWBs are separately responsive to toluene (ADPWH_Tol), salicylate (ADPWH_lux), alkanes (ADPWH_alk), benzoate (ADPWH_BenM) and genotoxicity ADPWH_RecA. The result showed that the BWBs array correctly indicated the contaminants present in the mixture without cross-sensing (Fig. 8.6). It suggests that the construction of a high-throughput BWBs array could be feasible in near future.

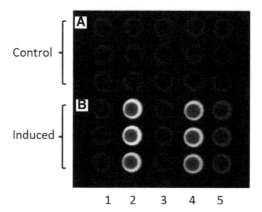

Figure 8.6. A BWBs array for the detection of a contaminant-mixed sample, which contained 100 µM sodium salicylate, 10 µM sodium benzoate and 100 nM mitomycin C. Panel **(A)** is BWBs were added with water as a control and panel **(B)** is BWBs were added with the multi-contaminant sample. Lane 1 is ADPWH_Tol for toluene; lane 2 is ADPWH_lux for salicylate; lane 3 is ADPWH_alk for alkanes; lane 4 is ADPWH_BenM for benzoate; and lane 5 is ADPWH_RecA for toxicity. BWBs array correctly responded to the contaminants in the mixture without cross-talking.

APPLICATION IN TOXICITY, BIOMARKERS AND PATHOGENS DETECTION

BWBs for cytotoxicity detection were usually constructed by fusing a reporter gene with a constitutively expressed promoter, which leads to a continuous expression of reporting signal. This type of biosensors was widely used in detection of acute toxicity or other cell growth inhibition chemicals. One of the most used promoters is *lac*, due to its high transcription level in plasmids. Usually cytotoxicity BWBs give a quick response to heavy metals and some small molecular organic compounds (Paton et al. 2006) in less than 30 min after being exposed to a substrate.

However, the cytotoxicity BWBs are unable to detect genotoxins which may not have an immediate enzymatic inhibition effect. Traditionally, genotoxicity assessment methods include animal tests and the Ames *Salmonella typhimurium*/microsome mutagenicity assay (Ames test). Animal testing is less desirable as it is laborious and costly, and it requires the sacrifice of animals. The Ames test requires a large quantity of the test compound (1 g) and at least a 72 hr incubation period, which hampers its widespread use for either online or in situ monitoring. Since DNA damage can have many different forms (e.g. point mutation, deletion, insertion, break, dimer formation, cross-linking, rearrangement), the Ames test must use a cocktail of *his*- Salmonella tester strains to encompass different types of DNA damage, each for a specific DNA target (Mortelmans and Zeiger 2000).

A few BWBs have been constructed for genotoxicity detection. The "SOS" mechanisms in *E. coli* has been well-established and the "SOS" promoters have been frequently used to construct genotoxicity BWBs (Davidov et al. 2000). *Salmonella typhimurium* has become another favourable host choice for genotoxicity BWBs as the toxicity data obtained from *Salmonella* BWBs could be compared with normalized Ames results (Mortelmans and Zeiger 2000). Current commercialized BWBs kits for genotoxicity test, such as *umu*/SOS chrometest, Vitotox, *sulA*-test, all use *Salmonella typhimurium* as host cell (vanderLelie et al. 1997). Since most genotoxicity BWBs are "light on" system, it may cause false negative results due to high concentrations of contaminants. In an attempt to monitor radioactivity and genotoxicity in a highly radioactive environment, it is proposed that the genetically modified radio-resistant bacteria *Deinococcus radiodurans* could serve as a good host while other former biosensors may have failed to finish the task due to decrease of cell viability (Gao et al. 2008). *Acinetobacter baylyi* strain ADP1 was genetically modified to create a *recA::luxCDABE* fusion on the chromosome of the host cell, while high baseline *recA* expression was used to circumvent the false negative problem (Song et al. 2009). A recent review paper provides a list of genetically engineered genotoxicity BWBs (Biran et al. 2010).

BWBs are also be used to sense biomarkers for diseases. Glucose is a biomarker for diabetes. It is required to constantly monitor glucose level in a patient's blood and subsequently injecting insulin to control sugar level in blood. Glucose-sensing circuit was designed to sense glucose and activates insulin synthesis (Yang et al. 2011). High concentration of glucose inhibits cAMP synthesis, leading to a high level phosphorylation of *OmpR* that triggers a promoter of *OmpC* gene (Yang et al. 2011). Yang and colleagues fused two genes encoded peptides α and β subunits of insulin with an *OmpC* promoter to achieve glucose controlled insulin synthesis in *E. coli*. Such BWBs provided a platform to test and debug glucose controlled insulin synthesis system (designated as GlucOperon) before the sense-synthesis system can be used in therapeutic application. Bacteria are able to activate the responding regulatory system in the presence of drugs and drug-specific BWBs has been successfully used for the detection and discovery of novel drugs (e.g. ferrimycin A1) (Urban et al. 2007). Due to a different regulatory response, a soil bacterium *Bacillus subitlis* has been employed as a chassis for the construction of antibiotic BWBs (Urban et al. 2007). Five *Bacillus subtilis* promoters were fused with firefly luciferase reporter luc, which are related to DNA (*yorB*), RNA (*yvgS*), protein (*yheI*), cell wall (*ypuA*) and fatty acids (*fabHB*) biosynthesis separately. Those BWBs are a light-off system (Fig. 8.1) and new antibiotics can be discovered due to their inhibitive effect of DNA, RNA, protein, cell wall or fatty acids synthesis. Bacillus based BWBs were used as a high-throughput approach to screen 14,000 pure natural products with potential antibiotic activities (Urban et al. 2007). Quorum sensing is a bacterial cell-cell communication mechanism that uses signal molecules (autoinducers-AIs) to sense bacterial population density and triggers expression of some genes including virulent genes when the AIs concentration reaches a threshold. Some pathogens such as *Pseudomonas aeruginosa, Erwinia* spp., *Burkholderia cepacia, Serratia liquifaciens, Yersinia enterocholitica,* and *Vibrio* spp secrete small, self-generated signal molecules (N-acyl homoserine lactones-AHL in most gram-negative bacteria) to achieve intercellular communication. Different bacteria produce different AHLs and have responding AHL-controlled promoters. Pathogen BWBs consist of an AHL-controlled promoter fused to a reporter gene (e.g. *lacZ* or the *luxCDABE*), which is able to detect the presence of AHLs (at pM level) as indicators of specific pathogens (Massai et al. 2011). To simplify and explore the clinical application, BWBs have been immobilized and dried on filter paper strips to the detection of AHLs in human saliva with sensitivity down to 10 nM (Struss et al. 2010). BWBs in medical and clinical application are showing great promise. However there are limited publications of the application of BWBs on human serum and urine samples. Future work will be needed to explore more clinical applications.

APPLICATIONS TO AREAS OF HEALTH AND DISEASE

The function of BWBs is biologically detecting substances of interest. One of the practical applications of the BWBs is to detect specific biomarkers related to human diseases, such as uric acid, glucose, antibodies, hormones and cancer-related factors including chemical agents, virus, and DNA damage caused by smoking and UV radiations. The role of BWBs for human health and disease is firstly to provide a way to measure and assess the potential risk that the substances would pose on humans and on the environment when humans are exposed. The substances include pesticides, herbicides, solvents, petrochemicals, heavy metals, antibiotics and drugs, endocrine disruptors, pharmaceutical waste and radiation. Chemicals and contaminants in the environment may exert two types of toxicity effects on human and animal cells: cytotoxicity and genotoxicity. Cytotoxicity involves inhibition of enzyme activity and cellular respiration, resulting in an acute and immediate effect; while genotoxicity involves deleterious actions on cellular genetic materials (DNA or RNA) and is usually related to cancers and other genetic diseases. Since chemical compounds are continually entering the environment, it is essential for safety assessments to effectively identify contaminants that could potentially damage DNA and lead to cancer or other genetic diseases.

The second role of BWBs for human health and disease is to provide a way to diagnose and estimate the health status of a person by detecting biomarkers when the person is being medical examined. This may include saliva, serum, urine or skin-surface testing for disease biomarker, pathogen and chemicals.

KEY FACTS

- The first whole-cell biosensor assay for toxicity detection was reported in 1981 by Bulich (Bulich and Isenberq 1981). It used bioluminescent bacteria to indicate the toxicity through the decrease of bioluminescence upon the exposure to contaminated aquatic samples.
- Most biosensors used in environmental and health monitoring cost US$1–15 per analysis. Their sensitivity could reach as much as parts per million to part per billion.
- Cytotoxicity is related to inhibition of enzyme and cellular activity, which usually results in an acute and immediate effect. Compounds responsible for cytotoxicity can be enzyme inhibitors such as cyanide and heavy metal ions.
- Genotoxicity involves deleterious actions on cellular genetic materials (DNA or RNA) and is usually related to cancers and other genetic diseases. High concentrations of genotoxic compounds may also result in cytotoxic effects.
- PAHs are of concern because some compounds have been identified as carcinogenic and mutagenic. One PAH compound, benzo[a] pyrene, is notable for being the first chemical carcinogen to be discovered. However it is acknowledged that PAHs compounds

donot damage the DNA until they are activated to their oxidized metabolic intermediate by mammal cytochrome P450 in the liver.

- Dioxins are a group of most toxic organic compounds. They are from industrial processes (e.g. incineration) and natural events (e.g. volcano eruptions). About 30 dioxin-related compounds are significantly toxic, among them 2,3,7,8- tetrachlorodibenzo-*p*-dioxin (TCDD) is the most toxic.

DEFINITIONS

- *Ames Test*: The Ames test is a biological assay to assess the mutagenic potential of chemical compounds. A positive test indicates that the chemical might act as a carcinogen.
- *Bioavailability*: In microbiology, bioavailability describes the extent to which a chemical is accessed by an organism as well as the ability for the chemical to cross an organism's cellular membrane from the environment.
- *Cell array*: A biosensor cell array consists of several biosensors with different functions, so that a property matrix of chemicals or samples being tested can be achieved in one analysis.
- *Chassis*: Chassis is the bacterial host of BWBs, in which the genetic modified elements is expressed.
- *Förster (or Fluorescence) resonance energy transfer (FRET)*: is a distance-dependent interaction between two chromophores are located in near-field (usually < 10 nm), in which excitation is transferred from a donor molecule to an acceptor molecule. The FRET-based biosensor is able to report local protein configuration changes induced by ions and small molecules.
- *Induction ratio*: Induction ratio is a very important concept in biosensing which refers to the ratio of signal given by an induced biosensor with a uninduced biosensor in the "light on" system. A higher induction factor of different chemicals or samples to certain biosensors implies a stronger effect of that chemical. And a higher induction factor of the same chemical towards different biosensors implies the sensitivity of the biosensor. For "light off" system, inhibition ratio is used accordingly.
- *Bionanotechnology*: Or Nanobiotechnology is to bring nanotechnology to biological research. It uses nanomaterials to modify biological elements to achieve novel functions.
- *Quorum sensing*: Bacterial cell-cell communication by releasing and sensing of small diffusible signal molecules, which trigger some gene expression when it reaches a threshold.

- *Riboswitch*: A part of an mRNA molecule that can directly bind a small target molecule which affects gene expression.
- *Synthetic biology*: Reprogram cells to carry out novel tasks by modifying their software (e.g. nucleic acids, DNA/RNA) and hardware (cellular machinery system, e.g. ribosome, RNA ploymerase).

SUMMARY POINTS

- BWBs have unique advantages over other sensing technologies in toxicity and bioavailability assessment;
- Synthetic biology has been employed in BWBs design, construction, modelling, debugging and test and characterization, which may be able to push the technology forward by helping to develop more robust, reliable and novel sensors for broad application;
- Bionanotechnology will endow BWBs with novel functions (e.g. special catalysis) and make BWBs more compatible and manipulative.
- With advances in molecular technology, a broad range of host species (chassis) can be chosen to construct new BWBs that serve to different purposes;
- BWBs can be rapid, simple, cost-effective, sensitive and accurate enough to be a semi-quantitative tool, complementary to more precise chemical analysis;
- Significant improvements in the biosensor characteristics are still necessary, particularly the ability to sense contaminants in a complex milieu containing multiple contaminants alongside many other non-toxic compounds such as is found in natural soil or water samples;
- The use of BWBs for human health applications involving patients also faces challenges with regards to patient safety. Future work will be needed to explore more clinical applications.

ACKNOWLEDGEMENTS

We thank the Royal Society Research 2009/R2 and NERC (NE/F011938/1) fund to W.E.H. We also thank China National Natural Science Foundation (Project 40730738), and Tsinghua Research Funding (Project 20121080049).

ABBREVIATIONS

2,4-D	:	2,4-Dichlorophenol
AHL	:	N-acyl homoserine lactones
BaP	:	Benzo[a]pyrene
BTEX	:	Benzene, toluene, ethylbenzene, and xylene
BWBs	:	Bacterial whole-cell biosensors
CYP	:	cytochrome-P450-dependent monooxygenase
FMNH$_2$	:	reduced flavin mononucleotide
FPs	:	fluorescent proteins
GFP	:	green fluorescent protein
YFP	:	yellow fluorescent protein
CFP	:	cyan fluorescent protein
RFP	:	red fluorescent protein
GC/MS	:	gas chromatography/mass spectrometry
HPLC	:	high-pressure liquid chromatography
MMC	:	mitomycin C
NPs	:	Nanoparticles
MNPs	:	Magnetic nanoparticles
PAHs	:	polycyclic aromatic hydrocarbons
PCB	:	polychlorinated biphenyls

REFERENCES

Alonso S, JM Navarro-Llorens, A Tormo and J Perera. 2003. Construction of a bacterial biosensor for styrene. Journal of Biotechnology 102: 301–306.

Applegate BM, SR Kehrmeyer and GS Sayler. 1998. A chromosomally based tod-luxCDABE whole-cell reporter for benzene, toluene, ethybenzene, and xylene (BTEX) sensing. Applied and Environmental Microbiology 64: 2730–2735.

Biran A, S Yagur-Kroll, R Pedahzur, S Buchinger, G Reifferscheid, H Ben-Yoav, Y Shacham-Diamand and S Belkin. 2010. Bacterial genotoxicity bioreporters. Microbial Biotechnology 3: 412–427.

Bulich AA and DL Isenberq. 1981. Use of the luminescent bacterial system for the rapid assessment of aquatic toxicity. ISA Transactions 20: 29–33.

Daunert S, G Barrett, JS Feliciano, RS Shetty, S Shrestha and W Smith-Spencer. 2000. Genetically engineered whale-cell sensing systems: Coupling biological recognition with reporter genes. Chemical Reviews 100: 2705–2738.

Davidov Y, R Rozen, DR Smulski, TK Van Dyk, AC Vollmer, DA Elsemore, RA LaRossa and S Belkin. 2000. Improved bacterial SOS promoter:: lux fusions for genotoxicity detection. Mutation Research-Genetic Toxicology and Environmental Mutagenesis 466: 97–107.

DeFraia CT, EA Schmelz and ZL Mou. 2008. A rapid biosensor-based method for quantification of free and glucose-conjugated salicylic acid. Plant Methods. 4.

Elad T, JH Lee, S Belkin and M Gu. 2008. Micorbial whole-cell array. Microbial Biotechnology 1: 137–148.

Gao GJ, L Fan, HM Lu and YJ Hua. 2008. Engineering Deinococcus radiodurans into biosensor to monitor radioactivity and genotoxicity in environment. Chinese Science Bulletin. 53: 1675–1681.

Gu M, R Mitchell and B Kim. 2004. Whole-Cell-Based Biosensors for Environmental Biomonitoring and Application. Advances in Biochemical Engineering/Biotechnology 87: 269–305.

Hansen LH and SJ Sorensen. 2000a. Versatile biosensor vectors for detection and quantification of mercury. Fems Microbiology Letters 193: 123–127.

Hansen LH and SJ Sorensen. 2000b. Detection and quantification of tetracyclines by whole cell biosensors. Fems Microbiology Letters 190: 273–278.

Hay AG, JF Rice, BM Applegate, NG Bright and GS Sayler. 2000. A Bioluminescent Whole-Cell Reporter for Detection of 2,4-Dichlorophenoxyacetic Acid and 2,4-Dichlorophenol in Soil Appl Environ Microbiol 66: 4589–4594.

Huang WE, H Wang, LF Huang, HJ Zheng, AC Singer, IP Thompson and AS Whiteley. 2005. Chromosomally located gene fusions constructed in Acinetobacter sp ADP1 for the detection of salicylate. Environmental Microbiology 7: 1339–1348.

Huang WE, LF Huang, GM Preston, M Naylor, JP Carr, YH Li, AC Singer, AS Whiteley and H Wang. 2006. Quantitative in situ assay of salicylic acid in tobacco leaves using a genetically modified biosensor strain of Acinetobacter sp ADP1. Plant Journal 46: 1073–1083.

Huang WE, AC Singer, AJ Spiers, GM Preston and AS Whiteley. 2008. Characterizing the regulation of the Pu promoter in Acinetobacter baylyi ADP1. Environmental Microbiology 10: 1668–1680.

Katz E and I Willner. 2004. Integrated nanoparticle-biomolecule hybrid systems: Synthesis, properties, and applications. Angewandte Chemie-International Edition 43: 6042–6108.

Kiel C, E Yus and L Serrano. 2010. Engineering Signal Transduction Pathways Cell 140: 33–47.

King JMH, PM Digrazia, B Applegate, R Burlage, J Sanseverino, P Dunbar, F Larimer and GS Sayler. 1990. Rapid, Sensitive Bioluminescent Reporter Technology For Naphthalene Exposure and Biodegradation. Science 249: 778–781.

Kitney R,. P Freemont and V Rouilly. 2007. Engineering a molecular predation oscillator. IET Synth Biol 1: 68–70.

Liu XM, KJ Germaine, D Ryan and DN Dowling. 2007. Development of a Gfp-based biosensor for detecting the bioavailability and biodegradation of polychlorinated biphenyls (PCBs). Journal of Environmental Engineering and Landscape Management 15: 261–268.

Massai F, F Imperia, S Quattruccib, E Zennaroa, P Viscaa and L Leonia. 2011. A multitask biosensor for micro-volumetric detection of N-3-oxo-dodecanoyl-homoserine lactone quorum sensing signal. Biosensors and Bioelectronics 26: 3444–3449.

Mioni CE, AM Howard, JM DeBruyn, NG Bright, MR Twiss, BM Applegate and SW Wilhelm. 2003. Characterization and field trials of a bioluminescent bacterial reporter of iron bioavailability. Marine Chemistry 83: 31–46.

Mortelmans K and E Zeiger. 2000. The Ames Salmonella/microsome mutagenicity assay. Mutation Research-Fundamental and Molecular Mechanisms of Mutagenesis 455: 29–60.

Palmer AE, Y Qin, JG Park and JE McCombs. 2011. Design and application of genetically encoded biosensors. Trends in Biotechnology 29: 144–152.

Paton GT, W Cheewasedtham, IL Marr and JJC Dawson. 2006. Degradation and toxicity of phenyltin compounds in soil. Environmental Pollution 144: 746–751.

Quillardet P, O Huisman, R Dari and M Hofnung. 1982. SOS Chromotest, a direct assay of induction of an SOS function in Escherichia-coli K-12 to measure genotoxicity. Proceedings of the National Academy of Sciences of the United States of America-Biological Sciences 79: 5971–5975.

Schwarzenbach RP, BI Escher, K Fenner, TB Hofstetter, CA Johnson, U von Gunten and B Wehrli. 2006. The challenge of micropollutants in aquatic systems. Science 313: 1072–1077.

Sinha J, SJ Reyes and JP Gallivan. 2010. Reprogramming bacteria to seek and destroy an herbicide. Nature Chemical Biology 6: 464–470.

Song YZ, GH Li, SF Thornton, IP Thompson, SA Banwart, DN Lerner and WE Huang. 2009. Optimization of Bacterial Whole Cell Bioreporters for Toxicity Assay of Environmental Samples. Environmental Science & Technology 43: 7931–7938.

Sorensen SJ, M Burmolle and LH Hansen. 2006. Making bio-sense of toxicity: new developments in whole-cell biosensors. Current Opinion in Biotechnology 17: 11–16.

Struss A, P Pasini, CM Ensor, N Raut and S Daunert. 2010. Paper Strip Whole Cell Biosensors: A Portable Test for the Semiquantitative Detection of Bacterial Quorum Signaling Molecules. Analytical Chemistry 82: 4457–4463.

Tyo KEJ, KP Ajikumar and R Stepanauskas. 2009. Stabilized gene duplication enables long-term selection-free heterologous pathway expression. Nature Biotechnology 27: 760–765.

Urban A, S Eckermann, B Fast, S Metzger, M Gehling, K Ziegelbauer, H Rubsamen-Waigmann and C Freiberg. 2007. Novel whole-cell antibiotic biosensors for compound discovery. Applied and Environmental Microbiology 73: 6436–6443.

vanderLelie D, L Regniers, B Borremans, A Provoost and L Verschaeve. 1997. The VITOTOX(R) test, an SOS bioluminescence Salmonella typhimurium test to measure genotoxicity kinetics. Mutation Research-Genetic Toxicology and Environmental Mutagenesis 389: 279–290.

van der Meer JR and S Belkin. 2010. Where microbiology meets microengineering: design and applications of reporter bacteria. Nature Reviews Microbiology 8: 511–522.

VanEngelenburg SB and AE Palmer. 2008. Fluorescent biosensors of protein function. Current Opinion in Chemical Biology 12: 60–65.

Virta M, J Lampinen and M Karp. 1995. A luminescence-based mercury biosensor. Analytical Chemistry 67: 667–669.

Wei H and E Wang. 2008. Fe3O4 magnetic nanoparticles as peroxidase mimetics and their applications in H2O2 and glucose detection. Analytical Chemistry 80: 2250–2254.

Whitesides GM. 2005. Nanoscience, nanotechnology, and chemistry. Small. 1: 172–179.

Wu CH, D Le, A Mulchandani and W Chen. 2009. Optimization of a Whole-Cell Cadmium Sensor with a Toggle Gene Circuit. Biotechnology Progress 25: 898–903.

Yang C, D Lin, L Yen, W Liao, S Wen and C Chang. 2011. Designer Biosystem with Regulated Insulin Expression and Glucose Auto-Sensing for Diabetes. International Journal of Systems and Synthetic Biology 1: 143–153.

Zhang D, RF Fakhrullin, M Özmen, H Wang, J Wang, VN Paunov, G Li and WE Huang. 2011. Functionalization of whole-cell bacterial reporters with magnetic nanoparticles. Microbial Biotechnology 4: 89-97.

9

Environmental Toxicology Monitoring with Polyazetidine-based Enzymatic Electrochemical Biosensors

Franco Mazzei[1,a,*] and Gabriele Favero[1,b]

ABSTRACT

Polyazetidine prepolymer (PAP) has been used since two decades as an immobilizing agent for proteins in biosensors development; some of them are suitable for determination of toxics and pollutants either in inhibition or direct analysis. PAP immobilization procedure is a combination of physical entrapment and chemical cross linking towards several moieties in the proteins; the most interesting features displayed by the obtained layer are the permeability towards both analytes and electrochemical mediators and the formation of an aqueous microenvironment suitable to preserve a native-like structure of immobilized enzymes. Furthermore the PAP layer ensures an optimal contact time between the enzyme and the analytes to be determined especially important for the inhibition based biosensor configuration. Examples of determination of pesticides, herbicides, phycotoxins, biogenic amines and polyphenols with PAP-based biosensors, are described herein.

[1]Department of Chemistry and Drug Technologies, Sapienza University of Rome, P.le Aldo Moro, 5 – 00185 Roma, Italy.
[a]E-mail: franco.mazzei@uniroma1.it
[b]E-mail: gabriele.favero@uniroma1.it
*Corresponding author

List of abbreviations after the text.

INTRODUCTION

The last years have been characterized by an increasing number of pollutants released in the environment, especially by human activities in water. For this reason there is a great need for disposable systems for screening and real time monitoring, possibly "on-site", of the several contaminants as cheaply as possible. A wide number of analytical methods, based on the most commonly employed physico-chemical techniques for the detection of several pollutants (mainly HPLC, GC, MS or various combinations of them), in environmental matrices are available (Barcelò 1993). These traditional techniques are extremely powerful in terms of sensitivity and selectivity, but, apart from economic consideration, they also present some drawbacks, because of the need for qualified staff and to carry out extensive pretreatment processes on the sample to be assayed. More important, these techniques cannot be applied on the spot, thus making the continuous monitoring of a risky area virtually impossible. These reasons have led to the development of alternative analytical devices and methods to be applied for screening and monitoring of various contaminants in environmental matrices, possibly minimizing the pretreatment of samples, reducing the cost and time of analysis. Nonetheless, an ideal screening method should satisfy the basic requirement of ensuring the detection of the different contaminants without the risk of false negatives, and with a percentage of false positive as low as possible.

In this direction, the use of alternative analytical methods, such as those based on biosensors, would be very welcome, since they could represent one of the most practical and inexpensive analytical devices, especially for the preliminary screening of huge numbers of samples.

Biosensors are based on the coupling of a biochemical component (usually an enzyme, an antibody or an aptamer) to a physic-chemical transducer (electrochemical, optical, acoustic and so on). A particular subclass of biosensors is represented by the electrochemical biosensors or bioelectrodes which show a great potential in recent years and thus are proposed as analytical tools for effective monitoring of the toxicological risk (Palchetti et al. 2009). Electrochemical biosensors are cheap, fast, sensitive and can be miniaturized so they can be used for on-site analysis to provide enough information for routine testing and screening of samples. In this work we focused our attention on the employment of enzymatic electrochemical biosensors for toxicological risk assessment. Electrochemical enzymatic biosensors for environmental monitoring are mainly based on two operational mechanisms: the first mechanism involving the biocatalytic reaction of the pollutants as enzymatic substrates and the second mechanism involves the detection of pollutants that inhibit the enzymatic activity.

These biosensors, can be classified into first, second and third generation: a) first generation electrochemical biosensors, are based on the measurement of the product of the enzymatic reaction, (i.e. hydrogen peroxide, NADH) or on the monitoring of the consumption of the co-factor (i.e. oxygen); b) second generation biosensors use an electron transfer mediator between the enzyme redox center and the electrode surface; c) third-generation biosensors are characterized by the direct electron transfer between the electrode surface and the active site of the enzyme (Chaubey and Malhotra 2002). Nowadays, the most widely used electrochemical biosensors are the electron transfer ones, which are based on the use of enzymes catalyzing redox reactions.

One of the most important aspects that have to be taken into account in the development of electrochemical enzymatic biosensors is represented by the proteins immobilization on the electrode surface. An ideal immobilization procedure should guarantee the stable presence of the proteins at the electrode surface, maintaining their catalytic and biochemical properties and, at the same time, allowing a good accessibility to the target analytes, inhibitors, co-factors and in the case of second generation electrochemical biosensors also to redox mediators. In the case of electrochemical third generation biosensors the immobilization procedure should ensure also intact electrocatalytic properties to the enzymes and promote an efficient electron transfer communication with the electrode surface (Armstrong and Wilson 2000; Katz and Willner 2004; Wang 2008). For these reasons the protein immobilization procedure and its improvement are of paramount importance in order to enhance biosensor performance.

The immobilization procedure based on the use of polyazetidine uses both a chemical immobilization and a physical entrapment: this is possible due to the peculiar features characterizing the polyazetidine prepolymer. Indeed, PAP acts as a cross-linking agent as it is able to react with several different organic moieties (thiolic, oxydril, carboxyl, and amino groups), thus increasing the likelihood of chemical bonds being created with the enzyme and thereby enhancing immobilization efficiency (see Scheme 1).

This scheme of possible reactions involving PAP and some of the most important organic moieties emerges from the characterization performed by Hercules, Inc., and was confirmed by us in several previously published papers (Mascini and Mazzei 1987a, Mazzei et al. 1996b, Mazzei et al. 2004; Botrè et al. 1992). In these works, polyazetidine prepolymer (PAP) has also been extensively used to develop first generation biosensors. In the last few years, the employment of PAP as an immobilizing agent for redox proteins on electrodes aiming to second and third generation biosensors development is rapidly increasing thanks to the ability of PAP to preserve the enzymatic electrocatalytic properties in both direct and mediated electron transfer phenomena.

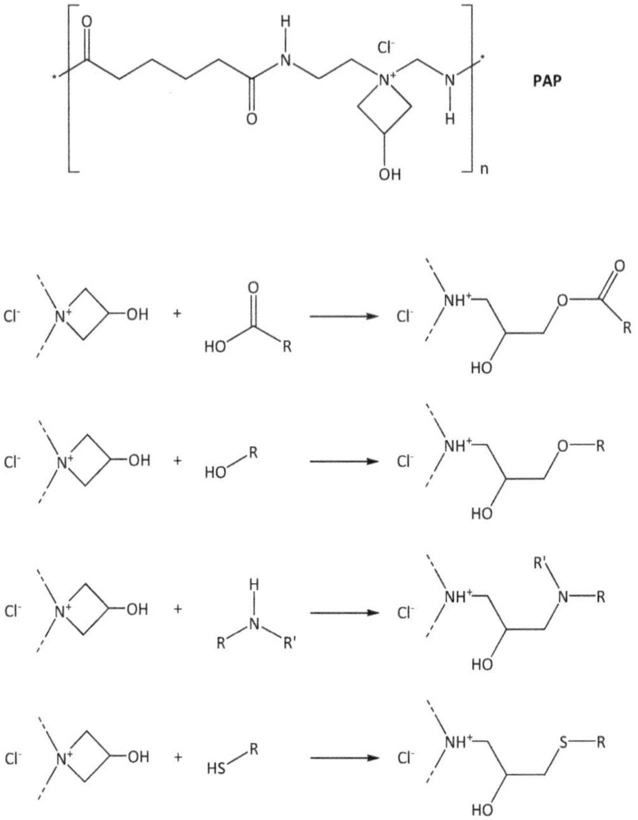

Scheme 1

The use of PAP as an immobilizing agent of redox proteins for use as chemical sensors, transducers of biosensors, electrodes of biofuel cells, and other devices allows to overcome several drawbacks occurring with other procedures; in fact, one common method used for this purpose is to spread polymeric membranes over the electrode surface. Such membranes can be divided into two categories: redox polymers in which the redox active sites are included in the polymer backbone or in the side chain, and ion-exchange polymers in which the redox sites are immobilized by electrostatic attraction (Cracknell et al. 2008). Most of the recent research in this field has been focused on the ion-exchange polymeric films: the most extensively studied of these is Nafion, a perfluorinated polysulfonate material (Tsai et al. 2005). Nevertheless, a problem associated with many ion exchange polymers, including Nafion, is that they yield small diffusion coefficients for the immobilized species, specifically, a reduction of up to four orders of magnitude with respect to the same species free in solution. For certain applications requiring rapid charge transfer, such as electrocatalysis, these small diffusion coefficients are undesirable. On the other hand, hydrogels have a number of characteristics that make them desirable electrode coatings, especially for bioelectrochemical applications: high water content, biocompatibility, low interfacial tension between hydrogel surface and aqueous solution, excellent diffusion characteristics for small molecules and ions, and optical transparency (Ratner and Hoffman 1976). PAP, once spread onto an electrode surface without the addition of other compounds to form a pure polymeric film, acts as a hydrogel and may thus be described as a single-phase "aqueous" matrix with excellent diffusional characteristics, representing an interesting alternative as an immobilizing agent in second generation electrochemical biosensor development.

The special features of PAP become particularly important also in the development of third generation biosensors (Shan et al. 2007; Zhang et al. 2004). These biosensors are characterized by several advantages over the traditional electrochemical biosensors: indeed, the possibility of having direct electron transfer between the protein and the electrode surface enables in constructing biosensors without the use of redox mediators thus obtaining a higher selectivity, because they can operate in a potential window closer to the redox potential of the enzyme leading to a lower sensitivity towards possible interfering reactions (Gorton et al. 1999). Despite these advantages, direct electrical communication between redox proteins and electrode supports is uncommon as it is often hindered by the inaccessibility of the protein redox center, which thus limits the development of this type of biosensor (Lu et al. 2006). The considerable progress made in this field is evident by the modification of electrodes using appropriate support providing a favorable micro-environment for the protein to exchange its electron directly with the underlying electrode, entailing a new opportunity for the detailed study of enzyme electrochemistry (De Groot et al. 2007). In the last few years several immobilization procedures suitable for studying direct electron transfer have been developed (Lu et al. 2007; Wang et al. 2003; Zhang et al. 2007). Most of them are characterized by both advantages and drawbacks, the greatest of which generally consists in (i) the difficulty to ensure an adequate protein mobility in order to correctly orientate its redox center, and thus to achieve a better electron transfer with the electrode surface and (ii) the possible denaturation of the proteins when they are directly adsorbed on to the electrode, leading to the loss of any electrochemical activity (De Groot et al. 2005).

In view of the possible use of PAP as an immobilizing agent for electron transfer based biosensors for toxicological studies, either the ability to retain in a native-like structure of proteins and the polymer permeability to redox mediator were evaluated.

(Mazzei et al. 2004; Frasconi et al. 2009; Di Fusco et al. 2010; Tortolini et al. 2010). More recently, a thorough investigation of the permeability of PAP toward redox mediators when used as an immobilizing agent in second-generation biosensor development was performed by determining the diffusion coefficient of different redox mediators using the same electrode surface before and after PAP deposition.

PERMEABILITY OF PAP IN BIOSENSORS DEVELOPMENT

The major problem to be addressed in the chronoamperometric measurement of the diffusion coefficient using a membrane coated electrode is the unknown analyte concentration in the hydrogel; to work around this the data needs to be treated in accordance with the normalized Cottrell equation (Denuault et al. 1991). The latter consists of a rearrangement of the classical Cottrell equation describing the chronoamperometric response of an electroactive species at a microdisk electrode (I): the rearrangement consists in dividing by the steady state current (I_d) thus obtaining that the dependence on the analyte concentration in the hydrogel layer has been elided.

$$\frac{I}{I_d} = \frac{r \cdot \sqrt{\pi}}{4 \cdot \sqrt{D \cdot t}} + 1$$

The plot of I/I_d vs $t^{-1/2}$ (where t is the time elapsing after the application of an appropriate potential step) has the form of a straight line with an intercept equal to unity and a slope $S = (r \cdot \pi^{1/2})/(4 \cdot D^{1/2})$, from which it is possible to evaluate D if the radius (r) of the microdisk is known. While potential-step chronoamperometry allows a simple evaluation of the diffusion properties, scan voltammetry can be used to investigate the efficiency of heterogeneous electron transfer. Because of the steady-state nature of the diffusion layer at a microdisk electrode, a sigmoidal voltammogram should be obtained at a slow scan rate (Howell and Wightman 1984). The equation for the voltammetric response of such a quasireversible redox reaction under steady-state conditions at a microdisk electrode was proposed by Galus et al. (Galus et al. 1988): The logarithmic form of this equation can be used to determine the standard heterogeneous rate constant (k_s):

$$E - E^{0'} \frac{1}{(1-\alpha)nf} \cdot \ln \frac{4D}{\pi k_s r} - \frac{1}{(1-\alpha)nf} \bullet \ln \left(\frac{I_d - I}{I} - \frac{I_d - I_r}{I_r} \right)$$

Where I_d is the steady-state current, and I_r is the calculated reversible current. For a given value of the radius r, the first term on the right-hand side is constant; therefore, a plot of $E-E^{0\prime}$ vs $\ln\{[(I_d-I)/I]-[(I_d-I_r)/I_r]\}$ should be linear with slope $1/(1-\alpha)nf$ and intercept $[1/(1-\alpha)nf]\ln(4D/\pi k_s r)$ from which the k_s value may be easily calculated, provided that r and D are known.

Measurement of Apparent Diffusion Coefficients and Standard Heterogeneous Rate Constants

The apparent diffusion coefficients D_{app} of seven simple redox molecules (2,2'-azino-bis(3-ethylbenzothiazoline-6-sulfonic acid) diammonium salt (ABTS), catechol, dopamine, ferrocenecarboxylic acid (FcCOOH), ferricyanide, ferrocyanide, osmium complex bis(2,2-bipyridyl)-4-aminomethylpyridine chloride hexafluorophosphate ($Os[(bpy)_2$ 4-AMP $Cl]^+$)) were obtained (Di Fusco et al. 2011) by means of potential-step chronoamperometry at different temperatures either in absence and in presence of PAP; in the first case the probe concentration is known, because of the lack of the membrane, so the data were fitted (using a nonlinear fit algorithm) according to the Cottrell equation and the diffusion current obtained at the steady-state I_d was used to calculate D_{app} values. On the other hand, the heterogeneous rate constants k_s of the compounds at different temperatures were also determined using steady-state voltammetry. The averaged results of all compounds are listed in Table 9.1. The heterogeneous rate constants k_s of the compounds at different temperatures were also determined using steady-state voltammetry observing that both the heterogeneous rate constants and diffusion coefficients increase with increasing temperature as result of the strict dependence of k_s on D_{app}.

Table 9.1. Calculated apparent diffusion coefficients and heterogeneous electron transfer constants for several compounds at 298 K using a platinum microelectrode (Di Fusco et al. 2011).

Compound	D_{app} (cm^2 s^{-1})	RSD (%)	k_s (m s^{-1})	RSD (%)
ABTS	3.4×10^{-6}	0.5	5.5×10^{-5}	0.5
Catechol	1.3×10^{-5}	0.7	1.7×10^{-4}	0.5
Dopamine	7.0×10^{-6}	0.1	9.1×10^{-5}	0.1
FcCOOH	6.2×10^{-6}	0.1	8.4×10^{-5}	0.1
$Fe(CN)_6^{4-}$	5.2×10^{-6}	0.2	7.2×10^{-5}	0.1
$Fe(CN)_6^{3-}$	7.5×10^{-6}	0.1	9.8×10^{-5}	0.1
$Os[(bpy)_2$ 4-AMP $Cl]^+$	1.8×10^{-6}	0.5	2.9×10^{-5}	0.3

The apparent diffusion coefficients D_{app} of the same compounds were determined at different temperatures also in the presence of a PAP membrane. Considering that in this case the unmodified Cottrell equation cannot be employed because the presence of the polymeric layer produces a variation in the concentration of the electroactive species near the electrode surface, the data were fitted according to the normalized Cottrell equation (described above), which allowed the diffusion coefficient to be calculated. The results obtained (apparent diffusion coefficients and concentrations) for the considered substrates are listed in Table 9.2.

Table 9.2. Calculated apparent diffusion coefficients and heterogeneous electron transfer constants for several compounds at 298 K using a platinum microelectrode in the presence of PAP membrane (Di Fusco et al. 2011).

Compound	D_{app} (cm^2 s^{-1})	RSD (%)	C (mmol L^{-1})	k_s (m s^{-1})	RSD (%)
ABTS	2.4×10^{-6}	1.3	1.97	4.3×10^{-5}	1.5
Catechol	3.0×10^{-6}	2.1	4.58	1.2×10^{-4}	2.2
Dopamine	1.5×10^{-6}	1.2	5.04	5.6×10^{-5}	1.3
FcCOOH	1.9×10^{-6}	1.4	1.83	3.4×10^{-5}	1.3
Fe(CN)$_6^{4-}$	5.3×10^{-7}	1.4	25.13	1.1×10^{-5}	1.6
Fe(CN)$_6^{3-}$	5.3×10^{-6}	1.5	2.05	9.0×10^{-5}	1.5
Os[(bpy)$_2$4-AMP Cl]$^+$	9.3×10^{-7}	3.8	1.96	2.9×10^{-5}	3.7

From the experimental data, a decrease in D_{app} values in the presence of the coated electrode is clearly evident compared to those obtained with the bare electrode. In particular, it can be observed that on going from substrates free to diffuse from the solution to substrates forced to cross a membrane, the percent decrease of the apparent diffusion coefficient values is different for the compounds investigated and is partly related to their strength as electrolytes and their charge status. While ABTS and Fe(CN)$_6^{3-}$ are only slightly hindered and the apparent diffusion coefficients decrease by about 30%, the positively charged osmium complex is more strongly affected and the apparent diffusion coefficient decreases by about 50%; on the other hand, the weak electrolytes FcCOOH, catechol, and dopamine are hindered to a much greater extent and the apparent diffusion coefficients are disrupted by 70–80%. Finally, it is interesting to note that Fe(CN)$_6^{4-}$ shows a completely different behavior with respect to the other negatively charged compounds since its D_{app} value is the most affected and decreases by almost 90%. This clearly suggests that the polymeric structure of PAP is charged and this charge plays a crucial role in disrupting the diffusion of the selected compounds toward the electrode surface. Indeed, the membrane structure is characterized by the presence of protonatable moieties such as nitrogen atoms and amide groups that can exert a variable influence on the apparent diffusion coefficients. For instance, the positively charged polymeric structure of PAP probably causes repulsion toward the osmium complex, significantly reducing the corresponding apparent diffusion coefficient value. Likewise, the substrates that are weak electrolytes could be hindered by electrostatic interactions between the protonated nitrogen atom of PAP and the -COOH and -OH groups present in their side chains. Although this tentative approach based on electrostatic repulsion provides some explanation of the different behaviors observed, it is far from being exhaustive: indeed, $Fe(CN)_6^{4-}$ shows a different behavior with respect to the other negatively charged compounds which seems to be inexplicable, particularly in comparison with the behavior of $Fe(CN)_6^{3-}$. One explanation that could satisfactorily fit all this experimental evidence is the tendency of $Fe(CN)_6^{4-}$ to form a stable ionic couple with quaternary ammonium ions while, conversely, $Fe(CN)_6^{3-}$ shows no effect at all under the same conditions (Cohen and Plane 1957).

Even if the D_{app} and k_s values in the presence of PAP are smaller than in solution, this reduction is small enough to indicate that the PAP membrane possesses excellent diffusion and electron-exchange properties with respect to other membranes that are frequently used and reported in the literature. The results indicate a very good permeability of the PAP layer to classical electrochemical mediators, except for $Fe(CN)_6^{4-}$, the performance of which is probably disrupted by the formation of a relatively stable ionic couple with the polymeric structure of the hydrogel. This is of great importance in view of the use of PAP as an immobilizing agent in electron transfer biosensor development.

Key facts for diffusion

- Diffusion describes the spread of particles through random motion (flux) from regions of higher concentration to regions of lower concentration (concentration gradient);
- This phenomenon is described by the two Fick's laws (considering one dimensional motion of particles along the x axis, first Fick's law: $J = -D\dfrac{\partial c}{\partial x}$; second Fick's law: $\dfrac{dc}{dt} = D\dfrac{\partial^2 c}{\partial x^2}$)
- Diffusion coefficient is proportionality constant between the molar flux due to molecular diffusion and the gradient in the concentration of the species (or the driving force for diffusion).

PAP BASED ENZYMATIC BIOELECTRODES FOR ENVIRONMENTAL ANALYSIS

Inhibition Based Biosensors

Over the past 15 years, several examples of classical enzymatic biosensors based on PAP immobilization have been reported; among them some have been proposed for pollutants determination in environmental analysis. Most of these biosensors are based on the inhibition effect of selected toxic compounds towards target enzymes. The principle of operation of these biosensors is based on the interaction that occurs between specific chemical and biological agents (inhibitors), present in the sample, and the biocatalyst (an enzyme, a polyenzymatic sequence and/or even a whole tissue) immobilized on the biosensor itself. The response of the biosensor is therefore a function of the reduced rate of the enzymatic reaction which takes place at the sensor interface. The inhibition process is generally diffusion-limited; in fact, the inhibitor concentration is normally low with respect to the enzyme loading on the membrane. Another aspect that has to be taken into account is the incubation time; in fact, there is a strict relationship between the contact time of the inhibitor with the enzyme and the inhibition response of the biodevices.

PAP based biosensors for pesticides and herbicides analysis

As far as inhibition biosensors are concerned, most of them are suitable for pesticides determination and rely on the inhibition of cholinesterase (AChE) (Botrè et al. 1994; Vastarella et al. 2007; Campanella et al. 1996a) or more recently on the inhibition of either acid (Mazzei et al. 1996; Croci et al. 2001) or alkaline phosphatase (Botrè et al. 2000; Mazzei et al. 2004).

In fact, pesticides and nervine gases are among the best known cholinesterase inhibitors; indeed organophosphorous compounds phosphorylate hydroxyl groups of serine residues present an the enzyme macromolecule with consequent changes of the enzyme structure and a permanent inhibition of its activity (O'Brien 1967; Corbett 1974; Jury et al. 1987). Several biosensors either electrochemical, optical or piezoelectric are based on the activity or inhibition of cholinesterase and are usually based on the combined activity of the enzymes acetyl- or butyril-cholinesterase and choline oxidase (ChOx) which catalyze the following reactions, respectively:

$$acetylcholine + H_2O \xrightarrow{\text{AChE}} choline + acetic\ acid$$

$$choline + 2O_2 + H_2O \xrightarrow{ChOx} betaine + H_2O_2$$

Both enzymes are entrapped and eventually crosslinked using PAP while the electrochemical transducer is usually an amperometric hydrogen peroxide sensing electrode which detects the production of hydrogen peroxide (Botrè et al. 1994; Vastarella et al. 2007). An alternative determination has been proposed (Campanella et al. 1996) based on the use of the enzyme butyril-cholinesterase alone immobilized with PAP and the detection of pH change produced by the enzymatic reaction using an ion selective FET as transducer; the use of a single enzyme entails a simpler assembly and a more robust and cheaper device.

In one of these works a new integrated bioreactor-biosensor based system has been proposed, for the determination of AChE inhibitors (Botrè et al. 1994), in particular organophosphorous compounds. The innovative aspects of this device with respect to other systems described in the literature are the following:

i.   The two enzymes (AChE and ChOx) are not co-immobilized on the same physic-chemical transducer; on the contrary, an AChE–chitin complex is loaded on a column reactor, while ChOx is immobilized onto an amperometric H_2O_2-sensing electrode.

ii. AChE is immobilized after extraction from a whole plant tissue (i.e. the inner portion of the grapefruit shell), thus drastically reducing the overall costs of the all system.

iii. The acetylcholine required for the detection of AChE inhibition is continuously recycled into the reactor, thus enhancing the lower detection limit of the system.

In Table 9.3 the main electroanalytical characterization of the AChE-ChOx integrated device in the analysis of two of the most used pesticides have been reported.

The AChE plant tissue bioreactor, coupled with the ChOx sensor, resulted to be very effective for the detection of acetylcholine and of acetylcholinesterase inhibitors. Particularly, the detection limits of 0.5 μM (for ACh) and about 1 ppb (for malathion) are sufficient to match most of the needs required by preliminary assays in various biological and environmental studies. The advantages of the method proposed are: i) the extremely easy preparation of both the AChE-chitin column reactor and the ChOx-H_2O_2 bioelectrode; ii) the markedly reduced overall cost of the system, in comparison with "traditional" systems involving the use of purified AChE; iii) the suitability for "on line" measurement; and iv) the possibility of automation of the procedure.

Table 9.3. Analytical characterization of the system constituted by the AChE-chitin reactor and the ChOx-H$_2$O$_2$ sensor in pesticides standard solutions. Measurement performed in phosphate buffer 0.1M pH=7.5 at 30°C (Botrè et al. 1994).

	Malathion	Paraoxon
Sample recycling time (incubation time):	40 min	40 min
Equation of the calibration graph:	Y = 0.50X+12.45 Y= % of inhibition; X= [Malathion] in ppb	Y=0.29X+10.70 Y= % of inhibition; X= [Paraoxon] in ppb
Linearity range:	10–100 ppb	32–240 ppb
Correlation coefficient:	0.9967	0.9951
Lower detection limit:	1 ppb	10 ppb
Pooled Standard Deviation (in the linearity range):	2.8 %	3.0%

Even if the enzymatic systems that have been most extensively employed for the realization of pesticide-sensitive biosensors are the cholinesterases and especially the acyl-cholinesterase, some alternative systems have been described. In fact, the major drawback of those biosensors is the irreversible inhibition of the enzymatic activity; to overcome this limitation a new class of pesticide-selective biosensors based on the reversible inhibition of acid phosphatase enzyme have been presented. The proposed biosensors (Botrè et al. 2000; Mazzei et al. 2004) are basically glucose-6-phosphate (G6P) sensitive biosensors, based on the combination of two enzymes, acid phosphatase (AcP) and glucose oxidase (GOx), catalyzing the following reactions respectively:

$$glucose\text{-}6\text{-}phosphate + H_2O_2 \xrightarrow{\ AcP\ } glucose + H_2PO_4^-$$

$$glucose + O_2 \xrightarrow{\ GOx\ } gluconolactone + H_2O_2$$

AcP and GOx are coupled to an amperometric hydrogen peroxide sensitive electrode, so that the current measured in the amperometric detection of H$_2$O$_2$ is proportional to the concentration of G6P in the sample.

While GOx has always been employed in the form of purified enzyme, physico-chemically immobilized on a suitable inert support, AcP has been used either as a purified enzyme, coimmobilized with GOx, or in the form of a whole plant tissue. In this latter case, a thin slice of potato (*Solanum tuberosum*) tissue has been coupled to the GOx-based sensor. The resulting bienzymatic biosensor, containing a biocatalytic membrane with immobilized and purified enzyme as well as a whole plant tissue, is usually defined as a "hybrid" biosensor. The determination of pesticides (malathion, methyl parathion, paraoxon) (Mazzei et al. 1996a) has been carried out by measuring their inhibition of the catalytic activity of AcP by either one of the two G6P-selective biosensors.

Even alkaline phosphatase (AlP) instead of acid phosphatase (AcP) has been employed for the same purpose either in the same configuration of AcP (i.e. coupled to GOx to form a bienzymatic biosensor) (Botrè et al. 2000) or in a more reliable monoenzymatic configuration using 3-indoxyl phosphate as the substrate.

This reaction sequence allows to use only AlP (without the need to couple it to GOx) because the obtained enzymatic reaction product spontaneously oxidates forming hydrogen peroxide and thus allowing direct amperometric transduction.

The analytical performances of all these types of PAP based inhibition first generation biosensors for pesticides and herbicides analysis are summarized in the Table 9.4.

Table 9.4 comparison of analytical performances for inhibition PAP-based biosensors.

Enzyme(s) used	Transducer	Compounds detected	Linearity range (ppb)	LOD (ppb)	Ref.
AChE + ChOx	Pt electrode for H_2O_2	Malathion Paraoxon Aldicarb	10–100 32–240 400–1300	1 10 100	(Botrè et al. 1994)
AChE + ChOx	Graphite SPE for H_2O_2	Paraoxon	5.2–103.3	4.5	(Vastarella et al. 2007)
BuChE	Ion selective FET	Paraoxon	3.3–12 17–82	3.0 10	(Campanella et al. 1996)
AcP + GOx	Pt electrode for H_2O_2	Malathion Methyl parathion Paraoxon	4.4–15 0.7–12 6.2–18.3	3 0.5 5	(Mazzei et al. 1996a)
AlP + GOx	Pt electrode for H_2O_2	2,4-dichlorophenoxyacetic	15–4200	10	(Botrè et al. 2000)
AlP	Pt electrode for H_2O_2	Malathion 2,4-dichlorophenoxyacetic	0.2–45.0 1.5–60.0	0.1 0.5	(Mazzei et al. 2004)

AChE: acetylcholinesterase; AcP: acid phosphatase; AlP: alkaline phosphatase; BuChE: butyrylcholinesterase; ChOx: choline oxidase; GOx: glucose oxidase

PAP based biosensors for pesticides and phycotoxins determination

The massive marketing and global distribution of marine organisms makes the monitoring of seafood for toxin content a critical issue of food control activity. Mussels filter approximately 20 liters of sea water per hour on average, thus accumulating a great variety of chemical compounds in their tissues. Among them, the most relevant as food contaminants are undoubtedly the so-called algal toxins, i.e. toxins produced by various algae, which are responsible for various diseases in man when ingested. These toxins are classified according to their toxic effects. Usually, monitoring procedures for algal toxins consist of sampling molluscan shellfish at the harvesting sites and analyzing the samples before harvesting occurs. When toxins are found in concentrations close to the safety tolerance threshold established by law, the area of sampling is immediately closed to harvesting. The area is re-opened when the values of toxin concentrations are again below the safety threshold. During these periods, the frequency of sampling can become as frequent as daily sampling. It is clear that a rapid, simple, inexpensive, but nonetheless reliable and specific screening test for each toxin (or group of toxins) would be invaluable for such needs.

For this aim a AcP/GOx biosensor previously described, based on the inhibition effect of okadaic acid towards acid phosphatase has been employed for the detection of one of the most important algal toxins (Croci et al. 2001); okadaic acid (OA) a diarrhetic shellfish poisoning (DSP) which is a lipophilic compound produced by several marine dinoflagellates belonging to the genera *Dinophysis* and *Prorocentrum.*

The concentration of OA in the sample is easily determined by following the reduced rate of H_2O_2 production that takes place as a consequence of the inhibitory capacity of OA towards AcP catalytic activity, making this system suitable for the preliminary screening of real samples (see Tables 9.5 and 9.6).

The use of this bienzymatic inhibition electrode presented can be recommended for the routine determination of OA in real samples of mussels, due to its combination of good analytical features (response time, lowest detection limit and range of linearity), reduced costs of operation and maintainance, prolonged lifetime of operation of the bienzymatic layer, and facility of use.

All the cited PAP based biosensors are based on the inhibition effect of toxic compounds towards different enzymes; despite inhibition methods being most widely employed for toxics and pollutants analysis, it has also been reported that a PAP based biosensor based on different amine oxidase able to directly detect the concentration of diamines are growing considerably as important environmental compounds (Botrè and Mazzei 1999).

Table 9.5. Analytical characterization of the hybrid AcP/GOx inhibition biosensor for the determination of OA in Citrate buffer 0.1M pH=6.0 at 37°C (Croci et al. 2001).

Response Time	20 min
Equation of the calibration graph: Y = ΔI (nA); X = [okadaic acid concentration] (ppb)	Y = 6.8 + 1.8 X
Linearity range (ppb):	2–22
Correlation coefficient:	0.9948
Lower detection limit (ppb):	0.5
Repeatibility of the measurements (as pooled standard deviation in the linearity range):	2.5 %
Lifetime of operation (expressed as total number of assays)	60–80
Long term stability of the AcP/GOx biocatalytic layer at -18°C (days)	75–90

Table 9.6. Comparison between the results obtained by the proposed AcP/GOD-based inhibition bioelectrode and by a reference HPLC technique. Data refer to the assays carried out on acetonic extract of the hepatopancreas of naturally contaminated (n.c. 1–2) and artificially contaminated (a.c. 1–2) mussels. Blank 1–2: control acetonic extract; n.d.: not detectable. Values of OA concentration are given in ppm (Croci et al. 2001).

Sample n°	Inhibition biosensor (a)	HPLC (b)	(a-b)/b (%)
Blank 1	n.d.	n.d.	-
Blank 2	n.d.	n.d.	-
n.c. 1	0.58	0.53	+9.4
n.c. 2	0.58	0.54	+7.4
a.c. 1	2.2	2.0	+10
a.c. 2	1.6	1.5	+6.7

Key facts for inhibition based biosensors

- The inhibition process is diffusion-limited;
- A major drawback of inhibition biosensors is the irreversible inhibition of the enzymatic activity; to overcome this limitation a new class of pesticide-selective biosensors based on the reversible inhibition has been developed;
- There is a strict relationship between the contact time of the inhibitor with the enzyme and the inhibition response of the biodevices.
- The methods proposed here are characterized by: i) easy preparation; ii) reduced overall cost; iii) sustainable for "on line" measurement and automation of the procedure.

PAP based Enzymatic Electrodes for the Direct Determination of Environmental Analytes

Biogenic amines determination

Biogenic amines (BAs) are nitrogen-containing compounds produced during fermentation, from the decarboxylation of amino acids by yeast and bacteria. The main biogenic amines are tyramine, histamine, putrescine, cadaverine, tryptamine, spermine and spermidine. Formation of biogenic amines can occur during food processing and storage as a result of bacterial activities. Consequently, higher levels of these amines may be found in food as a consequence of the use of poor quality raw materials, microbial contamination and inappropriate food processing conditions or microbial contamination. We have developed a first generation electrochemical biosensors based on plant tissue diamino oxidase (DAO) PAP bioelectrode (Botrè and Mazzei 1999) for the determination of biogenic amines. DAO is an enzyme which catalyzes the oxidation of agmatine, cadaverine, and putrescine with the production of NH_3 and H_2O_2 which can be detected amperometrically.

$$putrescine + O_2 + H_2O \xrightarrow{DAO} \Delta' - pyrroline + H_2O_2 + NH_3$$

$$cadaverine + O_2 + H_2O \xrightarrow{DAO} \Delta' - piperidine + H_2O_2 + NH_3$$

Based on these reactions, it is possible to detect putrescine in the range 0.5-320 µM with a LOD of 0.25 µM and cadaverine in the range 0.5–200 µM with a LOD of 0.25 µM, respectively. This aspect, yet not fully developed, opens to the possibility of direct measurement of diamines in environmental samples by a PAP based biosensor.

Polyphenols determination

Recently described PAP based biosensors also offer the opportunity of directly measuring the concentration of substrates for laccase enzymes, namely polyphenols or catecholamines which may cause environmental issues. In this study (Tortolini et al. 2010) either *Trametes versicolor* Laccase (TvL) or *Trametes hirsuta* Laccase (ThL) was immobilized by means of PAP on screen printed electrodes based on multi-walled carbon nanotubes (MWCNTs) thus enabling increased conductivity, high surface area matrix, flexibility and reactivity thereof. In order to explore the possibility of applying the proposed laccase biosensor to polyphenols determination in real samples, typical calibration curves for the considered compounds were carried out by chronoamperometry measurements at a fixed potential

(E=−100 mV vs Ag/AgCl/KCl$_{sat}$). The non-phenolic ABTS shows a good sensitivity with respect to both Laccases as well as catechol among the phenolic compounds. Catecholamines show similar sensitivity for TvL based biosensor, on the contrary for ThL based one, methyldopa shows a different behavior. The limits of detection (LODs) of the proposed biosensors toward the phenolic compounds under investigation are ranged 0.01–0.32 mg L^{-1} for TvL-based biosensor and 0.003–0.24 mg L^{-1} for ThL one. Moreover, in the case of catecholamines analysis the LODs of TvL and ThL-modified electrode are within 0.24–0.75 mg L^{-1} and 0.001–0.53 mg L^{-1}, respectively.

Key facts for PAP based biosensors for the direct determination of environmental analytes

- The linearity range obtained for the determination of putrescine is 0.5–320 µM with a LOD of 0.25 µM and for cadaverine is 0.5–200 µM with a LOD of 0.25 µM.
- The LODs obtained for several phenols taken into account are ranged 0.01–0.32 mg L^{-1} for TvL-based biosensor and 0.003–0.24 mg L^{-1} for ThL one.
- For catecholamines analysis the LODs of TvL and ThL-modified electrode are within 0.24–0.75 mg L^{-1} and 0.001–0.53 mg L^{-1}, respectively.

DEFINITIONS

- *Biosensor*: A biosensor is an analytical device for the detection of an analyte that combines a biological component with a physicochemical detector component.
  It consists of: a) the biorecognition element (biological material (e.g. tissue, microorganisms, organelles, cell receptors, enzymes, antibodies, nucleic acids, etc.), a biologically derived material or biomimic); b) the physicochemical transducer (optical, piezoelectric, electrochemical, etc.) that transforms the signal resulting from the interaction of the analyte with the biological element into an electrical signal.
- *Polyazetidine prepolymer*: it helps in realizing both a chemical and a physical immobilization: this is possible due to its peculiar features. Indeed, polyazetidine acts as a cross-linking agent as it is able to react with several different organic moieties (thiolic, oxydril, carboxyl, and amino groups), thus increasing the likelihood of chemical bonds being created with the enzyme and thereby enhancing immobilization efficiency.

- *Pesticides*: are substances or mixture of substances intended for preventing, destroying, repelling or mitigating any pest.
- *Herbicides*: An herbicide, commonly known as a weedkiller, is a type of pesticide used to kill unwanted plants
- *Polyphenols*: are a structural class of natural, synthetic, and semisynthetic organic chemicals characterized by the presence of large multiples of phenol units (phenol is an organic compound with the chemical formula C_6H_5OH).
- *Marine toxins*: are naturally occurring chemicals that can contaminate certain seafood. The seafood contaminated with these chemicals frequently looks, smells, and tastes normal, but can make people sick if they eat it.

APPLICATION TO OTHER AREAS OF HEALTH AND DISEASE

Electrochemical biosensors represent low-cost disposable devices for a variety of application areas related to human health diagnosis. A number of biosensors based on different enzymes are available for monitoring clinically important parameters such as glucose, pyruvate (Mascini and Mazzei et al. 1987a), and lactate for whole blood in extracorporeal experiments with an endocrine artificial pancreas (Mascini et al. 1987b), as well as for contaminants in food matrices (toxins, heavy metals, etc.).

SUMMARY POINTS

- The suitability of PAP as an enzyme immobilizing agent for the development of both second- and third generation electron transfer based biosensors is well documented in literature; the main key facts of PAP for these applications are: (i) a good permeability of the PAP layer to classical electrochemical mediators; (ii) the efficient entrapment of proteins therein without loss of bioelectrochemical properties, suggesting instead the maintenance of the native-like structural properties and (iii) the effective heterogeneous electron transfer at the interface PAP-electrode surface have been demonstrated.
- The use of PAP resulted in a simplified and cost-effective immobilization procedure that also ensured a good stability and reproducibility of the enzymatic–polymeric film.

- PAP offers a biocompatible micro-environment for confining biomacromolecules and foreshadows the great potentiality of this immobilizing agent not only in theoretical studies of protein direct electron transfer but also from the point of view of possible application to the development of electrochemical biosensors.

ABBREVIATIONS

ABTS	:	2,2'-azino-bis(3-ethylbenzothiazoline-6-sulfonic acid) diammonium salt
AChE	:	acetylcholinesterase
AcP	:	acid phosphatase
AlP	:	alkaline phosphatase
BuChE	:	butyrylcholinesterase
ChOx	:	choline oxidase
FcCOOH	:	ferrocenecarboxylic acid
GOx	:	glucose oxidase
LOD	:	Limit of detection
PAP	:	polyazetidine prepolymer
Os[(bpy)$_2$ 4-AMP Cl]$^+$	:	osmium complex bis(2,2-bipyridyl)-4-aminomethylpyridine chloride hexafluorophosphate

REFERENCES

Armstrong FA and GS Wilson. 2000. Recent developments in faradaic bioelectrochemistry. Electrochim. Acta 45: 2623–2645.

Barcelò D. p 149. In: D Barcelò [ed]. 1993. Official Methods of analysis of priority pesticides in water using gas chromatographic techniques. Environmental Analysis: Techniques, Applications and Quality Assurance Elsevier Science Publishers, Amsterdam, The Netherlands.

Botrè C, F Botrè, M Galli, G Lorenti, F Mazzei and F Porcelli. 1992. Determination of glutamic acid decarboxylase activity and inhibition by an H_2O_2-sensing glutamic acid oxidase biosensor. Anal Biochem 201: 227–232.

Botrè C, F Botrè, F Mazzei and E Podestà. 2000. Inhibition-based biosensors for the detection of environmental contaminants: determination of 2,4-dichlorophenoxyacetic acid. Environ Toxicol Chem 19: 2876–2881.

Botrè F and F Mazzei. 1999. Interactions between carbonic anhydrase and some decarboxylating enzymes as studied by a new bioelectrochemical approach. Bioelectrochem Bioenerg 48: 463–467.

Botrè F, G Lorenti, F Mazzei, G Simonetti, F Porcelli, C Botrè and G Scibona. 1994. Cholinesterase based bioreactor for determination of pesticides. Sens Actuat 18–19: 689–693.

Campanella L, C Colapicchioni, G Favero, MP Sammartino and M Tomassetti. 1996. Organophosphorus pesticide (Paraoxon) analysis using solid state sensors. Sens Actuat B 33: 25–33.

Chaubey A and BD Malhotra. 2002. Mediated biosensors. Biosens Bioelectron. 17: 441–456.

Cohen SR and RA Plane. 1957. The Association of Ferrocyanide Ions with Various Cations. J Phys Chem 61: 1096–1100.

Corbett JR. 1974. The Biochemical Mode of Action of Pesticides, Academic Press, London.

Cracknell JA, KA Vincent and FA Armstrong. 2008. Enzymes as Working or Inspirational Electrocatalysts for Fuel Cells and Electrolysis. Chem Rev 108: 2439–2461.

Croci L, A Stacchini, L Cozzi, G Ciccaglioni, F Mazzei, F Botrè and L Toti. 2001. Evaluation of rapid methods for the determination of okadaic acid in mussels. J Applied Microb 90: 73–77.

De Groot MT, M Merkx and TM Koper. 2005. Heme Release in Myoglobin–DDAB Films and Its Role in Electrochemical NO Reduction. J Am Chem Soc 127: 16224–16232.

De Groot MT, TH Evers, M Merkx and TM Koper. 2007. Electron transfer and ligand binding to cytochrome c' immobilized on self-assembled monolayers. Langmuir 23: 729–736.

Denuault G, MV Mirkin and AJ BardJ. 1991. Direct determination of diffusion coefficients by chronoamperometry at microdisk electrodes. J Electroanal Chem 308: 27–38.

Di Fusco M, C Tortolini, D Deriu and F Mazzei. 2010. Laccase-based biosensor for the determination of polyphenol index in wine. Talanta 81: 235–240.

Di Fusco M, G Favero and F Mazzei. 2011. Polyazetidine-Coated Microelectrodes: Electrochemical and Diffusion Characterization of Different Redox Substrates. J Phys Chem B 115: 972–979.

Frasconi M, G Favero, M Di Fusco and F Mazzei. 2009. Polyazetidine-based immobilization of redox proteins for electron-transfer-based biosensors. Biosens Bioelectron 24: 1424–1430.

Galus Z, J Golas and J Osteryoung. 1988. Determination of kinetic parameters from steady-state microdisk voltammograms. J Phys Chem 92: 1103–1107.

Gorton L, A Lindgren, T Larsson, FD Munteanu, T Ruzgas and I Gazaryan. 1999. Direct electron transfer between heme-containing enzymes and electrodes as basis for third generation biosensors. Anal Chim Acta 400: 91–108.

Howell JO and RM Wightman. 1984. Ultrafast voltammetry of anthracene and 9,10-diphenylanthracene. J Phys Chem 88: 3915–3918.

Jury AW, AM Winer, WF Spencer and DD Focht. 1987. Transport and transformations of organic chemicals in the soil-air-water ecosystem. Rev Environ Contamin Toxicol 99: 119–164.

Katz E and I Willner. 2004. Integrated nanoparticle-biomolecule hybrid systems: Synthesis, properties and applications. Angew Chem Int Ed 43: 6042–6108.

Lu Q, C Hu, R Cui and S Hu. 2007. Direct Electron Transfer of Hemoglobin Founded on Electron Tunneling of CTAB Monolayer. J Phys Chem B 111: 9808–9813.

Lu X, Z Wen and J Li. 2006. Hydroxyl-containing antimony oxide bromide nanorods combined with chitosan for biosensors. Biomaterials 27: 5740–5747.

Mascini M and F Mazzei. 1987a. Amperometric sensor for pyruvate with immobilized pyruvate oxidase. Anal Chim Acta 192: 9–16.

Mascini M, F Mazzei, D Moscone, G Calabrese and MM Benedetti. 1987b. Lactate and pyruvate electrochemical biosensors for whole blood in extracorporeal experiments with an endocrine artificial pancreas. Clin Chem 33: 591–3.

Mazzei F, F Botrè and C Botrè. 1996a. Acid phosphatase/glucose oxidase-based biosensors for the determination of pesticides. Anal Chim Acta 336: 67–75.

Mazzei F, F Botrè, G Lorenti and F Porcelli. 1996b. Peroxidase based amperometric biosensors for the determination of γ-aminobutyric acid. Anal Chim Acta 328: 41–46.

Mazzei F,.F Botrè, S Montilla, R Pilloton, E Podestà and C Botrè. 2004. Alkaline phosphatase inhibition based electrochemical sensors for the detection of pesticides. J Electroanal Chem 574: 95–100.

O'Brien RD. 1967. Insecticides: Action and Metabolism, Academic Press, New York.

Palchetti I,.S Laschi and M Mascini. 2009. Electrochemical biosensor technology: application to pesticide detection. Methods Mol Biol 504: 115–26.

Ratner BD and AS Hoffman. Hydrogels for Medical and Related Applications. pp 1–36. *In:* JD Andrade [ed.] 1976. ACS Symposium Series 31, American Chemical Society:

Washington, DC, USA, Shan D, E Han, H Xue and S Cosnier. 2007. Self-Assembled Films of Hemoglobin/Laponite/Chitosan: Application for the Direct Electrochemistry and Catalysis to Hydrogen Peroxide. Biomacromolecules 8: 3041–3046.

Tortolini C, M Di Fusco, M Frasconi, G Favero and F Mazzei. 2010. Laccase-polyazetidine prepolymer-MWCNT integrated system: biochemical properties and application to analytical determinations in real samples. Microchem J 96: 301–307.

Tsai Y-C, S-C Li and J-M Chen. 2005. Cast Thin Film Biosensor Design Based on a Nafion Backbone, a Multiwalled Carbon Nanotube Conduit, and a Glucose Oxidase Function. Langmuir 21: 3653–3658.

Vastarella W, V Rosa, C Cremisini, L Della Seta, MR Montereali and R Pilloton. 2007. A preliminary study on electrochemical biosensors for the determination of total cholinesterase inhibitors in strawberries. Int J Environ Anal Chem 87: 689-699.

Wang J. 2008. Electrochemical Glucose Biosensors. Chem Rev 108: 814–825.

Wang J, M Musameh and Y Lin. 2003. Solubilization of Carbon Nanotubes by Nafion toward the Preparation of Amperometric Biosensors. J Am Chem Soc 125: 2408–2409.

Zhang M, A Smith and W Gorski. 2004. Carbon Nanotube-Chitosan System for Electrochemical Sensing Based on Dehydrogenase Enzymes. Anal Chem 76: 5045–5050.

Zhang Q,.L Zhang, X Liu and J Li. 2007. Assembly of quantum dots-mesoporous silicate hybrid material for protein immobilization and direct electrochemistry. Biosens Bioelectron 23: 695–700.

Polycyclic Aromatic Hydrocarbon (PAH) Sensitive Bacterial Biosensors in Environmental Health

Mona Wells

ABSTRACT

Polycyclic aromatic hydrocarbon (PAH) exposure leads to adverse effects on most systems in the human body. A primary concern is carcinogenicity, but prenatal developmental mental impairment extending into early childhood and endocrine disruption are emerging concerns. PAHs are one of the most prevalent environmental contaminants, released primarily by anthropogenic/human activity. Bioavailability is the key —PAHs that are not bioavailable are not harmful. Presently, however, environmental PAHs are often characterized by total chemical load. Bioavailbility, when investigated, is often approximated using abiotic methods. Whole-cell bacterial biosensors, known as bioreporters, offer a unique approach to the determination of bioavailability, and can convey additional information, e.g. via subpopulation responses, visualization of response as a function of heterogeneous matrix, etc. The functional mechanism of bioreporters covered here depends on a target compound's passing through the cell membrane and binding to

Helmholtz Centre for Environmental Research, Department of Environmental Microbiology, Permoserstraße 15, Leipzig, 04318, Germany.
E-mail: mona.wells@ufz.de; mona@kimsey-wells.com

List of abbreviations after the text.

a regulatory protein, which activates transcription of a reporter gene. Subsequent translation produces an optically active reporter molecule. An overview of important developments of bioreporter applications to environmental PAHs is given, including notable examples of using bioreporters to understand PAH provenance and bioavailability from oil spills. Areas for future development are briefly summarized, including development of 1) reporters for higher molecular weight PAHs, 2) improved bioreporter-specific assays, and 3) improved assays for bioavailability.

INTRODUCTION

Polycyclic aromatic hydrocarbons (PAHs) are a class of compounds that consist of fused aromatic rings and are one of the most widespread environmental pollutants. PAH-bearing materials usually contain a number of different individual PAHs, many of which are known or suspected to pose an environmental health threat, inclusive of being carcinogenic, mutagenic, and teratogenic. Measurement of PAHs is a prerequisite to risk assessment, and the common approach is chemical analysis. Chemical analysis has the fundamental shortcoming of not addressing bioavailability, which is important since that which is not bioavailable, is not toxic. Due to their environmental prevalence and chemical behavior, wherein there is a large variation of PAH affinity for various natural materials, bioavailability is arguably even more important to understand for PAHs than for other classes of environmental pollutants.

Whole-cell bacterial biosensors, or bioreporters, combine advantages of both chemical and ecotoxicological approaches to environmental measurement and are suited for assessment of PAH environmental health issues. This chapter will describe environmentally important PAHs, their health effects, their chemical properties and important aspects of these that relate to environmental health. Subsequently the chapter details measurement, a review of progress to date in using bioreporters to assess environmental PAH bioavailability, including notable examples of bioreporter applications to oil spill studies, differentiating aspects of bioreporters, and a future outlook for technology.

ENVIRONMENTALLY IMPORTANT PAHs, SOURCES, SINKS, AND OVERVIEW OF EFFECTS

PAHs are highly prevalent in the environment, and anthropogenic and/or anthropogenically labilized sources constitute the majority of bioavailable PAHs. Important environmental sources of PAHs are summarized as follows (IPCS 1998):

- Adventitious mobilization of tars and pitch products;
- Oil spills, and spills/release of refined fossil fuel products that contain high quantities of PAHs;
- Burning to clear agricultural fields;
- Exhausts from mobile sources (e.g. road vehicles, ships, and aircraft);
- Industrial operations (e.g. aluminum foundries, incinerators, other stationary industrial combustion);
- Combustion from domestic heating/fireplaces;
- Burning coal for energy production;
- Burning wastes;
- Smoking tobacco;
- Spreading of contaminated sewage on agricultural fields; and
- Forest fires and volcanic eruptions.

The above list reflects three basic processes that lead to PAH release in the environment; these are 1) release from materials that inherently contain large amounts of PAHs, 2) release from high-temperature chemical transformations associated with burning/combustion, and 3) release from contaminated materials. Crude oil and coal deposits fall into the first class, inherently containing substantial amounts of PAHs that arise from diagenesis of natural organic matter and subsequent processing alterations thereof (Roy 1995). Most of the remaining sources are in the second class, involving induced high-temperature transformations. The third class of PAH source, contamination, is typically an outcome of PAH chemical properties —for instance, sewage in itself is not likely to contain PAHs. However, PAHs are lipophilic, i.e. more like oil than water, and not very soluble in water (solubility decreases approximately one order of magnitude for each additional ring fusion). As a result, PAHs in the environment "stick" to soil, sediment, aerosol particles in air, or any other solid rather than remaining in water or air itself. Sewage sludge is thus readily contaminated from, for example, PAHs in stormwater runoff containing automobile exhaust particulates, engine oil, etc. This stickiness also entails that vegetables/produce and livestock coming into contact with contaminated water or soil could serve as a human exposure route via food.

In recognition of environmental health concerns, in 1979 the US EPA classed 16 PAHs as EPA Priority PAHs (US EPA 1979). Though additional PAHs have since been identified as posing environmental concern, these still constitute the suite that are most frequently analyzed in evaluation of environmental samples. In the EU, the Scientific Committee on Food (SCF) has identified 15 PAHs of concern with respect to carcinoginicity, subsequent to which an additional PAH has been added as probably carcinogenic (EFSA 2008). This group of 16 PAHs is commonly referred to as 15+1 EU priority PAHs for which there is EU legislation to ensure food safety (EU 2005). The US Agency for Toxic Substances and Disease Registry (ATSDR 1995) has classed yet a third grouping of 17 PAHs as concern with respect to health effects. Although there are additional groupings of PAHs that have been categorized as posing environmental health threats, these three groupings, listed in Table 10.1, are very commonly used as food safety and environmental assessment indicators.

Because the genotoxic activity of PAH benzo[*a*]pyrene (BaP) has been studied the most, it is the basis for the US EPA (1993) relative potency factors (RPF) for cancer, having a RPF = 1. In contrast, of the other US EPA Priority PAHs, dibenz[*ah*]anthracene has an RPF of 5, and the PAHs benzo[a]anthracene, benzo[*b*]fluoranthene, benzo[*k*]fluoranthene, and indeno[1,2,3-*cd*]pyrene have RPFs of 0.1. Figure 10.1 shows the structures of the PAHs listed in Table 10.1, with the PAH bay region being indicated for dibenzo[*ah*]pyrene (Melendez-Colon et al. 2000). The US EPA RPFs are under review, with a draft document issued, though citation of same is proscribed at this time.

PAH toxicity varies greatly, depending on the structure of the individual PAH, and can range from relatively nontoxic to very toxic. Since PAHs typically present as a mixture, it is difficult to assess the relative

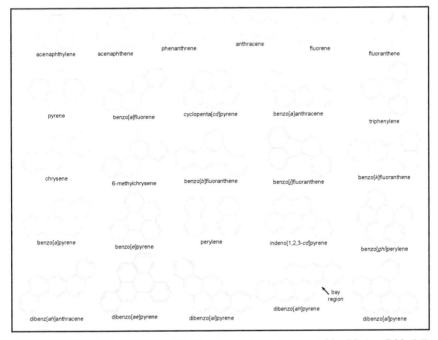

Figure 10.1. Chemical structures of PAHs relevant to environmental health (see Table 9.1). Unpublished figure.

Table 10.1. Summary of PAHs important to environmental health, by molecular weight.[a]

PAH	Molecular Weight (a.m.u.)	EPA Priority[b]	EU 15+1	ATSDR
Acenaphthylene	152	✓		✓
Acenaphthene	154	✓		✓
Fluorene	166	✓		✓
Phenanthrene	178	✓		✓
Anthracene	178	✓		✓
Fluoranthene	202	✓		✓
Pyrene	202	✓		✓
Benzo[a]fluorene	216		✓	
Cyclopenta[cd]pyrene	226		✓	
Benz[a]anthracene	228	✓	✓	✓
Triphenylene	228			
Chrysene	228	✓	✓	✓
6-Methylchrysene	242		✓	
Benzo[b]fluoranthene	252	✓	✓	✓
Benzo[j]fluoranthene	252		✓	✓
Benzo[k]fluoranthene	252	✓	✓	✓
Benzo[a]pyrene	252	✓	✓	✓
Benzo[e]pyrene	252			✓
Perylene	252			
Indeno[1,2,3-cd]pyrene	276	✓	✓	✓
Benzo[ghi]perylene	276	✓	✓	✓
Dibenzo[ah]anthracene	278	✓	✓	✓
Dibenzo[al]pyrene	302		✓	
Dibenzo[ai]pyrene	302		✓	
Dibenzo[ae]pyrene	302		✓	
Dibenzo[ah]pyrene	302		✓	

[a]Luch (2005) also classes the PAHs coronene and ovalene of important human health/environmental concern.
[b]The original group of 16 US EPA Priority PAHs includes naphthalene, though it is not technically a PAH; see main text for details.
Unpublished table.

contributions that individuals will have in mediating effects, particularly with the potential for synergistic effects. Table 10.1 PAHs relate most closely to carcinoginicity, but concerns are not limited to cancer. There is emerging evidence about how PAHs affect children, infants, and embryos. Studies indicate that high prenatal exposure to PAH is associated with low IQ, low birth weight, heart malformations, DNA damage and early childhood developmental delays (Perera et al. 2009 and references therein). The EU Commission Regulation (EC) No 208/2005 acknowledges the concerns for infant health in that it contains four subsections (in Section 7.1.2) with specific limits on levels permissible for infants, and these limits are at minimum a factor of two lower than any others in the regulation (EU 2005). PAHs have also recently been recognized as potential environmental endocrine-disrupting chemicals (EDCs) acting as antiestrogens by blocking the activation of the estrogen receptor or by binding the aryl hydrocarbon (Ah) receptor leading to induction of Ah-responsive genes that lead to antiestrogenic responses (Rodriguez-Mozaz et al. 2004).

The world human population is still growing, and the anthropogenic provenance of PAHs ensures that the prevalence of PAHs in the environment will increase, and with greater concerns regarding PAH exposure and impact on environmental health. This entails an increasing requirement for measurement of PAHs and different approaches to measurement in order to assess environmental health risks.

MEASUREMENT OF PAHs—CHEMICAL LOAD *vs.* BIOAVAILABILITY

The traditional approach to PAH measurement is chemical analysis. Most environmental samples have a complex matrix, i.e. the material containing the mixture of PAHs requiring analysis is itself a very complex mixture of chemicals. Such a matrix binds PAHs in analytically inaccessible forms. As a result, one of the first steps in chemical analysis is total digestion wherein a sample is typically macerated in a strong organic solvent, such as methylene chloride or hexane, effectively extracting PAHs, followed by purification/ concentration and instrumental chemical analysis, e.g. see Deepthike et al. (2009). Due to the nature of the digestion, results obtained in this way are typically referred to as representing the total chemical load of PAHs in a material. Digestion methods used for determination of total chemical load represent a process of liberating the PAH from the matrix that is entirely artificial and not at all equivalent to biological processes. Hence chemical load, in some cases, will be a gross overestimation of the fraction of PAHs that are bioavailable or even bioaccessible.

Recognizing the limitations of total chemical load, the last two decades have witnessed an increasing level of interest in methods to assess bioavailability and bioaccessibility. The word bioavailable is most often taken to indicate availability to biota. Traditional ecotoxicological methods involve biota and measure the consequent effects of bioavailability—i.e. mortality may result from springtails or fat head minnow exposure to a compound that is both bioavailable and toxic, but the net result depends on both. Perhaps counterintuitively, if not perversely, there are many chemical methods that do not employ biota of any sort that are presently used to measure bioavailability. There are many examples of such non-biotic approaches, and these rely on chemical partitioning. For instance, semi-permeable membrane devices (SMDs), e.g. Short et al. (2008) function by PAH's diffusing from surrounding water or soil/sediment through a membrane and preferentially partitioning into a lipophilic interior (with subsequent extraction and chemical analysis). This type of approach may well mimic biological systems in a static sense—i.e. when there is no complex biological dynamic or series of feedback loops present, and all biological uptake occurs via unperturbed partitioning. However, it is a misnomer to label such an approach as measuring bioavailability since no biota are involved and instead biota are simulated via chemical proxy. As such, it is more appropriate to recognize that such approaches measure "availability to chemical proxy", i.e. chemoavailability, a term whose use acknowledges that in some cases measurements involving chemical proxy differ from bioavailability, as explained in more detail below.

Bacterial whole-cell biosensors, or bioreporters, as living organisms measure bioavailability (van der Meer et al. 2004). On exposure to an analyte or mixture of analytes, bacterial bioreporters respond by producing a measurable signal, not always, but usually, optical (via a chromophore, fluorophore, or luminophore reporter molecule). Bioreporters are usually genetically modified organisms whose cellular response is based on a change in the physicochemical condition (pH, temperature, etc.) or the molecular recognition of a chemical compound (e.g. PAHs), the said response being reported by the reporter gene that leads to synthesis of reporter molecules (Fig. 10.2). A motif of the reporters discussed herein is that their function is based on sensor and regulator proteins of a metabolic pathway, i.e. not an intracellular pathway for toxicity response. A reporter that senses a PAH in the process of utilizing the same as a carbon source is clearly engaged in a process that is not equivalent to chemical partitioning (chemoavailability). Most bioreporters that have been reported for PAHs demonstrate response to more than one PAH, which is a disadvantage if the goal is quantitation of each individually, but may be an advantage in terms of detecting groups of PAHs having similar properties, particularly given the Bay Region Hypothesis of PAH carcinoginicity. The basic paradigmatic differences between approaches to measurement of total chemical load, chemoavailability, bioavailability (e.g. via bioreporters), and ecotoxicological response is shown in Fig. 10.3.

A review of bioreporters distinguishes three separate categories—Class I, II, and III, depending on how bioreporter molecular recognition/response takes place (van der Meer 2004). Bioreporters that react specifically to a single compound or class of compounds via an increase in the output

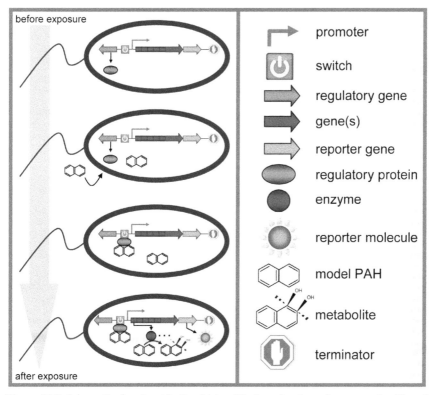

Figure 10.2. Schematic showing idealized/simplified mechanism of response for Class I bacterial bioreporters. Unpublished figure.

Color image of this figure appears in the color plate section at the end of the book.

signal are Class I, those that react to more generic stimuli (e.g. heat shock, oxidative stress, etc.) via an increase in the output signal are Class II, and those that react to a single or mixture of compounds and/or stresses via a decrease in the output signal are Class III. Class III reporters are, in terms of the measurement paradigms illustrated in Fig. 10.3, ecotoxicological, and Class I reporters are most relevant to understanding bioavailability of PAHs (compound/compound-class specific). The next section of this chapter covers an overview of selected applications of Class I bioreporters to the determination of environmental PAHs. The penultimate section details aspects of why bioavailability (e.g. as determined with bioreporters) is a fundamentally different measurement than chemoavailability, and the chapter will conclude with an outlook for the future.

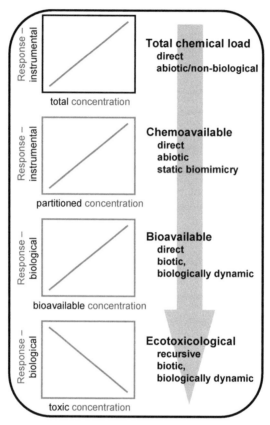

Figure 10.3. Paradigmatic differences between measurement approaches for PAHs. Unpublished figure.

OVERVIEW OF BIOREPORTER STUDIES ON PAHs

Class I bioreporters were in development in the 1990s, but the technology is still not in wide use compared with methods to determine, e.g. chemical load, or chemoavailability. Regarding PAHs, some notable early papers were published on the bioluminescent strain *Pseudomonas fluorescens* HK44, which responds to naphthalene. While naphthalene is not technically a PAH according to the International Union of Pure and Applied Chemists (IUPAC), because it is only bicyclic (2 fused rings), it is often referred to as such and used as a model PAH. One early paper involving strain HK44 (Matrubutham et al. 1997) looked at response over 35 dy in simulated groundwaters and for alginate immobilized cells. This study found that there was no bioluminescent response from cells incubated in groundwater samples with pH below 6, and pHs below 6 would certainly not be unusual for the intended application, i.e. the strain would be useful for monitoring in very mildly acidic to neutral conditions. The study also found that strain response was much better in nutrient (and carbon) deprived conditions, such as would be normal for long-term field studies.

A year later, in 1998, the BBIC (bioluminescent bioreporter integrated circuit) was demonstrated, wherein a bioreporter was immobilized to a CMOS (complementary metal-oxide semiconductor) integrated circuit to produce a chip-based sensor with a bioreporting transducer element. Simpson et al. (1998) demonstrated the principle using a strain sensitive to toluene, but a later expanded report utilized strain *Pseudomonas fluorescens* 5RL (Simpson et al. 2001), sensitive to naphthalene, and more recently a study has been published that reports an improved performance BBIC (Vijayaraghavan et al. 2007). The effector molecule for the strains HK44 and 5RL in these studies is salicylate, i.e. the strains are responsive to a metabolite of naphthalene, and induction of response was achieved in early studies using salicylate. Vijayaraghavan et al. (2007) specifically demonstrate applicability of air analysis to detect naphthalene in an air stream. While all of these studies indicated that the strains used were specific to naphthalene (salicylate), none explicitly stated details of response testing on other PAHs.

In the first notable demonstration of reporter technology for a true PAH, the authors mutated a fluorene-degrading strain to produce a modification, strain *Sphingomonas* sp. LB132, wherein fluorene degradation was impaired and exposure instead resulted in luminescence (Bastiaens et al. 2001). Strain LB132 demonstrated that, as with chemical analysis (though probably different in causation), there is an analyte concentration range below detection (no response), an optimal detection range with linear response, and a concentration range at which response is uniform/saturated. Response was tested for a number of other PAHs, but none demonstrated a detectable signal-to-noise ratio (greater than three). Diesel fuel, on the other hand, which contains a complex mixture of PAHs, demonstrated a clearly detectable response (see Fig. 10.4). The strain was also responsive to four fluorene metabolites, suggesting that, as with *Pseudomonas* strains HK44 and 5RL discussed above, the reporter is sensitive to metabolites rather than the analyte of interest per se.

In 2006, Tecon et al. reported the development and use of a phenanthrene bioreporter, *Burkholderia* sp. strain RP037, which produces a green fluorescent protein (GFP) signal. One issue that led to the popularity of using naphthalene as a model PAH is aqueous solubility, which is higher for naphthalene than for any of the PAHs. Solubility is naphthalene > fluorene > phenanthrene, with the difference being around an order of magnitude between each, hence this paper represented a significant advance. Unlike studies of reporters showing a surfeit of carbon source can sometimes

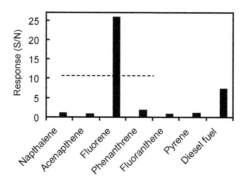

Figure 10.4. Average response of fluorene reporting strain *Sphingomonas* sp. LB132 to individual PAHs and a mixture of PAHs in diesel fuel. Raw data from Bastiaens et al. (2001). Unpublished figure. Dashed line represents limit of detection (units are S/N, signal to noise ratio).

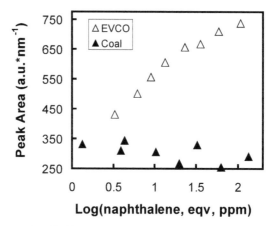

Figure 10.5. Response of *Burkholderia* sp. strain RP037 to a mixture of PAHs (units expressed as naphthalene equivalents) in Gulf of Alaska coals and in Exxon Valdez crude oil (from oil spill). Coal response is same as blank, i.e. PAHs from coal are biounavailable, whereas PAHs from crude oil are fully bioavailable. Response is expressed in fluorimetric peak areas; raw data from Deepthike et al. (2009). Unpublished figure.

adversely affect response (e.g. *vida supra*), the RP037 response precision was higher with a secondary carbon source. The strain was also found to be differentially (more) responsive to naphthalene and produced a strong positive response to a lampblack soil contaminated with a complex PAH assemblage. The authors proposed that the cumulative PAH response might be referred to in terms of phenanthrene equivalents, though this has since been changed to naphthalene equivalents in recognition of naphthalene being the model compound. The most notable aspect of this work involved its consideration of the unique contribution that bioreporters have to make in the measurement of bioavailability, described later.

An important source of environmental PAHs is oil spills. The recent catastrophe of the Deepwater Horizon demonstrates the unfortunate validity for concern with respect to environmental health, given the known PAH load in crude oil (Deepthike et al. 2009). While no bioreporter studies have yet been published relating to this disaster, it did occur in the immediate wake of publication of two reports demonstrating the use of bioreporters for PAHs from oil (Deepthike et al. 2009; Tecon et al. 2010). The latter examined the response of a multi-strain platform to artificial (i.e. lab simulated) oil spills. The strains on the platform were responsive to alkanes, BTEX (benzene, toluene, ethyl benzene, xylene), hydroxybiphenyl, DNA damaging agents (Class II response), and PAHs in the form of naphthalene, dimethylnaphthalene, and phenanthrene (luminescent strain *B. sartisoli* RP007). This study found that, unlike alkanes and BTEX, for PAHs the appearance of naphthalene equivalent aqueous concentrations started later and resembled a saturation-type dissolution. The results indicated that volatilization and microbial degradation of PAHs proceed slowly, with the implication being that this contributes to the late toxicity of oil spills. Chemical analysis showed that the most abundant dissolved PAHs were naphthalene, methylnaphthalenes, dimethylnaphthalenes, and phenanthrene, and the bioreporter response observed was significantly higher than what would be expected for the level of naphthalene alone, hence verifying that the PAH reporter was responding to an assemblage of bioavailable PAHs.

Another oil spill bioreporter study was published by Deepthike et al. (2009) and involved a 20th year retrospective analysis of PAH provenance with respect to the Exxon Valdez Oil Spill. The premise of the study was simple regarding PAH provenance. Authochthonous coals and related geomaterials from around the Gulf of Alaska have been implicated as responsible for persistent PAH signatures there rather than crude oil released from the Exxon Valdez spill. The study used the *Burkholderia sartisoli* sp. RP037 strain of Tecon et al. (2006) to assess bioreporter response to a suite of relevant coals, and compared the coal response to that of Exxon Valdez Crude Oil (EVCO) soiled kaolinite. Since lipophilic materials do not measurably partition to kaolinite, this represents material from which the bioavailability of EVCO PAHs can be assessed in a geomorphic form (Kohlmeier et al. 2008). Coal and EVCO/kaolinite sample suites had a similar concentration range of PAHs, from low to high. The results showed that PAHs in the Alaskan coals were biounavailable, with bioreporter response being the same, to within uncertainty, as that for non-PAH containing controls. Conversely, the EVCO soiled kaolinite showed a clear dose-related response. As with the Tecon et al. (2010) study, the bioreporter response to EVCO was approximately equivalent to the naphthalene equivalent PAH load (i.e. greater than for naphthalene alone).

The Deepthike et al. (2009) study, despite its simplicity, generated substantial controversy (Deepthike et al. 2010; Page et al. 2010). While it is impossible to separate the technical from the inherent conflict of interest in the energy industries' funding of scientific research, the extent of which has been the topic of a report entitled *Big Oil U.*, published by the Center for Science in the Public Interest's Integrity in Science Project (CSPI 2008). Recalling that the vast majority of sources for PAHs in the environment are anthropogenic, and that as a class of compounds the known health threats from PAH exposure are increasing, the need for disinterested objectivity *and transparency* will be increasingly important as well.

This section concludes with a brief consideration of the potential versatility of bioreporters. The Tecon et al. (2006) study cited above also assessed live vs. dead cells via staining and noted that dead cells were low in GFP, whereas live cells were high in GFP. Their use of GFP also demonstrated another facet of bioreporter application—the reporter gene used can be exploited to different effect. Bioluminescent reporters are perhaps most commonly used since the absence of optical background is useful in obtaining the highest possible detection limits. On the other hand, the use of fluorescent reporters allows possibilities that are either absent or more difficult to exploit with luminescence. One of these is exploitation of subpopulation effects/analysis, as demonstrated implicitly by Tecon et al. (2006), and explicitly discussed in terms of the effect on detection by Kohlmeier et al. (2007). Another is single cell analysis (to date only demonstrated for an arsenic reporter, Wells et al. 2005). Finally, visualization of response in heterogeneous media is an exciting prospect that has been demonstrated for PAH detection by Tecon et al. (2009) wherein a double tagged reporter, *Burkholderia sartisoli* RP037-mChe, was used to track PAH release from complex environmental matrices and PAH diffusion through media over time. Unifying several of these themes (subpopulation analysis, simultaneous study of live and dead cells, with potential to monitor bioavailability and toxicity at once, visualization), Shin et al. (2011) published a study wherein two fluorescent phenanthrene reporters, *Sphingomonas paucimobilis* EPA505 strains D and S, were exposed to phenanthrene in a sand/water mixture and the response monitored with confocal laser scanning microscopy. The first strain was engineered to produce GFP upon degradation of phenanthrene. The second was designed to die on exposure to phenanthrene, hence being amenable to quantification of toxic response. The manner in which the latter is implemented will require further results, but the approach shows great promise.

CHEMOAVAILABILITY *vs.* BIOAVAILABILITY REVISITED

Chemoavailability of PAHs is based on chemical partitioning or sequestration from an aqueous into a /lipophilic environment, mimicking partitioning in the environment between aqueous and solid phases of various characters. The process is inherently governed by thermodynamic equilibrium, however, since equilibrium is an asymptotic condition, non-equilibrium conditions may often presage the final equilibrium state.

Bioavailability, in its simplest form, may also represent chemical partitioning to a living organism. For this reason, a number of studies have noted a correlation between bioaccumulation (for which bioavailability is the prerequisite), i.e. as determined with biota, and chemoavailability. For instance, a study utilizing a naphthalene bioreporter found a correlation between bioreporter response, chemoavailability (as measured by extraction using the lipophilic material Tenax), and partitioning coefficients that serve as an indicator of the strength of attraction of various geomaterials for naphthalene (Kohlmeier et al. 2008).

Numerous studies (bioreporters inclusive) have demonstrated that bioavailability is not always the product of chemical partitioning, and the work of Tecon et al. (2006) is an illustrative case involving a bioreporter wherein response was not related to phenanthrene concentration but rather to cumulative flux. For chemoavailability, the proxy will approach, or arrive at equilibrium (e.g. Kohlmeier et al. 2007; Short et al. 2008) and the amount of analyte collected will be determined by same. The reporter that Tecon et al. used is capable of utilizing phenanthrene as a carbon source, as many PAH reporters are, and hence no saturation behavior based on thermodynamics will be achieved, nor is such in danger of being approximated. Different factors will govern the overall system behavior, and these are a concatenation of multiple effects that are, in sum, much more complex than those based on partitioning. This difference in fundamental performance could also explain why Tecon et al. (2010) found that bioreporter response to simulated oil spills was not in agreement with results from chemical analysis.

Chemoavailability is fundamentally a more proscribed measurement than that of bioavailability, i.e. it represents a simplest-case limit for the latter. This can be an advantage or disadvantage for either, depending on the needs of analysis. Au fond, it is highly useful to have diverse instruments for measurement in order to obtain a complete understanding of the environmental health consequences of PAHs. It is important however to recognize the distinction between what is meant by these two terms in order to exploit the potential differences. Chemoavailability is, relatively speaking, better understood, more convenient, and more advanced in terms of its developmental maturity. Bioavailability is less understood and does not have an accepted definition within the field of environmental science, but clearly involves methods of measurement that involve biota, from the etymological source of the term, and then as a result of more complex sets of biological processes is capable of reflecting more than chemical partitioning.

LOOKING FORWARD

The state of development of bioreporter technology is immature relative to chemical analysis for total load, ecotoxicological analysis, biomonitoring of bioaccumulation, and analysis of chemoavailability. There are three major issues with regard to bioreporters for environmental health and PAHs. First, there is an urgent need for bioreporters that report PAHs of higher molecular weight than phenanthrene. Secondly, enhanced uptake of this technology requires development of bioreporter-optimized assays and methods. There are a wide variety of inventive and efficient methods for chemoavailability measurements (only SMDs and Tenax mentioned here), and accordingly the methods available for making different types of bioreporter measurements require diversification. Thirdly, bioreporter studies must increasingly focus on the fundamental differences that bioreporters have to offer in measuring bioavailability.

The work reviewed herein illustrates how researchers in the area are making efforts to bring bioreporters into a context of relevance to environmental health by developing applications that utilize complex matrix samples. Such work can serve to facilitate uptake of the technology, and there is considerable future scope and need in this domain. Of particular note, since primary exposure pathways to PAHs are via airborne particles and food, it is of interest to develop methods that demonstrate the applicability of bioreporters in measuring the bioavailability of PAHs from these matrices.

APPLICATIONS TO AREAS OF HEALTH AND DISEASE

Most of what is known about the *in vivo* effects of PAHs on human health and disease results from studies on non-human animals. Such studies indicate that PAH exposure results in adverse effects including respiratory and dermal effects (cancer and noncancer), promotion of arterial plaque formation, gastrointestinal toxicity, effects on rapidly proliferating tissues (attack on bone marrow and blood forming elements), increased liver weight, dose-related increase in liver enzymes, nephropathy, immunological and lymphoreticular effects, reproductive impairment, developmental impairment, mutagenic and tumorogenic activity, and death (ATSDR 1995). Results from these studies, taken with results from studies on humans with known occupational or other collateral exposure and on *in vitro* effects, then lead to weight of evidence assessments. The genotoxicity/carcinogenicity of PAHs is of primary concern and is addressed by the Bay Region Hypothesis, a hypothesis that is reasonably consistent with experimental findings and which predicts that PAH structures with bay regions (*vide infra*) that are strongly reactive are more genotoxic and carcinogenic (ATSDR 1995). In this hypothesis, the mutagenic and tumorigenic activity of PAHs arises from the formation of bay region diol epoxides, which then form DNA adducts that ultimately lead to DNA damage/cancer. Because some PAHs have been studied much more than others, the United States Environmental Protection Agency (US EPA 1993) has developed a system of model equivalency (cancer potency factors), though the approach is presently being refined.

KEY FACTS OF PAH ENVIRONMENTAL HEALTH

- Carcinoginicity of some PAHs is a key concern.
- One route of exposure to PAHs is air; urban airborne particles and tobacco smoke can have very high PAH concentrations, posing a cancer risk.
- A second exposure route is ingestion via food, for which the European Union has amended regulated permissible levels due to cancer risk (EU 2005).
- Cooked meat and processed meats and fish have been shown to contain carcinogenic PAHs well above EU regulated limits. Independently, studies by the World Cancer Research Fund (WCRF 2007, 2009) show a significant cancer risk for meat eating leading to recommended elimination of dietary processed meats and limiting other meat intake.
- Most forms of black carbon, including coal, are high in PAHs, but many of these materials contain biounavailable PAHs, and are therefore not of environmental concern, e.g. Kohlmeier et al. 2008; Deepthike et al. 2009, and references therein.
- Crude oil and many petroleum products contain large amounts of PAHs that are bioavailable. That coal PAHs are biounavailable and crude oil PAHs are highly available has generated controversy regarding provenance (Deepthike et al. 2010; Page et al. 2010), and with economic interests entailed in energy, this will foreseeably continue (CSPI 2008). Therefore, understanding bioavailability is inherent to understanding provenance, risk, and damage of/from PAHs.
- Some chemical tests proxy biota in the measurement and assessment of PAH bioavailability, but these are more accurately described as measuring chemoavailability, as discussed herein.

DEFINITIONS OF KEY TERMS

- *Bioavailable*: though often used, even by regulatory agencies, as pertains to environmental studies, there is not a single widely accepted definition of this term; Semple et al. (2004) give a definition in keeping with the etymological origin of the word as the amount of a chemical that is at hand or readily available to biota. Since this definition implies an instantaneous quantity, which often is not practically relevant, the term is used here to indicate bioavailability as measured over a finite amount of the measurement or observation time.
- *Bioaccessible*: as per the definition of Semple et al., the amount of a chemical that is not bioavailable, but can *potentially* become bioavailable.
- *Bioaccumulation*: the biological sequestering of a substance or chemical at a higher concentration than that at which it occurs in the surrounding environment or medium.
- *Carcinogenic*: a chemical or substance with the ability or tendency to produce cancer.
- *Chemoavailable*: the chemical analogue of bioavailable, only measured with a chemical proxy to simulate biota.
- *Genotoxic*: a chemical or substance that is damaging to DNA, and thereby capable of causing mutations or cancer.
- *Mutagenic*: a chemical or substance which has the property of being able to induce genetic mutation.
- *Teratogenic*: a chemical or substance that causes birth defects.
- *Tumorogenic*: a chemical or substance causing formation or production of tumors.

SUMMARY POINTS

- Most major sources of environmental PAHs are released from anthropogenic activity. Black carbon/coals can be high in PAHs but these are, generally, biounavailable, whereas crude oil and petroleum products contain large amounts of bioavailable PAH. Primary exposure routes include airborne particles, tobacco smoke and ingestion via cooked and processed meat and fish. Bay and fjord PAHs are thought to be most strongly implicated in carcinogenesis. Information available suggests that integrated environmental health effects of PAHs can be severe and future increases in population ensure that PAH exposure will be of growing concern.
- PAH measurement by chemical analysis measures total load. In many instances, however, the bioaccessible fraction of the total

load may be small. Risk assessment based on loads then may be overly conservative, and bioavailability is the relevant metric. Many approaches to bioavailability employ a chemical proxy or biomimetic extraction based on partitioning to estimate bioavailability, hence these correctly speaking measure chemoavailability (availability of a substance to chemical proxy) and not bioavailability. Bioreporters are organisms that, on exposure to an analyte, respond by producing a measurable signal, and as living organisms, measure bioavailability.

- Class I bioreporter technology is still not in wide use compared to methods that determine chemical load or chemoavailability. Early papers described naphthalene bioreporters; while naphthalene is not technically a PAH, it is often used as a model for same. Bioreporters detecting fluorene and phenanthrene have since been reported. Two articles have used PAH bioreporters to assess bioavailability of PAHs from oil spills, one assessing Exxon Valdez Oil Spill provenance. Other bioreporter papers seek to exploit analysis of differing subpopulation responses, simultaneous study of live and dead cells (with potential to monitor bioavailability and toxicity at once), and visualization.

- In circumstances wherein bioavailability is dictated only by chemical partitioning, chemoavailability will approximate bioavailability. Hence, some studies have noted a correlation between bioaccumulation (bioavailability prerequisite) and chemoavailability. The PAH bioreporters detailed herein have active metabolic processes that govern a regulated response to PAHs. Thus, their mechanism of detection is not based solely on partitioning. This is why bioreporter response does not always correlate to chemical load or chemoavailability analysis.

- Bioreporter technology is immature relative to chemical analysis for total load, ecotoxicological analysis, biomonitoring of bioaccumulation, and analysis of chemoavailability. Three areas for future development with respect to environmental health applications and PAHs are noted. These are 1) development of bioreporters that report PAHs of higher molecular weight than phenanthrene, 2) development of bioreporter optimized assays, and 3) development of methods that exploit bioreporter measurement of bioavailability. Additionally, applications that address primary PAH exposure routes for humans (airborne particles and food) are of interest.

ABBREVIATIONS

Ah	:	Aryl hydrocarbon
ATSDR	:	Agency for Toxic Substances and Disease Registry
BaP	:	Benzo[a]pyrene
BBIC	:	Bioluminescent bioreporter integrated circuit
BTEX	:	Benzene toluene ethylbenzene xylene
CMOS	:	Complementary metal-oxide semiconductor
CSPI	:	Center for Science in the Public Interest
DNA	:	Deoxyribonucleic acid
EDC	:	Endocrine disrupting chemical
EFSA	:	European Food Safety Authority
EU	:	European Union
EVCO	:	Exxon Valdez Crude Oil
GFP	:	Green fluorescent protein
IPCS	:	International Programme on Chemical Safety
IUPAC	:	International Union of Pure and Applied Chemists
PAH	:	Polycyclic aromatic hydrocarbon
RPF	:	Relative potency factor
SCF	:	Scientific Committee on Food
SMD	:	Semi-permeable membrane device
US EPA	:	United States Environmental Protection Agency
WCRF	:	World Cancer Research Fund

REFERENCES

ATSDR. Agency for Toxic Substances and Disease Registry, U.S. Department of Health and Human Services, Public Health Service. 1995. Toxicological Profile for Polycyclic Aromatic Hydrocarbons (PAHs). Atlanta, Georgia, USA.

Bastiaens L, D Springael, W Dejonghea, P Wattiauc, H Verachtert and L Dielsa. 2001. A transcriptional *luxAB* reporter fusion responding to fluorene in *Sphingomonas* sp. LB126 and its initial characterisation for whole-cell bioreporter purposes. Res Microbiol 152: 849–859.

CSPI. Center for Science in the Public Interest. 2008. Big Oil U. Washington DC, USA.

Deepthike HU, R Tecon, G van Kooten, JR van der Meer, H Harms, M Wells and J Short. 2009. Unlike PAHs from Exxon Valdez Crude oil, PAHs from Gulf of Alaska coals are not readily bioavailable. Environ Sci Technol 43: 5864–5870.

Deepthike HU, R Tecon, G van Kooten, JR van der Meer, H Harms, M Wells and J Short. 2010. Response to Comment on Unlike PAHs from Exxon Valdez Crude Oil, PAHs from Gulf of Alaska Coals are not Readily Bioavailable. Environ Sci Technol 44: 2212–2213.

EFSA. European Food Safety Authority 2008. Scientific Opinion of the Panel on Contaminants in the Food Chain on a request from the European Commission on Polycyclic Aromatic Hydrocarbons in Food. The EFSA Journal.

EU. European Union, Office Journal of the European Union 2005. Commission Regulation (EC) No 208/2005 amending Regulation (EC) No 466/2001 as regards polycyclic aromatic hydrocarbons, Brussels.

IPCS. International Programme on Chemical Safety, World Health Organization 1998. Environmental Health Criteria 202. Selected Non-heterocyclic Polycyclic Aromatic.

Kohlmeier S, M Mancuso, R Tecon, H Harms, J van der Meer and M Wells. 2007. Bioreporters: gfp versus lux revisited and single-cell response. Biosens Bioelectr 22: 1578–1585.

Kohlmeier S, M Mancuso, U Deepthike, R Tecon, J van der Meer, H Harms and M Wells. 2008. Comparison of naphthalene bioavailability determined by whole-cell biosensing and availability determined by extraction with Tenax. Environ Pollut 153: 803–808.

Luch A. 2005. The Carcinogenic Effects of Aromatic Polycyclic Hydrocarbons, Imperial College Press, London, UK.

Matrubutham U, JE Thonnard and GS Sayler. 1997. Bioluminescence induction response and survival of the bioreporter bacterium Pseudomonas fluorescens HK44 in nutrient-deprived conditions. Appl Microbiol Biotechnol 47: 604–609.

Melendez-Colon VJ, A Luch, A Seidel and WM Baird. 2000. Formation of stable DNA adducts and apurinic sites upon metabolic activation of bay and fjord region polycyclic aromatic hydrocarbons in human cell cultures. Chem Res Toxicol 13: 10–17.

Page DS, PD Boehm and JM Neff. 2010. Comment on Unlike PAHs from Exxon Valdez Crude Oil, PAHs from Gulf of Alaska Coals are not Readily Bioavailable. Environ Sci Technol 44: 2210–2211.

Perera FP, L Zhigang, R Whyatt, L Hoepner, S Wang, D Camann and V Rauh. 2009. Prenatal Airborne Polycyclic Aromatic Hydrocarbon Exposure and Child IQ at Age 5 Years. Pediatrics 124: 195–202.

Rodriguez-Mozaz S, M-P Marco, MJ Lopez de Alda and D Barcelo. 2004. Biosensors for environmental monitoring of endocrine disruptors: a review article Anal Bioanal Chem 378: 588–598.

Roy GM. 1995. Activated Carbon Applications in the Food and Pharmaceutical Industries, CRC Press, Boca Raton, Florida, USA.

Semple KT, KJ Doick, KC Jones, P Burauel, A Craven and H Harms. 2004. Defining bioavailability and bioaccessibility of contaminated soil and sediment is complicated. Environ Sci Technol 38: 228A–231A.

Shin D, HS Moon, C-C Lin, T Barkay and K Nama. 2011. Use of reporter-gene based bacteria to quantify phenanthrene biodegradation and toxicity in soil. Environ Pollut. 159: 509–514.

Short JW, KR Springman, MR Lindeberg, LG Holland, ML Larsen, CA Sloan, C Khan, PV Hodson and SD Rice. 2008. Semipermeable membrane devices link site-specific contaminants to effects: Part II—A comparison of lingering Exxon Valdez oil with other potential sources of CYP1A inducers in Prince William Sound, Alaska Mar Environ Res 66: 487–498.

Simpson ML, GS Sayler, BM Applegate, S Ripp, DE Nivens, MJ Paulus and GE Jellison, Jr. 1998. Bioluminescent-bioreporter integrated circuits form novel whole-cell biosensors. Trend Biotech 16: 332–338.

Simpson ML, GS Sayler, G Patterson, DE Nivens, EK Bolton, JM Rochelle, JC Arnott, BM Applegate, S Ripp and MA Guillorn. 2001. An integrated CMOS microluminometer for low-level luminescence sensing in the bioluminescent bioreporter integrated circuit. Sensors Act B 72: 136–140.

Tecon R, M Wells and JR van der Meer. 2006. A new green fluorescent protein-based bacterial biosensor for analysing phenanthrene fluxes. Environ Microbiol 8: 697–708.

Tecon R, O Binggeli and JR van der Meer. 2009. Double-tagged fluorescent bacterial bioreporter for the study of polycyclic aromatic hydrocarbon diffusion and bioavailability. Environ Microbiol 11: 2271–2283.

Tecon R, S Beggah, K Czechowska, V Sentchilo, P-M Chronopoulou, TJ McGenity and JR van der Meer. 2010. Development of a multistrain bacterial bioreporter platform for the monitoring of hydrocarbon contaminants in marine environments. Environ Sci Technol 44: 1049–1055.

US EPA. United States Environmental Protection Agency 1979. Toxic Pollutants Code of Federal Regulations 40 CFR 401.15, Washington DC, USA.

US EPA. United States Environmental Protection Agency, Office of Health and Environmental Assessment, Environmental Criteria and Assessment Office 1993. Provisional Guidance for Quantitative Risk Assessment of Polycyclic Aromatic Hydrocarbons, Cincinnati, Ohio, USA.

WCRF. World Cancer Research Fund (with the American Institute for Cancer Research) 2007. Food, Nutrition, Physical Activity, and the Prevention of Cancer: A Global Perspective, Washington, DC, USA.

WCRF. World Cancer Research Fund (with the American Institute for Cancer Research) 2009. Policy and Action for Cancer Prevention. Food, Nutrition, and Physical Activity: A Global Perspective, Washington, DC, USA.

Wells M, M Gösch, R Rigler, H Harms, T Lasser and JR van der Meer. 2005. Ultrasensitive reporter protein detection in genetically engineered bacteria. Anal Chem 77: 2683–2689.

van der Meer JR, D Tropel and M Jaspers. 2004. Illuminating the detection chain of bacterial bioreporters. Environ Microbiol 6: 1005–1020.

Vijayaraghavan R, SK Islam, M Zhang, S Ripp, S Caylor, ND Bull, S Moser, SC Terry, BJ Blalock and GS Sayler. 2007. A bioreporter bioluminescent integrated circuit for very low-level chemical sensing in both gas and liquid environments. Sensors Act B. 123: 922–928.

11

Quorum Sensing in Microbial Biosensors

Swati Choudhary[1,a] and Claudia Schmidt-Dannert[1,b,*]

ABSTRACT

Small signaling molecules are utilized by several unicellular organisms to assess the cell density of other organisms of the same species in their vicinity. Together, the biochemical pathways involved in the production, secretion and recognition of these diffusible signals are known as quorum sensing (QS). Upon establishing that their local concentration has reached a threshold, the unicellular organisms collectively undertake a change in their transcriptional profiles, initiating complex activities which benefit the group as a whole but would have had limited relevance at a lower population count. In bacteria, QS regulates diverse functions such as formation of biofilms, onset of virulence, competence and bioluminescence. Researchers have developed whole-cell microbial biosensors that detect the presence of QS signals in clinical and environmental isolates. These biosensors enhance our understanding of microbial ecosystems present in diverse locations including the mammalian gut and lake sediments. Components of bacterial QS machinery have found widespread application in the emerging field of synthetic biology for the engineering of complex genetic circuits with novel functionalities—for example, production of biochemicals, spatio-temporal control of gene expression and

[1]Department of Biochemistry, Molecular Biology and Biophysics, University of Minnesota, 140 Gortner Laboratory, St. Paul, MN 55108, USA.
[a]E-mail: swati@umn.edu
[b]E-mail: schmi232@umn.edu
*Corresponding author

List of abbreviations after the text.

creation of synthetic ecosystems. Engineered QS-based devices have been used to create microbial biosensors that localize to cancer cells or serve as improved live attenuated vaccines. In this chapter, we will discuss bacterial QS, and its usage in synthetic biology, followed by an overview of the applications of QS-based microbial biosensors in health and environment.

INTRODUCTION

It is now known that co-ordination of action is not restricted to cells constituting multi-cellular organisms. Several unicellular organisms use sophisticated signaling mechanisms to estimate the concentration of other cells belonging to their own species in their surroundings. The assessment of local population density occurs through secretion and detection of small signaling molecules, which are often specific for that species. Once the population is estimated to have reached a threshold, the unicellular organisms collectively initiate a change in their gene-expression profiles, resulting in co-ordinated forays into activities that would not have been profitable if commenced at a lower cell count. Assessing local cell density and employing it to change their actions allows unicellular organisms to function as multi-cellular systems. These cell-to-cell communication systems are collectively known as quorum sensing (QS). QS in various species regulates diverse activities including formation of biofilms, initiation of virulence, competence, mating, sporulation, formation of root nodules, synthesis of secondary metabolites and bioluminescence (Bassler and Losick 2006). Although QS has also been reported in unicellular eukaryotes, in this chapter we will focus on bacterial QS and their applications in biotechnology (Kruppa 2009).

Bacteria employ diverse small molecules as QS signals, including acyl homoserine lactones (AHLs, also called autoinducers due to positive feedback on their own expression) and translationally derived small oligopeptides known as auto-inducing peptides (AIPs). While many of the QS signals are species-specific, a few can be recognized by several bacterial species, and are hypothesized to facilitate inter-species communication (Bassler and Losick 2006). Additionally, several bacteria (including *Pseudomonas aeruginosa*) use more than one QS system to regulate their transcription profiles (Williams and Camara 2009). Quorum quenching—the disruption of heterologous QS—is practiced by several bacteria, and likely assists in reducing competition for nutrients in their ecological niche. Several eukaryotes have also developed strategies to counter bacterial QS (Joint et al. 2007). While QS is associated with large bacterial populations, confinement of a few bacterial cells in a small volume can also initiate this phenomenon (Carnes et al. 2010).

Components of bacterial QS systems have been widely used in engineering biological parts and devices, and have played an important role in the emerging field of synthetic biology (see Key facts of synthetic biology). QS-based parts can perform complex functions such as spatio-temporal control of gene expression, population control, maintenance of synthetic ecosystems, biocontrol, and prevention of biofouling (Purnick and Weiss 2009; Xiong and Liu 2010). Here, we present an overview of bacterial QS systems, and their applications in synthetic biology, followed by a discussion of the role of QS-based biosensors in identifying microbial signaling compounds in clinical and environmental isolates. We also present a review of recent progress in the development of novel whole-cell biosensors which incorporate QS-based artificial genetic circuits.

Bacterial Quorum Sensing systems

Secreted QS signals are recognized by an intracellular transcription factor, either through direct interaction, or via a transmembrane sensor histidine kinase. The portion of the QS machinery involved in production of the diffusible signal can be referred to as the "sender" module, and the signal recognition components as the "receiver" module. In several Gram negative bacteria, the sender module consists of an autoinducer synthase of the LuxI family (Fig. 11.1). These produce acyl homoserine lactone (AHL) signals which have a core homoserine lactone attached to a variable acyl side chain (Bassler and Losick 2006). AHLs produced by different LuxI

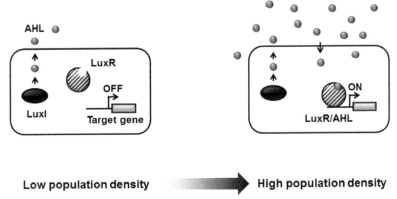

Figure 11.1. Acyl homoserine lactone (AHL)-mediated Quorum Sensing in Gram negative bacteria. Autoinducer synthases of the LuxI family produce AHL signals which diffuse freely through the Gram negative cell membranes. When the quorum is reached, the AHL signal molecule binds and activates transcription factors belonging to the LuxR family. The activated AHL/LuxR complex binds cognate promoters and regulates transcription of downstream genes.

homologs have side-chains with different lengths (four or more carbons), varying degrees of saturation and diverse R groups (Thiel et al. 2009). The unique acyl side-chain of each AHL signal is believed to impart signal specificity and exclusivity to its QS system. AHLs can freely diffuse in and out of Gram negative cells. In this case, the receiver module consists of intracellular receptors that are homologs of the *Vibrio fischeri* LuxR protein. At high concentrations, AHLs bind their cognate LuxR proteins, causing their activation and subsequent translocation to cognate promoters. This results in the transcriptional regulation of downstream genes.

As an example, the Lux R/I QS system of the marine Gram negative bacterium *Vibrio fischeri* is presented in Fig. 11.1. *V. fischeri* is a symbiont of marine animals like the Hawaiian bobtail squid *Euprymna scolopes*, and is responsible for producing bioluminescence which protects its host from detection by predators on moonlit nights. At high bacterial titers, the LuxI product *N*-(3-oxohexanoyl)-homoserine lactone (3-oxo-C6-HSL) activates LuxR, resulting in transcription of the *luxCDABE* operon, which contains genes involved in producing bioluminescence. The LuxR/AHL complex also creates a positive feedback loop by activating transcription of *luxI* and *luxR* genes. Homologs of the LuxR/I system have been discovered in several Gram negative bacteria. For example, in the opportunistic human pathogen *Pseudomonas aeruginosa*, two homologs of the LuxR/I system (named LasR/I and RhlR/I), control the expression of up to 10% of all genes, including those involved in production of the virulence factors pyocyanin, elastase, rhamnolipids, lectins and hydrogen cyanide (Williams and Camara 2009).

Instead of LuxR/I and AHLs, Gram positive bacteria utilize two-component histidine kinase response regulators systems which recognize extracellular short oligopeptides (auto-inducing peptides, AIPs). The sender module in Gram positive QS systems consists of genes encoding the AIP precursor and a dedicated transport machinery which facilitates its maturation and secretion. During this process, post-translational modifications such as formation of a cyclic thiolactone (AgrD, *Staphylococcus aureus*) or geranylation of certain amino acid residues (ComX, *Bacillus subtilis*) may be added to the AIP (Bassler and Losick 2006). Once the quorum threshold is achieved, the interaction of extracellular AIPs with a transmembrane histidine kinase (HK) triggers a phosphorelay which activates the downstream, cognate response regulator (RR). The activated RR then binds to responsive promoters and regulates transcription of QS-controlled genes. In contrast to the fairly promiscuous AHL-based systems, the AIP-based systems are extremely species- and even strain-specific.

The widely-studied *accessory gene regulator* (*agr*) QS system from the opportunistic pathogen *Staphylococcus aureus* is presented as a Gram positive QS example in Fig. 11.2. *agrD* encodes the AIP precursor, which is processed

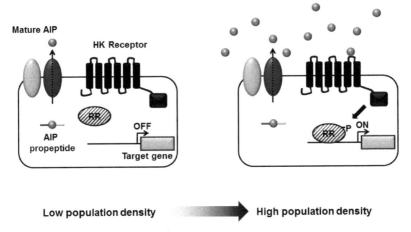

Figure 11.2. Auto-inducing peptide (AIP)-mediated Quorum Sensing in Gram positive bacteria. AIPs are translated as propeptides, with N-terminal signal sequences (shown as a dark gray line). A C-terminal tail may also be present (represented by a light gray line). During maturation, the oligopeptide may undergo post-translational modifications such as self-cyclization. This process is carried out by a dedicated transmembrane endopeptidase (light gray oval). The N-terminal leader is removed by a Type I signal peptidase (dark gray oval). At high AIP concentrations, the transmembrane histidine kinase receptor is activated, and it, in turn, initiates a phosphorelay resulting in activation of the cognate response regulator (RR). The activated RR directs transcription of QS-regulated genes.

by the transmembrane endopeptidase AgrB (Novick and Geisinger, 2008). During maturation, the C-terminal tail of the propeptide is removed, and a cyclic thiolactone is formed by a cysteine residue and the C-terminal end of the cleaved precursor. Secretion of the modified AIP involves cleavage of the N-terminal signal sequence by a Type I signal peptidase (Thoendel and Horswill 2010). The receiver module comprises the histidine kinase AgrC and a cognate response regulator named AgrA. At high AIP concentration, AgrC is activated and in turn activates AgrA. AgrA then induces transcription of downstream genes, including the *agr* operon and an adjacent regulatory RNA, known as RNAIII. In *S. aureus*, RNAIII modulates the expression of a large number of genes involved in pathogenesis. The production of RNAIII unleashes a transcriptional cascade which results in the production of virulence factors and biofilm formation, causing infection of the host organism. Figure 11.3 contains examples of AHLs and AIPs from Gram negative and Gram positive bacteria, respectively.

Quorum Quenching by Bacteria and Eukaryotes

Since QS triggers the production of virulence factors in several pathogenic bacteria, strategies for interfering with this cell-to-cell communication

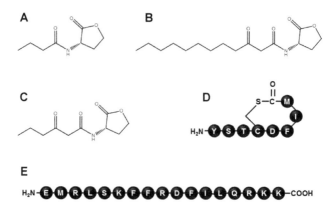

Figure 11.3. Bacterial QS signals. Many Gram negative bacteria use acyl homoserine lactones (AHLs) as Quorum Sensing signals. These share a core homoserine lactone ring. Signal specificity arises from the length of the acyl side-chain, and the presence of unsaturation or different substituent R-groups. **(A)** N-(butanoyl)-homoserine lactone (*Pseudomonas aeruginosa*). **(B)** N-(3-oxo-dodecanoyl)-homoserine lactone (*Pseudomonas aeruginosa*). **(C)** N-(3-oxo-hexanoyl)-homoserine lactone (*Vibrio fischeri*). Several Gram positive bacteria utilize auto-inducing peptides (AIPs) as QS signals. These may be unmodified, or carry post-translational modifications, for example, self-cyclization. **(D)** AIP-I (*Staphylococcus aureus*). **(E)** Competence Stimulating Peptide-1 (CSP-1) (*Streptococcus pneumoniae*).

would be useful in developing novel therapeutics for preventing and treating bacterial diseases. As QS affects pathogenicity but not survival, compounds that target QS would not elicit selection pressure, and are therefore expected to by-pass the development of drug-resistance.

Several bacteria secrete enzymes that degrade the AHL QS signals of heterologous bacteria. This phenomenon, known as quorum quenching, is believed to reduce colonization of the surrounding location by other species, and therefore reduce competition for common resources. Diverse quorum quenching enzymes that target various parts of the AHL signal have been reported. These include lactonases that degrade the core lactone ring, acylases/amidases that hydrolyze the amide bond between the lactone and the acyl side-chain, and oxidoreductases that modify the acyl side-chain (Chowdhary et al. 2007; Czajkowski and Jafra 2009). So far, bacterial enzymes that specifically target Gram positive AIPs have not been discovered.

Many plants, fungi, and animals have also developed strategies to "jam" bacterial QS communication. The seaweed *Delisea pulchra* secretes halogenated furanones that obstruct AHL-mediated QS and biofilm formation (Natrah et al. 2011). Synthetic brominated furanones have been demonstrated to disrupt biofilm formation in *Salmonella*, and are of considerable interest in developing anti-QS applications (Janssens et al. 2008). QS inhibitors penicillic acid and patulin are produced by fungi of the *Penicillium* genus (Rasmussen et al. 2005). Interestingly, the fungal secondary metabolite ambuic acid has been demonstrated to interfere with cyclic AIP production by Gram positive bacteria including *S. aureus*, *Enterococcus faecalis* and *Listeria innocua* (Nakayama et al. 2009). Mammalian paraoxonases degrade AHL signals, likely via hydrolysis of the central lactone ring, and are hypothesized to play an important role in host defense against invading Gram negative pathogens (Pacheco and Sperandio 2009).

APPLICATIONS OF QUORUM SENSING IN SYNTHETIC BIOLOGY

Synthetic biology is a new area of scientific study which melds biology with engineering (see Key facts of synthetic biology). The emphasis is on creating new biological systems with novel functionalities that did not exist naturally. In this approach, basic biological activities are considered from an engineering point-of-view, i.e. there is an effort to standardize biochemical functions, so that they may be combined to construct complex genetic circuits that perform predicted functions. Components of Gram negative QS systems have been widely used in engineering synthetic biological circuits (Choudhary and Schmidt-Dannert 2010). Their properties of signal recognition and signal amplification, along with the ability to regulate transcription of downstream genes, has been utilized to create genetic circuits that control diverse phenomena, for example, spatio-temporal control of gene expression, population control, bistable behavior, and pulse response (Purnick and Weiss 2009). In this section, we will describe a few synthetic biological circuits that incorporate QS. This will formulate our later discussion of how QS can be used to create microbial biosensors that perform complex functions.

Different input, processor and output modules can be combined in a "plug-and-play" strategy to engineer new functionalities. For instance, combining an AHL recognition module, a bistable switch module, and a fluorescent output module produces a biosensor which can detect AHL signals in its vicinity. Additionally, they can retain the memory of their interaction with AHLs, and therefore display a sustained response even after AHLs are cleared from their immediate surroundings (Kobayashi et al. 2004). These engineered devices can be of great use in designing biosensors for environment and health.

Components of the *V. fischeri* LuxR/I system have been utilized to engineer cell-to-cell communication that can be regulated with respect to both space and time (Fig. 11.4) (Basu et al. 2004). Here, two different types of synthetic circuits were devised. The "sender" circuit, which consisted of the *V. fischeri luxI* gene, was inserted into *E. coli* to create Sender cells which released 3-oxo-C6-HSL into their surroundings. The "receiver" circuit

E. coli Sender   E. coli Receiver

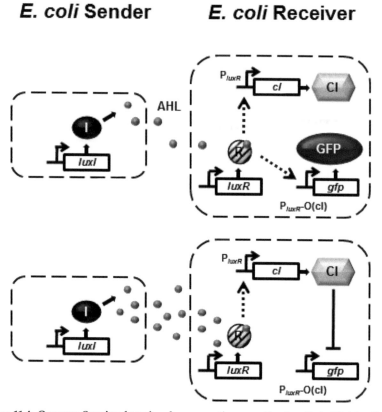

Figure 11.4: Quorum Sensing-based pulse generating genetic circuit. Artificial cell-cell communication engineered in *E. coli* using components of the *V. fischeri* LuxR/I QS system. *E. coli* cells with the Sender module produce the *V. fischeri* QS signal 3-oxo-C6-HSL. *E. coli* Receiver cells are equipped with a pulse-generating circuit which regulates transient expression of Green Fluorescent Protein (GFP). Upon activation, the LuxR/AHL complex directs transcription of both CI repressor and GFP, the former through the P_{luxR} promoter, and the latter via a hybrid promoter (P_{luxR}-O(cI)) with a CI binding site inserted at the +1 transcription start downstream of the P_{luxR} promoter. In time, concentration of the CI repressor reaches the threshold required to repress transcription of GFP, leading to a decrease in fluorescence. This scheme is modified from Basu et al. (2004) (QS: Quorum Sensing).

contained an AHL-inducible pulse-generating circuit made by interfacing the *luxR* gene with lux-inducible *cI* repressor and GFP regulated by both LuxR/AHL and CI. The genetic circuit was constructed so that an early GFP response to AHL was tamped down by a gradual increase in CI levels, thereby presenting a temporally-regulated output. The response was also controlled by the distance between the Sender and Receiver cells—while nearby Sender cells elicited a transient response from the Receivers, those that were further away failed to generate a strong output altogether. By incorporating spatio-temporal control into genetic circuits, one can design microbial biosensors that can identify the origin of the signal, which should be very helpful in locating, for example, the site of pathogen infection *in vivo* or in contaminated waste water.

Genetic circuits have also been designed using antibiotics, hormones, metabolites or volatile compounds to generate cell-density dependent responses (Bulter et al. 2004; Chen and Weiss 2005; Weber et al. 2007). These QS-like systems are very promising for designing biosensors that can sense a wide variety of inputs. QS has also been employed to create artificial communication between cells of different species—bacteria, fungi, mammalian cells and plants (Brenner et al. 2008; Weber et al. 2007). It is possible to engineer co-operative relationships between the participating species (Fig. 11.5). Mixed consortia can also be used to devise biosensors that sense environmental inputs and produce complex responses, which may not be possible by using a single biosensor species.

Broadly speaking, QS-based whole-cell microbial biosensors can be divided into two categories: those that recognize QS signals in clinical

Species A   Species B

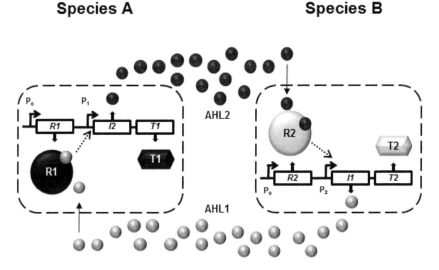

Figure 11.5. Quorum Sensing-based circuit for interspecies communication. Components of two QS systems from *P. aeruginosa* (Rhl and Las, represented here as 1 and 2, respectively) used to engineer bi-directional communication in *E. coli*. The response regulator R1 and autoinducer synthase I1 (responsible for production of the QS signal AHL1) constitute QS system 1. QS system 2 consists of the response regulator R2 and the autoinducer synthase I2 (producer of the QS signal AHL2). The transcription of target genes T1 and T2 is triggered at when the cell density of both communicating cell types reaches the threshold. The expression of I2 and T1 is directed by the R1/AHL1 complex. Similarly, the R2/AHL2 complex initiates transcription of I1 and T2. P_c = Constitutive promoter, P_1 = Promoter induced by QS system 1, P_2 = Promoter induced by QS system 2. This scheme is modified from Brenner et al. (2008) (QS: Quorum Sensing).

and environmental isolates, and those that incorporate QS machinery into synthetic biological circuits. In the following sections, we describe the applications of both varieties of QS-based biosensors in human health and environmental studies.

QUORUM SENSING-BASED MICROBIAL BIOSENSORS: APPLICATIONS TO AREAS OF HEALTH AND DISEASE

Here, we will discuss recent progress in utilizing bacterial QS for creating whole-cell microbial sensors for health (Fig. 11.6). These biosensors incorporate components of the QS machinery in sophisticated genetic circuits that perform diverse functions including recognition of pathogens, invasion of cancer cells, or serve as better versions of live attenuated vaccines.

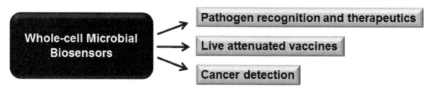

Figure 11.6. Quorum Sensing-based whole-cell microbial biosensors in health. Whole-cell microbial biosensors that incorporate QS systems have diverse applications in health, including pathogen diagnostics and therapeutics, vaccines, and detection of cancerous cells.

Pathogen Recognition and Therapeutics

Several whole-cell biosensors have been devised to recognize AHL signals produced by Gram negative bacteria (Kumari et al. 2008; Steindler and Venturi 2007). These biosensors are often outfitted with simple circuits that interface a LuxR homolog with a responsive promoter that drives the expression of a reporter gene. Researchers have also developed whole-cell biosensors that recognize the presence of *P. aeruginosa* 3-oxo-C12-HSL in clinical samples from cystic fibrosis patients (Massai et al. 2011). This provides a cost-effective and rapid diagnostic method for detecting infection by this common nosocomial pathogen. Liquid-drying of whole-cell AHL-recognizing bacteria on filter-paper has been used to create biosensors that can recognize the presence of C12-HSL in different clinical and environmental samples (Struss et al. 2010). Paper-strip biosensors are inexpensive, easy to store and use, and can be created to recognize a wide variety of AHL signals.

More complex circuits can be designed to recognize and respond to pathogens present in, for example, gut microbiota. QS-like circuits (mentioned previously) can be engineered to recognize secreted toxins or surface antigens presented by pathogenic bacteria. These input modules can feed into output modules that provide a readable output, and/or mount a concerted response against the pathogen. Probiotic biosensors can be engineered by inserting these genetic circuits into GRAS organisms—for example—lactic acid bacteria. Alternatively, directed evolution can be used to modify existing LuxR homologs to recognize the toxins produced by gut pathogens. It would be interesting to create "AND Gate" biosensors (activated by the presence of more than one input signal, i.e. more than one toxin), since they would be more robust, and less likely to give false positive readouts.

Towards Improved Live Attenuated Vaccines

Live genetically attenuated pathogenic bacteria are of considerable interest in providing immunity against several diseases including cholera, salmonellosis, shigellosis, listeriosis and tuberculosis (Silva et al. 2010). However, their widespread application is hindered by fears of insufficient attenuation, and the possibility that once released into the environment (by shedding of live bacteria by the inoculated host), they may regain virulence factors and become a major health hazard. A recent study provides an elegant solution to these issues (Silva et al. 2010). Researchers have created a synthetic biological device which interfaces the *Vibrio cholerae* QS input with a cell lysis gene. At high cell densities, the QS module is activated and results in death of the bacterial cells. In this way, excessive accumulation of attenuated cells within the host is avoided. This approach also addresses problems associated with possible shedding of live attenuated cells into the environment. It should be feasible to extend this strategy to create live attenuated vaccines against other bacterial pathogens as well.

Detection and Treatment of Cancer

3-oxo-C12-HSL, a major QS signal of the opportunistic pathogen *P. aeruginosa*, has been shown to reduce proliferation and promote apoptosis in human breast cancer cell lines (Li et al. 2004). Its mode of action is at present unclear. Since this QS signal may promote *P. aeruginosa* virulence in immune-compromised individuals, it is not feasible to administer it directly as an anti-cancer agent. However, this compound is under consideration as a spring-board for rational design of anti-cancer drugs that do not activate *P. aeruginosa* virulence (Oliver et al. 2009).

It has been shown that intravenously delivered *E. coli*, *Bifidobacterium longum*, and attenuated live strains of *V. cholerae*, *Listeria monocytogenes* and *Salmonella typhimurium* can selectively localize to murine solid tumors and metastases (Yazawa et al. 2000; Yu et al. 2004). These bacteria were able to proliferate within tumors in various locations, including mouse breast, bladder, and brain. It is hypothesized that poor immune surveillance inside the tumors contributes to the enhanced survival of bacteria within these environments. The ability of these bacteria to localize to and survive within tumors has been exploited to engineer whole-cell microbial biosensors that target cancer cells (Anderson et al. 2006). A genetic device which associated a *V. fischeri* LuxR/I QS system with the *Yersinia pestis* invasin gene was engineered into *E. coli*. The invasin gene product allowed attachment and entry of *E. coli* biosensors into human cancer cell lines. Including QS components into the synthetic circuit ensured that invasion of mammalian cells occurred only at high cell densities, which should ideally coincide with the presence of the tumor *in vivo*. The specificity of these devices can be enhanced by using AND gate input modules that require the presence of more than one cancer cell-surface antigen for activation. These bacteria can also be modified further to express anti-cancer drugs that target the tumor *in situ*. While further studies are required to assess the efficacy of anti-cancer bacterial biosensors *in vivo*, they certainly represent an exciting new approach in the fight against cancer.

APPLICATIONS OF QUORUM SENSING: MICROBIAL BIOSENSORS FOR THE ENVIRONMENT

QS-based microbial biosensors have multiple uses in detecting microbes present in the soil or other environment like contaminated waste water. Biosensors that recognize specific AHLs can also be applied to gain an understanding of the diverse microbial species present in an environmental niche. In this section, we present a few recent examples of QS-based environmental biosensors.

Savka and colleagues have reported *E. coli* biosensors that can recognize long-chain AHLs produced by Gram negative bacteria including *Agrobacterium vitis* (Savka et al. 2011). This biosensor device places bioluminescence genes under control of the *P. aeruginosa* LasR/I QS system. The natural product of LasI is 3-oxo-C12-HSL. However, the authors demonstrate that this biosensor is able to recognize a wide range of long-chain AHLs with 14-18 carbon side-chains. AHLs with oxygen-substituted or carbon-carbon double bond, containing side-chains, were also recognized efficiently. Long-chain AHLs are produced by several bacteria including *Agrobacterium vitis*, *Rhodobacter capsulatus*, *Paracoccus denitrificans*, *Rhizobium leguminosarum*, and *Sinorhizobium meliloti*. Therefore, the creation of this biosensor provides an important step in detection of these bacteria in environmental isolates.

Self-transmissible biosensor plasmids have been created by combining the broad host-range vector RP4 with receiver modules that respond to different AHLs (Lumjiaktase et al. 2010). By harnessing the conjugation and replication properties of the RP4 vector, these plasmids were able to distribute amongst microbial species forming lake sediment. In case they associated with a bacterium producing a recognizable AHL, a green florescent output was displayed. These biosensor plasmids present an interesting method of identifying AHL-secreting microbes in their native habitat.

Using a QS biosensor strain of the plant pathogen *Pseudomonas syringae*, researchers have demonstrated that bacterial epiphytes present on plant leaves produce excessive amounts of its cognate QS signal, thereby interfering with *P. syringae* QS and hindering the onset of virulence (Dulla and Lindow 2009). In this case, employment of a QS-based biosensor helped to narrow-down the mode of action used by other bacteria to reduce *P. syringae* pathogenicity. This study also suggests that exogenous application of QS signals of pathogens can be used to reduce crop disease.

The term "rhizosphere" refers to the thin region of soil around a plant's roots that contains compounds secreted by the plant as well as the soil micro-organisms present in this ecological niche. Quorum sensing-competent bacteria are a vital part of the rhizosphere microbial community. For instance, AHLs are produced by 8%, 12% and 24% of the bacterial isolates from wheat, tomato and wild oat rhizospheres, respectively (DeAngelis et al. 2008; Lumjiaktase et al. 2010). It has been established that several plant-associated bacteria produce AHLs. These include beneficial bacteria like *Rhizobium* and *Sinorhizobium*, as well as plant pathogens like *Erwinia*, *Pantoea*, and *Xanthomonas* (Cha et al. 1998). QS-based biosensors have also been used to detect cross-talk between bacteria constituting the rhizosphere (Lumjiaktase et al. 2010). Future design directions will include engineering beneficial rhizosphere bacteria to secrete large amounts of AHLs employed by pathogens. Application of these recombinant bacteria to farm soil is expected to enhance crop defense against bacterial pathogens. An alternative strategy is the introduction of quorum quenching enzymes into plants or rhizosphere-associated bacteria. However, since these broad-specificity enzymes would also disrupt QS between beneficial bacterial species, further research is warranted to establish the safety and efficacy of this approach.

CONCLUSIONS

Bacterial activities regulated by QS have a significant impact on human health, agriculture, and environmental safety. While a large amount of effort has been devoted towards understanding the biological functions of QS in diverse bacterial species, strategies to disrupt or modify QS are also of great interest. A better knowledge of the cross-talk between different bacterial species will formulate future strategies for controlling biofilms in surgical or environmental settings, and also provide solutions for crop disease.

Components of Gram negative QS systems have found widespread use in synthetic biological circuits. While Gram negative QS is somewhat promiscuous, the two component signaling systems of Gram positive bacteria are extremely specific in terms of the AIP recognized by them. Biosensors based on Gram positive QS systems would be extremely specific, and amenable to large-scale biotechnological applications.

So far, considerable work has gone into creating microbial biosensors that recognize AHL signals and produce a readable output. By incorporating QS components into the design of the genetic circuit, it should be possible to engineer microbial biosensors that undertake multiple activities—not only detect pathogens but also mount a strong defense against them. This goal will be assisted by the creation of microbial consortia biosensors that could undertake complex functions which cannot be performed by a single cell biosensor.

KEY FACTS OF SYNTHETIC BIOLOGY

- The goal of synthetic biology is to create novel biological systems by applying the engineering principles of standardization and hierarchical abstraction to biochemical reactions.
- This approach towards genetic engineering treats the flow of information across a biochemical pathway as analogous to the flow of information across an electronic circuit.
- Standard biological parts are DNA sequences with a defined function. These may include promoters, ribosome binding sites or open reading frames.
- Standard parts can be combined to produce a module that performs a specific function. For example, in the presence of 3-oxo-C6HSL and LuxR, an output module containing a lux-responsive promoter driving GFP will give a green fluorescent readout. In this case, the *luxR* gene and its promoter would be considered the "processor" module, while 3-oxo-C6HSL would be the "input".
- Different input, processor and output modules can be assembled to engineer a synthetic biological device, such as the "AHL detection and reporting" device described above.

DEFINITIONS

- *Quorum sensing*: Cell-to-cell communication systems used by unicellular organisms (including bacteria) to sense their local population density. Once the population is determined to reach a threshold (or quorum), the organisms collectively undertake a change in their transcriptional profiles, adopting activities that are advantageous at higher cell counts.
- *Autoinducer*: A chemical signaling molecule that is secreted and recognized by bacteria to sense their local population density. Several Gram negative bacteria use acyl homoserine lactone molecules as quorum sensing-signals, while many Gram positive bacteria rely on small peptides as signals for quorum sensing. Quorum sensing signals are also called autoinducers since their recognition usually triggers an increase in their own transcription.
- *Histidine kinases*: These are multifunctional proteins that frequently act as receptors for extracellular signaling molecules and facilitate signal transduction across the cell membrane. A conserved histidine residue within the kinase plays a key role in the phosphorelay.
- *Synthetic Biology*: A new area of scientific study which interfaces biology with engineering. The goal of synthetic biology is to create new biological systems with novel functionalities that do not exist in nature.
- *Bioluminiscence*: The generation and emission of light by living organisms. In the Gram negative marine symbiont *Vibrio fischeri*, the *lux* operon is responsible for creating bioluminescence once the quorum sensing threshold is reached.

SUMMARY POINTS

- Quorum sensing (QS) systems are used by unicellular organisms to assess their local concentrations, and accordingly change their activities to those that benefit the group, but would not be useful at lower cell counts.
- Phenomena regulated by QS include biofilm formation, onset of virulence, competence, mating, sporulation, root nodulation, bioluminescence and production of secondary metabolites.
- Many Gram negative bacteria use QS systems based on LuxR/I homologs and diffusible acyl homserine lactone (AHL) signals. LuxI homologs produce AHLs. When the quorum is achieved, the activated AHL/LuxR complex binds to cognate promoters and regulates transcription of downstream genes.

- Several Gram positive bacteria use extracellular auto-inducing peptides (AIPs) as QS signals, and employ two-component transmembrane histidine kinase—response regulator systems for AIP recognition. At high cell density, the interaction of AIP with the histidine kinase sets off a phosphorelay which activates the intracellular response regulator. In turn, the response regulator controls expression of downstream genes.
- Whole-cell microbial biosensors have been designed to recognize AHLs in clinical samples and environmental isolates. These are expected to enhance our knowledge of microbial ecosystems present in diverse niches. They also present an excellent avenue for devising diagnostic tests for the presence of pathogenic bacteria.
- Components of the bacterial QS systems have been incorporated into synthetic biological circuits to create biosensors that can perform complex functions. Examples include whole-cell biosensors that invade cancer cells and improved live attenuated vaccines.

ACKNOWLEDGEMENTS

The authors gratefully acknowledge support by the Office of Naval Research (grant #N00014-10-1-0157).

ABBREVIATIONS

AHL	:	Acyl Homoserine Lactone
AI	:	Autoinducer
AIP	:	Auto-Inducing Peptide
CSP	:	Competence Stimulating Peptide
GFP	:	Green Fluorescent Protein
GRAS	:	Generally Recognized As Safe
HK	:	Histidine Kinase
HSL	:	Homoserine Lactone
QS	:	Quorum Sensing
RR	:	Response Regulator

REFERENCES

Anderson JC, EJ Clarke, AP Arkin and CA Voigt. 2006. Environmentally controlled invasion of cancer cells by engineered bacteria. J Mol Biol 355: 619–627.
Bassler BL and R Losick. 2006. Bacterially speaking. Cell 125: 237–246.
Basu S, R Mehreja, S Thiberge, MT Chen and R Weiss. 2004. Spatiotemporal control of gene expression with pulse-generating networks. Proc Natl Acad Sci USA 101: 6355–6360.
Brenner K, L You and FH Arnold. 2008. Engineering microbial consortia: a new frontier in synthetic biology. Trends Biotechnol 26: 483–489.

Bulter T, SG Lee, WW Wong, E Fung, MR Connor and JC Liao. 2004. Design of artificial cell-cell communication using gene and metabolic networks. Proc Natl Acad Sci USA 101: 2299–2304.

Carnes EC, DM Lopez, NP Donegan, A Cheung, H Gresham, GS Timmins and CJ Brinker. 2010. Confinement-induced quorum sensing of individual *Staphylococcus aureus* bacteria. Nat Chem Biol 6: 41–45.

Cha C, P Gao, YC Chen, PD Shaw and SK Farrand. 1998. Production of acyl-homoserine lactone quorum-sensing signals by gram-negative plant-associated bacteria. Mol Plant Microbe Interact 11: 1119–1129.

Chen MT and R Weiss. 2005. Artificial cell-cell communication in yeast *Saccharomyces cerevisiae* using signaling elements from *Arabidopsis thaliana*. Nat Biotechnol 23: 1551–1555.

Choudhary S and C Schmidt-Dannert. 2010. Applications of quorum sensing in biotechnology. Appl Microbiol Biotechnol 86: 1267–1279.

Chowdhary PK, N Keshavan, HQ Nguyen, JA Peterson, JE Gonzalez and DC Haines. 2007. *Bacillus megaterium* CYP102A1 oxidation of acyl homoserine lactones and acyl homoserines. Biochemistry 46: 14429–14437.

Czajkowski, R and S Jafra. 2009. Quenching of acyl-homoserine lactone-dependent quorum sensing by enzymatic disruption of signal molecules. Acta Biochim Pol 56: 1–16.

DeAngelis KM, SE Lindow and MK Firestone. 2008. Bacterial quorum sensing and nitrogen cycling in rhizosphere soil. FEMS Microbiol Ecol 66: 197–207.

Dulla GF and SE Lindow. 2009. Acyl-homoserine lactone-mediated cross talk among epiphytic bacteria modulates behavior of *Pseudomonas syringae* on leaves. ISME J 3: 825–834.

Janssens JC, H Steenackers, S Robijns, E Gellens, J Levin, H Zhao, K Hermans, D De Coster, TL Verhoeven, K Marchal and et al. 2008. Brominated furanones inhibit biofilm formation by *Salmonella enterica* serovar Typhimurium. Appl Environ Microbiol 74: 6639–6648.

Joint I, K Tait and G Wheeler. 2007. Cross-kingdom signalling: exploitation of bacterial quorum sensing molecules by the green seaweed *Ulva*. Philos Trans R Soc Lond B Biol Sci 362: 1223–1233.

Kobayashi H, M Kaern, M Araki, K Chung, TS Gardner, CR Cantor and JJ Collins. 2004. Programmable cells: interfacing natural and engineered gene networks. Proc Natl Acad Sci USA 101: 8414–8419.

Kruppa M. 2009. Quorum sensing and *Candida albicans*. Mycoses 52: 1–10.

Kumari A, P Pasini and S Daunert. 2008. Detection of bacterial quorum sensing N-acyl homoserine lactones in clinical samples. Anal Bioanal Chem 391: 1619–1627.

Li L, D Hooi, SR Chhabra, D Pritchard and PE Shaw. 2004. Bacterial N-acylhomoserine lactone-induced apoptosis in breast carcinoma cells correlated with down-modulation of STAT3. Oncogene 23: 4894–4902.

Lumjiaktase P, C Aguilar, T Battin, K Riedel and L Eberl. 2010. Construction of self-transmissible green fluorescent protein-based biosensor plasmids and their use for identification of N-acyl homoserine-producing bacteria in lake sediments. Appl Environ Microbiol 76: 6119–6127.

Massai F, F Imperi, S Quattrucci, E Zennaro, P Visca, and L Leoni. 2011. A multitask biosensor for micro-volumetric detection of N-3-oxo-dodecanoyl-homoserine lactone quorum sensing signal. Biosens Bioelectron 26: 3444–3449.

Nakayama J, Y Uemura, K Nishiguchi, N Yoshimura, Y Igarashi and K Sonomoto. 2009. Ambuic acid inhibits the biosynthesis of cyclic peptide quormones in gram-positive bacteria. Antimicrob. Agents Chemother 53: 580–586.

Natrah FM, T Defoirdt, P Sorgeloos and P Bossier. 2011. Disruption of bacterial cell-to-cell communication by marine organisms and its relevance to aquaculture. Mar Biotechnol (NY) (in press).

Novick RP and E Geisinger. 2008. Quorum sensing in staphylococci. Annu Rev Genet 42: 541–564.

Oliver CM, AL Schaefer, EP Greenberg and JR Sufrin. 2009. Microwave synthesis and evaluation of phenacylhomoserine lactones as anticancer compounds that minimally activate quorum sensing pathways in *Pseudomonas aeruginosa*. J Med Chem 52: 1569–1575.

Pacheco AR and V Sperandio. 2009. Inter-kingdom signaling: chemical language between bacteria and host. Curr Opin Microbiol 12: 192–198.

Purnick PE and R Weiss. 2009. The second wave of synthetic biology: from modules to systems. Nat Rev Mol Cell Biol 10: 410–422.

Rasmussen TB, ME Skindersoe, T Bjarnsholt, RK Phipps, KB Christensen, PO Jensen, JB Andersen, B Koch, TO Larsen, M Hentzer and et al. 2005. Identity and effects of quorum-sensing inhibitors produced by *Penicillium* species. Microbiology 151: 1325–1340.

Savka MA, PT Le and TJ Burr. 2011. LasR receptor for detection of long-chain quorum-sensing signals: identification of N-acyl-homoserine lactones encoded by the avsI locus of *Agrobacterium vitis*. Curr Microbiol 62: 101–110.

Silva AJ, JA Benitez and JH Wu. 2010. Attenuation of bacterial virulence by quorum sensing-regulated lysis. J Biotechnol 150: 22–30.

Steindler L and V Venturi. 2007. Detection of quorum-sensing N-acyl homoserine lactone signal molecules by bacterial biosensors. FEMS Microbiol Lett 266: 1–9.

Struss A, P Pasini, CM Ensor, N Raut and S Daunert. 2010. Paper strip whole cell biosensors: a portable test for the semiquantitative detection of bacterial quorum signaling molecules. Anal Chem 82: 4457–4463.

Thiel V, B Kunze, P Verma, I Wagner-Dobler and S Schulz. 2009. New structural variants of homoserine lactones in bacteria. Chembiochem 10: 1861–1868.

Thoendel M and AR Horswill. 2010. Biosynthesis of peptide signals in gram-positive bacteria. Adv Appl Microbiol 71: 91–112.

Weber W, M Daoud-El Baba and M Fussenegger. 2007. Synthetic ecosystems based on airborne inter- and intrakingdom communication. Proc Natl Acad Sci USA 104: 10435–10440.

Williams P and M Camara. 2009. Quorum sensing and environmental adaptation in *Pseudomonas aeruginosa*: a tale of regulatory networks and multifunctional signal molecules. Curr Opin Microbiol 12: 182–191.

Xiong Y and Y Liu. 2010. Biological control of microbial attachment: a promising alternative for mitigating membrane biofouling. Appl Microbiol Biotechnol 86: 825–837.

Yazawa K, M Fujimori, J Amano, Y Kano and S Taniguchi. 2000. *Bifidobacterium longum* as a delivery system for cancer gene therapy: selective localization and growth in hypoxic tumors. Cancer Gene Ther 7: 269–274.

Yu YA, S Shabahang, TM Timiryasova, Q Zhang, R Beltz, I Gentschev, W Goebel and AA Szalay. 2004. Visualization of tumors and metastases in live animals with bacteria and vaccinia virus encoding light-emitting proteins. Nat Biotechnol 22: 313–320.

12

Biosensors Based on Immobilization of Proteins in Supramolecular Assemblies for the Detection of Environmental Relevant Analytes

Rosa Pilolli,[1,a] Maria Daniela Angione,[1,b] Serafina Cotrone,[1,c] Maria Magliulo,[1,d] Gerardo Palazzo,[1,e] Nicola Cioffi,[1,f] Luisa Torsi[1,g] and Antonia Mallardi[2,*]

ABSTRACT

Methods to immobilize proteins are of particular relevance for biosensing. In biosensors, proteins have to be assembled according to suitable architectures on solid surfaces such as electrodes, optical

[1]Department of Chemistry, University of Bari, Via Orabona 4, I-70126 Bari, Italy.
[a]E-mail: pilolli@chimica.uniba.it
[b]E-mail: angione@chimica.uniba.it
[c]E-mail: cotrone@chimica.uniba.it
[d]E-mail: magliulo@chimica.uniba.it
[e]E-mail: palazzo@chimica.uniba.it
[f]E-mail: cioffi@chimica.uniba.it
[g]E-mail: torsi@chimica.uniba.it
[2]CNR-IPCF, Istituto per i Processi Chimico-Fisici, Via Orabona 4, I-70126 Bari, Italy;
E-mail: a.mallardi@ba.ipcf.cnr.it
*Corresponding author

List of abbreviations after the text.

windows, or the organic semiconductor layer of electronic devices and their integrity and activity have to be preserved. This chapter deals with the use of immobilized proteins in several types of biosensors. The importance of molecular architecture control, particularly with synergistic combination of proteins and "other" material, is evidenced. Different methodologies for protein assembly are described, highlighting the environmental applications of the protein based biosensors. Results from the literature, grouped into large areas covering optical and electrochemical biosensors and also sensing exploiting field-effect transistors, are reported.

Layer by layer (LbL) immobilization of proteins on the transparent substrate of optical biosensors is proposed for its advantageous control of molecular architecture and versatility to accommodate layers having different functionalities.

Covalent immobilization is evaluated as an alternative process for the controlled incorporation of the recognition element on/into the electrode surface, in the case of electrochemical transducers. Finally, the integration of immobilized proteins in electronic devices is presented, especially in the context of using field-effect transistors (FETs) for biosensing.

It is hoped that this survey may assist researchers in choosing materials, molecular architectures, and detection principles, which may be tailored for specific applications.

INTRODUCTION

The detection and monitoring of contaminants in the environment, such as chemical compounds, toxins and pathogens is crucial to assess and avoid risks for environmental health. Highly sensitive and selective traditional analytical techniques exist, as liquid and gas chromatography combined with mass spectrometry, but they are time consuming, expensive and require highly trained personnel. To accomplish frequent monitoring, there is still the need for sensitive, simple, rapid, cost effective and portable detection methods. Biosensors meet all these requirements and are thus ideal for environmental monitoring.

A biosensor combines a recognition element with a suitable signal transduction method in such a way that the binding or reaction between the target and the recognition element is translated into a meaningful signal. Biosensors can be classified according to the type of recognition element (protein, whole cell or affinity-based biosensor) used. Proteins are important biological macromolecules with structural features and functionalities that make them a very attractive interfacing system for recognition in biosensing devices and historically were the first recognition elements included in biosensors.

The methods to immobilize proteins are of particular relevance for biosensing, as they have to be assembled according to suitable architecture and their activity must be preserved for long periods of time. To achieve this goal several strategies have been developed. A broad class of methods involves direct interactions of the proteins with the solid substrate encompassing physical, covalent or bioaffinity immobilization (Rusmini et al. 2007).

A convenient method for not-covalent protein immobilization is the electrostatic layer-by-layer (LbL) technique, based on the sequential self-assembly of different layers on a solid substrate (Decher et al. 1992). The LbL technique usually utilizes alternating layers of water soluble positively and negatively charged polyelectrolytes, a procedure suitable for proteins. The advantages of the LbL method lie on the easy control of molecular architecture yielding tailored sensing units, leading to films where the thickness and molecular architecture can be controlled accurately.

Alternatively, covalent immobilization of the biorecognition element onto electrode surfaces allows for tightly controlled molecular architecture to be exploited in electrochemical enzyme-based sensors. The affinity of thiols for some metal surfaces, particularly gold, makes alkanethiols ideal for the preparation of these modified electrodes (Chaki and Vijayamohanan 2002).

In this chapter different methodologies for protein assembly will be described. Results from the literature are grouped into large areas, covering LbL based optical biosensors, field-effect transistors and electrochemical sensors.

LBL ASSEMBLY OF PROTEINS

Films Fabrication

The alternate deposition of polyanions and polycations on a solid surface leads to the formation of films called polyelectrolyte multilayers (PEM). Such films are readily prepared by layer-by-layer (LbL) assembly, allowing for a suitable and versatile architecture. This simple process is based on the sequential adsorption, driven by electrostatic interactions, of cationic and anionic species on a charged substrate (for reviews see Decher 1997; Schonhoff 2003) as illustrated in Fig. 12.1. The immersion of a negatively charged surface into a solution of cationic polyelectrolyte results in the coverage of the surface by a positively charged polymer layer. The immersion of this new cationic surface in a solution of negatively charged polyelectrolyte results in its adsorption and parallel re-establishing of a negative charge density (Ariga et al. 2007). After deposition of each given

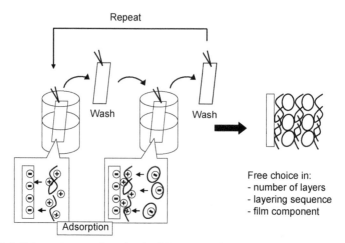

Figure 12.1. Schematic procedure for the fabrication of layer-by-layer films. The charged substrate is alternately dipped into solutions containing polycations and polyanions. The adsorption of materials is driven by electrostatic interactions.

layer, the substrate with the film deposited must be rinsed with water to remove weakly adsorbed molecules. The resulting films are thermally stable and resist to washing in aqueous solutions. Basically, the process can be repeated indefinitely, leading to multilayers with as many bilayers as desired, so the control of the thickness of the PEM is very simple. Moreover, since beakers and tweezers are the only apparatus required, it is a low cost procedure.

There are many advantages in using the water-based LbL assembly technique. First of all, multilayers of different thickness (from Å to nm) can be obtained. Because the method involves immersion of a surface in a charged species solution, multilayers conformally coat any substrate surface regardless of geometry or size. Moreover, several substrates can be used for the LbL adsorption: glass, quartz, silicon, gold, colloids or Teflon. The amount of material adsorbed and the film morphology depend on and can be controlled by several parameters, including pH, ionic strength, concentration of the material to be adsorbed and number of layers (Shiratori and Rubner 2000). Hybrid composites of varying complexity can be fabricated using both inorganic and organic materials. In addition to conventional polyelectrolytes such as poly(allylamine hydrochloride), poly(diallyldimethylammonium chloride), poly(ethyleneimine) and poly(sodium styrenesulfonate), any multiple charged species, like nanoparticles or dendrimers, can be incorporated, as well as bio-materials such as DNA and proteins.

Since the proteins themselves are polyelectrolytes, the LbL procedure has been used to prepare ordered films by alternating polyion layers with a number of different positively and negatively charged proteins (Campas and O'Sullivan 2003). Both water soluble proteins and detergent solubilized membrane proteins can be easily incorporated in polyelectrolyte multilayers that can be successfully employed in sensing devices.

Films Characterization Methods

Because of the characteristics of LbL assemblies, such as the number of layers, the interpenetration between layers and the surface roughness of the films, polyelectrolytes multilayers can be characterized by means of several techniques. Film deposition can be probed by UV-Vis and FT-IR spectroscopies if the adsorbed materials have specific absorption bands. The quartz crystal microbalance QCM (Marx 2003) successfully determines the mass of the films at the nanogram level and can be used to monitor in real time the growth of the polyelectrolyte layers (Campas and O'Sullivan 2003). The thickness of the adsorbed layers can be probed by ellipsometry (Ohlsson et al 1995) or by X-ray reflectivity (Li et al. 2000). The morphology of the films can also be analyzed using the various microscopies including atomic force microscopy (AFM) and scanning electron microscopy (SEM) (Caruso et al. 1998).

Due to their features, polyelectrolyte multilayers can be investigated by means of zeta-potential measurements and optical spectroscopy. The zeta-potential allows probing the inversion of surface charge with deposition of each layer (Ladam et al. 2000). On the other hand, when the solid support is transparent, the multilayer nature of polyelectrolyte multilayers furnishes an optical path long enough to measure the UV-visible optical density permitting also detailed studies on the spectral properties of the embedded proteins (Mallardi et al. 2007).

Optical Applications in Biosensing

In the last few years LbLs have found to be most successful in their employment in enzyme biosensors for several reasons: the sensing material is gently immobilized and maintains its biological activity; the incorporation of enzymes in these assemblies provides an ordered immobilization of the biorecognition element; analytes can easily diffuse through the assembled film and react with the immobilized enzyme while the corresponding product does not accumulate. LbL assembly leads to a wide range of combinations of charged species, which can be used for self-assembled ultrathin films each having different functionalities.

As glass, quartz and other transparent materials are well suited substrates for LbL assembly of PEM, optical detection techniques have been employed for signal transduction in LbL based sensors. Below, we focus our attention on LbL applications in optical sensors which use proteins (or enzymes) as bioactive sensing layer. In particular, attention will be devoted to optical biosensors used in detecting analytes of environmental interest.

The versatility of the LBL assembly to accommodate layers having different functionalities was exploited by Constantine et al. (2003). As shown in Fig. 12.2, the authors immobilized different charged components in the system, each having a different function. Several alternated layers of chitosan and thioglycolic acid—capped CdSe quantum dots were immobilized on a transparent support. In that assembly chitosan layer adopted a flat conformation that provided a homogeneous film on which quantum dots were adsorbed allowing an increase of their photoluminescent properties when compared to the ones in solution. This stable supramolecular film was then incorporated into a biosensing system by means of a further deposition of alternate layers of quantum dots and an enzyme: the organophosphorus hydrolase. This enzyme catalyzes the hydrolysis of a variety of organophosphorus compounds, quite extensively used in pesticides and insecticides. There are significant advantages associated with the immobilization of both quantum dots and enzyme in the same system. The enhanced optical properties of quantum dots allowed a direct fluorescence detection of paraoxon with high sensitivity. At the same time, as a result of the enzymatic activity of the organophosphorus hydrolase, the hydrolytic product *p*-nitrophenol could be released in solution and measured using UV-Vis spectroscopy.

The detection of aromatic compounds such pesticides and industrial pollutants has become of great interest, since these compounds withstand chemical oxidation and biological degradation, accumulating in the environment. The aerobic degradation of non-halogenated aromatic compounds, usually requires its initial conversion to a dihydroxybenzene derivative such as catechol, which is further degraded via ring cleavage by dioxygenases. Among the enzymes that act via ortho-cleavage yielding *cis–cis*-muconic acid, chlorocatechol 1,2-dioxygenase (CCD) has exhibited affinity to both halogenated and non-halogenated substrates. Zucolotto

⬭ OPH

⬤ Qdots

⬭ Chitosan

Figure 12.2. Sensing assembly for paraoxon. A top layer of the enzyme organophosphorus hydrolase (OPH), two bilayers of OPH/thioglycolic acid-capped CdSe quantum dots (Qdots) and five bilayers of chitosan/QDots have been immobilized.

and coworkers (2006) proposed a highly sensitive biosensor for detecting catechol, obtained with the immobilization of the enzyme CCD in layers alternated with poly(amidoamine) (PAMAM) dendrimer in a layer-by-layer fashion. Due to the mild immobilization conditions employed, CCD remained active in the films for periods longer than 3 wk. PAMAM/CCD films were employed in detecting catechol solutions using an optical approach. The detection was based on UV absorption experiments in which the formation of cis–cis muconic acid, resulting from the reaction between CCD and catechol, was monitored in solution. It was also evident that LbL films with the same organization were also deposited onto gold-interdigitated electrodes. Using impedance spectroscopy in a broad frequency range (1Hz–1kHz), catechol in solutions was detected at very low concentrations.

A biosensor for the detection of aromatic compounds like phenol and derivatives has also been proposed in the paper of Fiorentino et al. (2010). The authors developed a method to prepare an optical biosensor for detection of phenolic compounds by LbL deposition of the soluble enzyme mushroom tyrosinase and the cationic polymer poly(dimethyldiallylammonium chloride) (PDDA) on a quartz slide (Fig. 12.3). Tyrosinase catalyzes the oxidation of diphenolic compounds and this reaction is associated with drastic spectral changes, it has then been used in a simple optical bioassay. The analytical performances of this biosensor have been examined using the L-3,4-dihydroxyphenyl-alanine (L-DOPA) as archetype substrate but the biosensor was expected to work with others o-diphenolic compounds. The LbL deposition allowed a high loading of enzyme and biosensor sensitivity could be modulated varying the number of the immobilized enzyme layers. As many phenols are strongly fluorescent molecules, the same quartz-supported PEM could be used indifferently as the active element

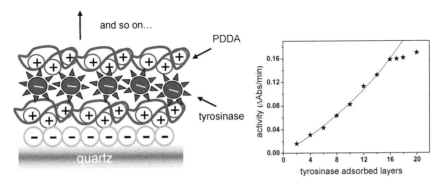

Figure 12.3. Sensing assembly for phenolic compounds. Different numbers of tyrosinase layers can be immobilized with the cationic polymer poly(dimethyldiallylammonium chloride) (PDDA) in a layer-by-layer film. The enzymatic activity of the multilayer increases with number of adsorbed layers.

for absorbance and fluorescence assays with an improvement in the sensing performance in passing to the fluorescence technique. The immobilized enzyme showed good time stability retaining its catalytic activity for weeks, allowing the reuse of the same multilayered support.

The work on biosensors made with immobilized proteins pointed to the importance of the scaffolding materials and film architecture in preserving their bioactivity. The first goal of LbL assembly is the gentle immobilization of biomaterials that can in this system maintain their structure and functionality intact for long periods. An excellent result in this respect is that obtained by Mallardi et al. (2007) with the immobilization of a photosynthetic membrane protein, the bacterial reaction centre (RC), in a LbL assembly, proposed as an optical herbicides biosensor. Photosynthetic herbicides, belonging to the classes of triazines, diazines, phenols, and ureas, represent toxic compounds that are still widely used in agriculture. They act inhibiting the initial steps of photosynthetic processes in plants, thus providing a low cost weed control. In the above mentioned work the RC (the protein target of herbicide) has been adsorbed onto a glass surface by alternating deposition with the polymer PDDA eventually obtaining a very simple and ordinate PEM where the protein retains its integrity and functionality over a period of several months (up to 7 mon). The RC protein is able to bind several kinds of herbicides and the RC-herbicides interaction can be optically monitored with ease since it alters the protein activity, such a variation resulting in a change of the optical properties of the protein. The RC/PDDA PEM revealed to be extremely sensitive to the presence of herbicides and several advantages could be evidenced in its use as a biosensor: the immobilized protein is more stable than in solution and the herbicide can be easily washed out, allowing the use of the same device for 100s of determinations and for several months without any loss in sensitivity or reproducibility of results.

COVALENT IMMOBILIZATION OF PROTEINS

Preparation and Characterization of Self-assembled Monolayers

Self-assembled monolayer (SAM) provides a simple route to functionalize electrode surfaces by means of organic molecules containing free anchor groups (e.g. thiols, disulphides, amines, silanes, or acids). The main advantages of such an immobilization approach lie in the tremendous versatility and enhanced reproducibility and robustness that have been successfully employed in several biosensing applications (Chaki et al. 2002). Despite the wide use of SAM, the functionalization procedure is usually tricky and often the importance of keeping its experimental parameters under a tight control is underestimated. The monolayer design step greatly influences the performance level of SAM based biosensors, regardless of the immobilization strategy and sensing mechanism. Infact, the affinity of thiols for metal surfaces (gold and silver, mainly) makes alkanethiols ideal for the realization of modified electrodes; long chain alkanethiols provide highly packed and ordered monolayers, resulting in a membrane-like microenvironment, useful for biomolecule immobilization (Love et al. 2005).

Two general approaches can be followed to functionalize surfaces: solution-based and gas phase derivatization. The substrate has to be either dipped in the required dilute solution or exposed to vapours of the adsorbent, for a specified time. In both cases, the cleanness of the surface is crucial, which is usually achieved by highly oxidizing pretreatments (e.g. piranha solution, electrochemical cycling in concentrated H_2SO_4 solutions, etc.). In addition, several experimental parameters affect the formation and packing density of monolayers such as the nature, crystallinity and roughness of substrate, nature and concentration of the adsorbate, etc (Chaki and Vijayamohanan 2002 and reference therein). Aiming at immobilizing large biomolecules, mixed monolayers are often preferred in such a way that steric hindrance between these bulky molecules and their target molecule can be avoided. For mixed SAM, the molar ratio of the mixture in solution is important as the same ratio in the mixed SAM can be obtained without any preferential segregation (Bain et al. 1989).

The monolayers can be characterized by a variety of methods including physical measurements (e.g. contact angle, wettability), electro-analytical techniques (e.g. cyclic voltammetry, impedance), spectroscopies (e.g. infrared, x-ray photoelectron spectroscopy), and microscopy techniques (e.g. scanning probe microscopies), each of them providing complementary information about the SAM properties (Chaki and Vijayamohanan 2002; Love et al. 2005).

Electrochemical methods are especially useful for monitoring the monolayer quality in biosensing devices, as they can provide information about the film surface density and/or the electrode degree of coverage, as well as the existence of coverage defects. For example, quantitative analysis of coverage can be performed by cyclic voltammetry in presence of a reversible redox couple, such as ferrocyanide/ferricyanide. The redox activity of the couple is inhibited as the monolayer becomes more and more dense and defect-free (Bethell et al. 1996; Chaki and Vijayamohanan 2002). Impedance measurements can provide mechanistic insights of the electron transfer processes by measuring the interfacial capacitance and resistance of these monolayers (Chaki and Vijayamohanan 2002).

Several spectroscopic and microgravimetric techniques have been exploited in order to provide useful information about SAM, discussion of the advantages and limitations of these methods are available elsewhere (Chaki and Vijayamohanan 2002; Love et al. 2005, and references therein).

Enzyme Inhibition-based Biosensors

Enzyme based biosensors can usually be classified into two main categories differentiated by the detection approach: i) investigation of the enzymatic product deriving from the analyte metabolism, ii) investigation of the decrease of enzymatic product formation on account of metabolism inhibition by the analyte (Amine et al. 2006).

Biosensors based on enzyme inhibition have found wide applications in the detection of environmental relevant analytes such as organophosphorous pesticides, organochloride pesticides, derivatives of insecticides, heavy metals and glycoalkaloids (Arduini et al. 2010).

Generally speaking, the development of these sensing systems relies on a quantitative comparison of the enzyme activity before and after exposure to a target analyte. Typically, the percentage of inhibited enzyme (I%) that results after exposure to the inhibitor is quantitatively related to the analyte concentration according to the following equation (Ivanov et al. 2003):

$$I\% = [(A_0 - A_i)/A_0] \times 100$$

where A_0 and A_i are the enzymatic activities in the absence and presence of the inhibitor, respectively. A linear correlation has usually been achieved between 20% and 80% of inhibition and the limit of detection is usually defined as the amount of inhibitor which decreases the enzyme activity by 20% (Arduini et al. 2010).

The aforementioned equation is used for both reversible and irreversible inhibition biosensors, but there is a substantial difference between these two kinds of systems. Reversible inhibition is characterized by non-covalent interaction between the inhibitor and the enzyme active centre. If the inhibitor is structurally related to the enzyme substrate, it competes with the substrate in the enzyme binding (competitive inhibition); alternatively, the inhibitor may bind the enzyme-substrate complex thus leading to a non-competitive inhibition. In both cases, after the inhibitor measurement the initial activity can be restored. Irreversible inhibition, on the contrary, results in a covalent bond between the enzyme active centre and the inhibitor; the decomposition of the enzyme-inhibitor complex results in the destruction of the enzyme and thus either a new biosensor or a reactivation procedure after the inhibitor measurement is required (Arduini et al. 2010).

Cholinesterase Inhibition-based Electrochemical Biosensors: A Case Study

The principal biological role of cholinesterase enzyme (AChE) is the termination of the nervous impulse transmission at cholinergic synapses by rapid hydrolysis of the neurotransmitter acetylcholine. Early investigations indicated that the active site of AChE contains two sub-sites, the esteratic and anionic sub-sites, corresponding to the catalytic site and the choline-binding pocket, respectively (Arduini et al. 2010 and reference therein). A serine residue of the esteratic site accounts for the substrate complexation but it may be also subjected to inhibition by some of the most common pesticides. In particular, organophosphate and carbamate pesticides are powerful neurotoxins that irreversibly inhibit the activity of AChE by modifying the catalytic serine residue. Taking into consideration this inhibition property, several innovative solutions for the sensitive detection of organophosphate and carbamate pesticides have been proposed. Some examples of these investigations will be reviewed, mainly focusing on biosensors obtained by covalent immobilization of AChE onto a SAM gold electrode.

In 2006 Somerset et al. reported on the development of a gold electrode modified with mercaptobenzothiazole (MBT) and either poly(o-methoxyaniline) (POMA) or poly(2,5-dimethoxyaniline) (PDMA) in the presence of polystyrene sulfonic acid (PSSA) (Somerset et al. 2006; Somerset et al. 2009). The biosensor performance was assessed in the aqueous phase detection of diazinon and carbofuran. Both biosensors exhibited low detection limits, calculated as percentage inhibition (Au/MBT/PDMA-PSSA/AChE: lower detection limits 0.07 ppb for diazinon and 0.06 ppb for carbofuran; Au/MBT/POMA-PSSA/AchE: lower detection limits 0.14 ppb for diazinon and 0.11 ppb for carbofuran). The average sensitivity was 4.2 µA/ppb, thus suggesting the feasibility in deploying polyaniline-based sensor systems as alarm devices for carbamate and organophosphate pesticides (Somerset et al. 2006).

An alternative AChE-based amperometric biosensor was developed by immobilization of the enzyme onto a 3-mercaptopropionic modified gold electrode by Pedrosa et al. (Pedrosa et al. 2007; Pedrosa et al. 2008). By an indirect approach involving the inhibition effect on the enzymatic reaction, parathion and carbaryl were quantitatively detected in spiked natural water and food samples without pretreatment or preconcentration steps (Pedrosa et al. 2008). The influence of several parameters such as pH, ionic strength, enzyme loading and concentration of glutaraldehyde on the biosensor response was investigated in order to evaluate the conditions for the best performance. The detection limits under optimized conditions were assessed at 9.3 mg/L for parathion and 9.0 mg/L for carbaryl (Pedrosa et al. 2007).

More complex architecture based on covalent immobilization of AChE was proposed in 2009: biosensors were constructed by means of gold nanoparticles and cysteamine self-assembled on glassy carbon paste (Du et al. 2009), or by vertically assembled carbon nanotubes wrapped by thiol terminated single strand oligonucleotide (ssDNA) on gold (Viswanathan et al. 2009).

In Du's paper, the biosensor was employed for quantitative determination of monocrotophos; the presence of gold nanoparticles not only provided an increased effective surface, but also promoted electron transfer reactions and catalyzed the electro-oxidation of thiocholine, thus improving the detection sensitivity; the detection limit achieved was lower than 0.3 ng/mL (Du et al. 2009).

The SAMs of single walled carbon nanotubes wrapped by thiol terminated ssDNA on gold was used to prepare a nano-sized polyaniline matrix for AChE enzyme immobilization (Fig. 12.4) (Viswanathan et al. 2009). The resulting electrochemical biosensor led to the determination of two of the most commonly used organophosphorous insecticides in

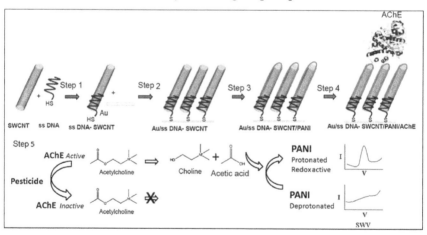

Figure 12.4. Representation of biosensor fabrication procedure and function. Step 1: thiol terminated single-stranded DNA (ssDNA) wrapped with single walled carbon nanotube (SWCNT); Step 2: ssDNA wrapped SWCNT self-assembled on gold electrode; Step 3: controlled electrochemical polymerization of aniline on the SWCNT ; Step 4: immobilization of acetilcolinesterase (AChE) enzyme by glutaraldehyde; Step 5: pesticide sensing mechanism. [Reprinted from Biosens Bioelectron, vol. 24, Viswanathan, S, Radecka, H and Radecki, J, "Electrochemical biosensor for pesticides based on acetylcholinesterase immobilized on polyaniline deposited on vertically assembled carbon nanotubes wrapped with ssDNA", pages 2772–2777, Copyright (2009), with permission from Elsevier].

vegetable crops: methyl-parathion and chlorpyrifos. The key feature of this biosensor was the small change of local pH in the vicinity of the electrode surface caused by AChE-acetylcholine enzymatic reaction. The dynamic range for the detection was found to be between 1.0×10^{-11} and 1.0×10^{-6} M.

IMMOBILIZED PROTEINS IN FETs

Electronic devices find attractive application in biosensing, as they provide a lot of potential advantages such as small size and weight, fast response time, high robustness and the possibility of low-cost fabrication. Nevertheless, the complete integration of a bioactive element with this kind of transducers is still a major challenge, particularly in terms of preserving its functioning and specificity while keeping the sensors electronic performance at an acceptable level.

Some example of strategies for interfacing proteins with different kinds of FET will be reported here.

Ion-sensitive field-effect transistor (ISFET) has found several applications for biosensing purposes, and a variety of devices differing in their design has been described (Lee et al 2009). In these biosensors, enzymes can be immobilized on the surface of the transducer allowing both direct and inhibitory analysis of enzyme activity.

A bioelectronic hybrid system for the detection of pesticides and nerve gas was assembled by way of covalent immobilization of the enzyme acetylcholine esterase (described in the above Section) to the gate surface of an ISFET (Fig. 12.5). AChE was immobilized using cyanuric-chloride

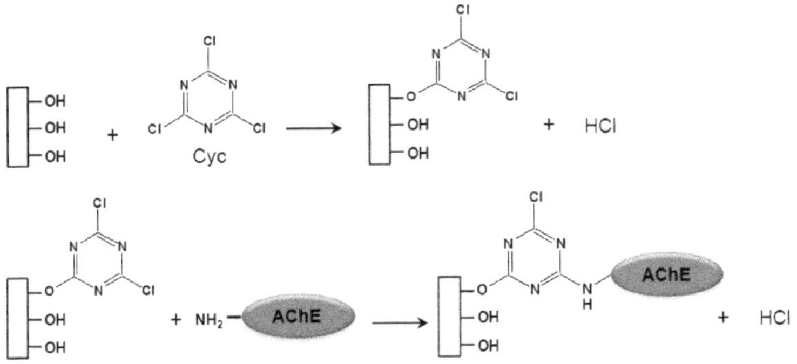

Figure 12.5. **Schematic representation of the immobilization procedure of Acetylcholine esterase onto the ion-sensitive field-effect transistor (ISFET) gate surface.** First step: under incubation at 70°C, the cyanuric-chloride (CyC) group condenses with the free hydroxyl group on the gate aluminum oxide surface, eliminating HCl and forming a covalent ether bound. Second step: condensation reaction (at room temperature) between the amine group of a lysine residue on the enzyme with the free cyanuric-chloride group of the immobilized CyC, with the elimination of HCl and the formation of a secondary amine covalent bond with the enzyme.

as short aromatic coupling molecule, allowing the formation of an AChE monolayer on silicon substrates (Hai et al 2006). The enzyme maintains its pharmacological properties, also if its Michaelis–Menten constant is moderately altered and the maximum reaction velocity is reduced by over an order of magnitude.

An alternative covalent attachment based on aminopropyl-triethoxysilane /glutaraldehyde was used to immobilize the organophosphorus hydrolase enzyme on a FET device for pesticide and insecticide detection (Simonian et al 2004). A sol-gel modified gate was also used in this case, providing a surface more suitable for enzyme immobilization and a better recovery of total enzyme activity.

A different strategy for immobilization of enzymes can be obtained by their entrapment in a membrane. Dzyadevych and coworkers (Dzyadevych et al 2002) formed a biologically active membrane on the sensitive transducer substrate by cross-linking of tyrosinase (for phenolic compounds and herbicides detection) with bovine serum albumin in saturated glutaraldehyde vapours. In this paper the influence of parameters such as enzyme loading in immobilization mixture and time of exposition to glutaraldehyde vapours was also investigated in order to optimize operating conditions and to avoid loss of the enzymatic activity.

Organic field-effect transistor (OFET) has been proposed as biosensing platforms (see section "Key fact of") and the integration of bio-species (proteins, cells, DNA) as active components in their realization has been obtained (Berggren and Richter-Dahlfors 2007).

A versatile method for immobilization of membrane proteins on top of the active layer of an OFET has been proposed by Torsi's group (Angione et al 2009). The immobilization has been obtained by incorporation of a suitable protein within liposomes (artificial lipid bilayers organized in vesicles) followed by their deposition onto the organic semiconductor of the OTFT. Spreading of lipid vesicles on a solid support allows their self assembly into fluid planar bilayers hosting the selected protein. By this method the authors immobilized the bacterial reaction centre protein (already mentioned in the previous-Optical Applications in Biosensing-Section), in view of an application of the device as a herbicide biosensor. A pictorial view of the obtained sensing device is shown in Fig. 12.6. The OFET was in a bottom gate-top contact configuration. The deposition of the lipid bilayer embedding the protein was performed via drop casting before evaporation of source and drain gold contacts. The lipid bilayer obtained facilitates a close association of the protein with the carriers accumulated in the channel at the semiconductor/gate dielectric interface.

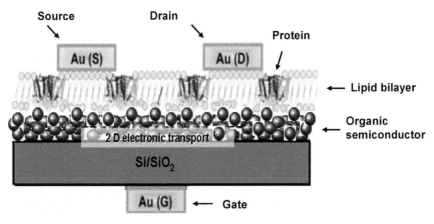

Figure 12.6. Schematic view of an organic field-effect transistor which integrates a lipid bilayer embedding a membrane protein. The device is constituted by a silicon/silicon dioxide substrate, an organic semiconductor layer, and on top of this the lipid bilayer embedding the reaction centre protein. Source and drain contacts are on top of the active layer; the gate electrode has been realized on the bottom part of the device.

APPLICATIONS TO OTHER AREAS OF HEALTH AND DISEASE

Improvement in the quality of life is one of the most important objectives of global research efforts. Of course, quality of life is closely linked to the control of environment, food safety and health. Protein based biosensors represent very promising tools in the fields of food and health control.

The greatest impact of protein biosensors is presently in the medical diagnostic field. Enzyme-based biosensors often meet the requirements of several biomedical applications. Starting from glucose biosensors that have became a commercial success, enzyme based sensors are also developed for several other compounds of biomedical interest. For example, the level of lactate can be determined in human saliva and serum, and that of cholesterol in serum. Urea can be determined in urine and fresh blood. Neurotransmitters like as glutamate can be measured, even if at this stage not in human samples but in cultured nerve cells or in rats. Enzyme-based amperometric sensors have successfully targeted all these substrates, and some of these have also been commercialized.

In the food industry, the quality of a product is evaluated through periodic chemical and microbiological analysis. These procedures use conventional analytical techniques that do not allow easy and continuous food monitoring. Enzymatic biosensors represent a promising tool for food analysis, especially for the determination of the composition, degree of contamination of raw materials and processed foods, allowing the detection of pathogens, pesticides, microorganisms and toxins. A large part of biosensor research for the food industry relies on enzyme-based electrochemical or optical biosensors but, in spite of the great number of publications, only few systems are commercially available.

SUMMARY POINTS

- Protein based biosensors are versatile analytical tools applied to different fields, such as environment monitoring, medicine, and food quality and safety control.
- The present work emphasizes the importance of scaffolding materials and film architecture to preserve the activity of biomolecules employed as recognition elements in biosensors.
- Among the various approaches, LbL deposition offers several advantages, such as precise control of composition and thickness of films, wide choice of materials and simplicity of procedure. Its mildness is one of its most pronounced characteristics and makes the method applicable to delicate biomaterials.
- Covalent immobilization represents a useful alternative, providing reproducibility, robustness and versatility as the main advantages. Nevertheless, tight control of the functionalization experimental parameters is required to obtain a good quality bio-active layer.
- The immobilization of different biomolecules, having an interesting role in environmental monitoring, in Field-Effect Transistors is also presented. Different immobilization techniques in electronic devices are specifically reported, in order to overcome issues related to the integration of biomolecules and electronic devices.

KEY FACTS OF ORGANIC FIELD EFFECT TRANSISTORS

- In recent years there is an increasing interest in thin film field effect transistors with an active layer made from an organic material, organic field-effect transistor (OFET). These devices have been developed to realize low-cost, large-area electronic products and/or biodegradable electronics.
- OFETs are three-terminal electronic devices consisting of a thin organic semiconducting film and an insulating (dielectric) layer along with three conductive terminals (source, drain and gate). OFETs have been fabricated with various device geometries. The

most commonly used geometry is bottom gate with top drain and source electrodes.

- The source and drain electrodes are fabricated to be directly in contact with the semiconductor. The semiconductor layer may consist of small molecules, polymers, or organic nanomaterials. The dielectric layer, which electrically isolates the semiconductor from the gate electrode, may be composed of inorganic oxides or properly tailored polymers or composites. Contacts are usually composed of gold. In such architecture, current amplification occurs at the interface between the gate-dielectric and the organic semiconductor where the two-dimensional field-effect transport occurs.
- OFET devices show very high repeatability, as well as fast and reversible responses. They allow for room temperature operation, and offer good selectivity towards a broad range of analytes. OFET have been proposed as sensing platforms in the field of environmental monitoring, military defense and medical care.
- Recent advancements in the field of biosensors have allowed for the integration of biomolecules as active components in the construction of organic transistors and have provided systems that are able to efficiently transduce biological events using electronic devices demonstrating the potential to be applied for the rapid screening of biological samples and point-of-care diagnostics.

DEFINITIONS

- *Acetylcholinesterase (AChE)*: enzyme that hydrolizes the neurotransmitter acetylcholine, producing choline and an acetate group. It is mainly found at neuromuscular junctions and cholinergic nervous system, where its activity serves to terminate synaptic transmission.
- *Dendrimer*: is a repeatedly branched, roughly spherical large molecule. The properties of dendrimers are dominated by the functional groups on the molecular surface. Dendrimers are classified by generation, which refers to the number of repeated branching cycles that are performed during their synthesis.
- *Enzyme*: catalyst produced by cells to speed up specific chemical reaction. Enzyme is a protein molecule with a characteristic sequence of amino acids that fold to produce a specific three-dimensional structure, which gives the molecule unique properties. Enzymes are usually classified and named according to the reaction they catalyze.
- *Pesticide*: any substance or mixture of substances intended for preventing, destroying, repelling or mitigating any pest. Although

there are benefits to the use of pesticides, there are also drawbacks, such as potential toxicity to humans and other animals.

- *Quantum dots*: semiconductor nanocrystals whose electronic characteristics are closely related to the size and shape of the individual crystal. Their band-gap energy increases with decrease in the particle size, resulting in a blue-shift of the electronic transitions and of the related features in optical spectra. Quantum dots are highly efficient fluorophores. Their colour emission can be tuned by size variation without altering their chemical properties.
- *Self assembled monolayer (SAM)*: organized layer of amphiphilic molecules which spontaneously give rise to absorption in a monomolecular layer on surfaces from either vapour or liquid phase. SAMs are created by the chemisorption of hydrophilic "head groups" onto the substrate, followed by a slow two-dimensional organization of hydrophobic "tail groups".

ABBREVIATIONS

AChE	:	acetylcholine esterase
AFM	:	atomic force microscopy
CCD	:	*cis–cis*-muconic acid, chlorocatechol 1,2-dioxygenase
FET	:	field-effect transistors
FT-IR	:	Fourier transform-infrared
ISFET	:	ion-sensitive field-effect transistor
LbL	:	layer by layer
L-DOPA	:	L-3,4-dihydroxyphenyl-alanine
MBT	:	mercaptobenzothiazole
OFET	:	organic field-effect transistor
PAMAM	:	poly(amidoamine)
PDDA	:	poly(diallyldimethylammonium chloride)
PDMA	:	poly(2,5-dimethoxyaniline)
PEM	:	polyelectrolyte multilayer
POMA	:	poly(o-methoxyaniline)
PSSA	:	polystyrene sulfonic acid
QCM	:	quartz crystal microbalance
RC	:	reaction center
SAM	:	self-assembled monolayer
SEM	:	scanning electron microscopy
ssDNA	:	single strand DNA
UV-Vis	:	ultraviolet-visible

REFERENCES

Amine A, H Mohammadi, I Bourais and G Palleschi. 2006. Enzyme inhibition-based biosensors for food safety and environmental monitoring. Biosens Bioelectron 21: 1405–1423.

Angione MD, A Mallardi, G Romanazzi, GP Suranna, P Mastrorilli, D Cafagna, E De Giglio, G Palazzo and L Torsi. 2009. Membrane Proteins Embedded in Supported Lipid Bilayers Employed in Field Effect Electronic Devices. Proceedings 3rd IEEE International Workshop on Advances in Sensors and Interfaces pp 218–221.

Arduini F, A Amine, D Moscone and G Palleschi. 2010. Biosensors based on cholinesterase inhibition for insecticides, nerve agents and aflatoxin B1 detection. Microchim Acta 170: 193–214.

Ariga K, JP Hill and Q Ji. 2007. Layer-by-layer assembly as a versatile bottom–up nanofabrication technique for exploratory research and realistic application. Phys Chem Chem Phys 9: 2319–2340.

Aslam M, K Bandyopadhyay, K Vijayamohanan and V Lakshminarayanan. 2001. Comparative behavior of aromatic disulfide and diselenide monolayers on polycrystalline gold using cyclic voltammetry, STM, and quartz crystal microbalance. J Colloid Interface Sci 234: 410–417.

Bain CD, J Evall and GM Whitesides. 1989. Formation of monolayers by the coadsorption of thiols on gold: variation in the head group, tail group, and solvent. J Am Chem Soc 111: 7155–7160.

Berggren M and A Richter-Dahlfors. 2007. Organic Bioelectronics. Adv Mat 19: 3201–3213.

Bethell D, M Brust, DJ Schiffrin and C Kiely. 1996. From monolayers to nanostructures materials: an organic chemist's view of self-assembly. J Electroanal Chem 409: 137–143.

Campas M and C O'Sullivan. 2003. Layer-by-layer biomolecular assemblies for enzyme sensors, immunosensing, and nanoarchitectures. Anal Lett 36: 2551–2569.

Caruso F, DN Furlong, K Ariga, I Ichinose and T Kunitake. 1998. Characterization of polyelectrolyte–protein multilayer films by atomic force microscopy, scanning electron microscopy, and Fourier transform infrared reflection–absorption spectroscopy. Langmuir 14: 4559–4565.

Chaki NK and K Vijayamohanan. 2002. Self-assembled monolayers as a tunable platform for biosensor applications. Biosens Bioelectron 17: 1–12.

Constantine C, KM Gatta´s-Asfura, SV Mello, G Crespo, V Rastogi, T Cheng, JJ DeFrank and RM Leblanc. 2003. Layer-by-Layer Biosensor Assembly Incorporating Functionalized Quantum Dots. Langmuir 19: 9863–9867.

Decher G. 1997. Fuzzy nanoassemblies: toward layered polymeric multicomposites. Science 277: 1232–1237.

Decher G, JD Hong and J Schmitt. 1992. Buildup of ultrathin multilayer films by a self-assembly process: III. Consecutively alternating adsorption of anionic and cationic polyelectrolytes on charged surfaces. Thin Solid Films 210: 831–835.

Du D, W Chen, J Cai, J Zhang, H Tu and A Zhang. 2009. Acetylcholinesterase biosensor based on gold nanoparticles and cysteamine self assembled monolayer for determination of monocrotophos. J Nanosci Nanotechnol 9: 2368–2373.

Dzyadevych SV, TMI Anh, AP Soldatkin, ND Chien, N Jaffrezic-Renault and JM Chovelon. 2002. Development of enzyme biosensor based on pH-sensitive field-effect transistors for detection of phenolic compounds. Bioelectrochem 55: 79–81.

Fiorentino D, A Gallone, D Fiocco, G Palazzo and A Mallardi. 2010. Mushroom Tyrosinase in Polyelectrolyte Multilayers as an Optical Biosensor for o-diphenols. Biosens Bioelectron 25: 2033–2037.

Hai A, D Ben-Haim, N Korbakov, A Cohen, J Shappir, R Oren, ME Spira and S Yitzchaik. 2006. Acetylcholinesterase–ISFET based system for the detection of acetylcholine and acetylcholinesterase inhibitors. Biosens Bioelectron 22: 605–612.

Hodak J, R Etchenique, EJ Calvo, K Singhal and PN Bartlett. 1997. Layer-by-layer self-assembly of glucose oxidase with a poly(allylamine)ferrocene redox mediator. Langmuir 13: 2708–2716.

Ivanov A, G Evtugyn, HC Budnikov, F Ricci, D Moscone and G Palleschi. 2003. Cholinesterase sensors based on screen-printed electrode for detection of organophosphorous and carbamate pesticides. Anal Bioanal Chem 377: 624–631.

Ladam G, P Schaad, JC Voegel, P Schaaf, G Decher and F Cuisinier. 2000. *In situ* determination of the structural properties of initially deposited polyelectrolyte multilayers. Langmuir 16: 1249–1255.

Lee C-S, SK Kim and M Kim. 2009. Ion-Sensitive Field-Effect Transistor for Biological Sensing Sensors 9: 7111–7131.

Li LS, R Wang, M Fitzsimmons and D Li. 2000. Surface electronic properties of self-assembled, oppositely charged macrocycle and polymer multilayers on conductive oxides. J Phys Chem B 104: 11195–11201.

Love JC, LA Estroff, JK Kriebel, RG Nuzzo and GM Whitesides. 2005. Self-Assembled Monolayers of Thiolates on Metals as a Form of Nanotechnology. Chem Rev 105: 1103–1169.

Lutkenhaus JL and PT Hammond. 2007. Electrochemically enabled polyelectrolyte multilayer devices: from fuel cells to sensors. Soft Matter 3: 804–816.

Mallardi A, M Giustini, F Lopez, M Dezi, G Venturoli and G Palazzo. 2007. Functionality of photosynthetic reaction centers in polyelectrolyte multilayers: toward an herbicide biosensor. J Phys Chem. B 111: 3304–3314.

Marx KA. 2003. Quartz crystal microbalance: a useful tool for studying thin polymer films and complex biomolecular systems at the solution–surface interface. Biomacromolecules 4: 1099–1120.

Ohlsson PA, T Tjarnhage, E Herbai, S Lofas and G Puu. 1995. Liposome and proteoliposome fusion onto solid substrates, studied using atomic force microscopy, quartz crystal microbalance and surface plasmon resonance. Biological activities of incorporated components. Bioelectrochem. Bioenerg 38: 137–148.

Pedrosa VA, J Caetano, SAS Machado, RS Freire and M Bertotti. 2007. Acetylcholinesterase immobilization on 3-mercaptopropionic acid self assembled monolayer for determination of pesticides. Electroanalysis 19: 1415–1420.

Pedrosa VA, J Caetano, SAS. Machado and M Bertotti. 2008. Determination of parathion and carbaryl pesticides in water and food samples using a self-assembled monolayer/acetylcholinesterase electrochemical biosensor. Sensors 8: 4600–4610.

Roberts ME, SCB Mannsfeld, N Queraltó, C Reese, J Locklin, W Knoll and Z Bao. 2008. Water-stable organic transistors and their application in chemical and biological sensors. PNAS 105: 12134–12139.

Rusmini F, ZY Zhong and J Feijen. 2007. Protein immobilization strategies for protein biochips. Biomacromolecules 8: 1775–1789.

Schonhoff M. 2003. Self-assembled polyelectrolyte multilayers. Curr Opin Colloid Interface Sci 8: 86–95.

Shiratori S and MF Rubner. 2000. pH Dependent thickness behavior of sequentially adsorbed layers of weak polyelectrolytes. Macromolecules 33: 4213–4219.

Shutava TG, DS Kommireddy and YM Lvov. 2006. Layer-by-layer enzyme/polyelectrolyte films as a functional protective barrier in oxidizing media. J Am Chem Soc 128: 9926–9934.

Simonian AL, AW Flounders and JR Wild. 2004. FET-Based Biosensors for The Direct Detection of Organophosphate Neurotoxins. Electroanalysis 16: 1896–1906.

Somerset VS, MJ Klink, MMC Sekota, PGL Baker and EI Iwuoha. 2006. Polyaniline-Mercaptobenzothiazole biosensor for organophosphate and carbamate pesticides. Anal Lett 39: 1683–1698.

Somerset VS, P Baker and EI Iwuoha. 2009. Mercaptobenzothiazole on-gold organic phase biosensor systems: 1.Enhanced organophosphate pesticide determination. J Environ Sci Health Part B 44: 164–178.

Viswanathan S, H Radecka and J Radecki. 2009. Electrochemical biosensor for pesticides based on acetylcholinesterase immobilized on polyaniline deposited on vertically assembled carbon nanotubes wrapped with ssDNA. Biosens Bioelectron 24: 2772–2777.

Zucolotto V, APA Pinto, T Tumolo, ML Moraes, MS Baptista, A Riul Jr, APU Araujo and ON Oliveira Jr. 2006. Catechol biosensing using a nanostructured layer-by-layer film containing Cl-catechol 1,2-dioxygenase. Biosens Bioelectron 21: 1320–1326.

13

Nanogravimetric and Voltammetric DNA-Biosensors for Screening of Herbicides and Pesticides

Maria Hepel,[1,*] Magdalena Stobiecka[1,2] and Anna Nowicka[1,3]

ABSTRACT

New methods for the assessment of DNA damage and alterations caused by herbicides and pesticides, based on the utilization of DNA biosensors, are presented. These methods enable evaluating the degree of DNA alterations or damage and offer simplicity of operation and scalability. We review DNA biosensors that can be applied for accelerated tests for DNA alterations by toxicants which could be subsequently applied for the development of assays for screening of manufacturer's herbicide and pesticide preparations. Particular attention is given to the investigations of interactions of herbicide atrazine as the model compound with double-stranded DNA using biosensors designed on gold-coated

[1]Department of Chemistry, State University of New York at Potsdam, Potsdam, New York 13676, USA; E-mail: hepelmr@potsdam.edu
[2]Permanent address: Department of Physics, Warsaw University of Life Sciences-SGGW, PL-02776 Warsaw, Poland; E-mail: magdalena_stobiecka@sggw.pl
[3]Permanent address: Department of Chemistry, University of Warsaw, PL-02093 Warsaw, Poland; E-mail: anowicka@chem.uw.edu.pl
*Corresponding author

List of abbreviations after the text.

quartz crystal piezoelectrodes with nanogravimetric and voltammetric transduction. The electrochemical quartz crystal nanobalance (EQCN), which has been employed in highly sensitive biosensor applications, is very well suited to monitor the immobilization of active sensory film components and analyzing the sensor response to DNA alterations and damage. The utility of the Nile Blue as the redox intercalator probe in studies of atrazine damage to DNA is illustrated using Au/aminothiol/DNA sensors and the enhanced-response gold-nanoparticle film sensors. The high responsiveness of this type of sensors to DNA alterations and damage by toxicants is due to the high sensitivity to the base-pair mismatches and to the degree of DNA underwinding. The interactions of herbicides and pesticides with nitrogen bases and DNA fragments were further investigated using molecular dynamics and quantum mechanical calculations, showing the role hydrogen bonding in these interactions.

INTRODUCTION

The observed rapid growth of environmental pollution has been attributed to expanding industrialization and application of intensive agricultural technologies. There is an urgent need to develop pollution monitoring (Del Carlo et al. 2004; Wang et al. 2009) and novel methods for fast screening of harmful compounds and the development of inexpensive and field-deployable biosensors responding to various pollutants. Of a particular importance are bioactive pollutants, interacting with DNA and causing DNA damage (Ashby et al. 2002; Clements et al. 1997; Freeman and Rayburn 2004; Greenlae et al. 2004), carcinogenesis (Alavanja et al. 2003; Hopenhayn-Rich et al. 2002; Schroeder et al. 2001), and a number of different diseases. Herbicides and pesticides are notorious and very serious environmental pollutants (Clements et al. 1997). There is mounting evidence for their highly acute toxicity and a wide range of biological activities (Fan et al. 2007; Ribas et al. 1995; Surralles et al. 1995). It is believed that active herbicides/pesticides in combination with adjuvants used in agricultural preparations are the main source of genotoxicity (Zeljezic et al. 2006). The Codex Alimentarius Commission of the Food and Agriculture Organization (FAO) and the World Health Organization (WHO) have established maximum residue limits for pesticides in food. Standard procedures, based on liquid chromatography and gas chromatography, are reliable techniques for routine laboratory detections of pesticides. However, there is also a need for innovative methods for rapid, inexpensive, and field-deployable testing. Recently, highly sensitive piezoimmunosensors for the detection of herbicides, atrazine (Pribyl et al. 2003) and 2,4-D (Halamek et al. 2001), have been developed. The affinity electrochemical biosensors for pollution control have been proposed by Mascini (Mascini 2001). Disposable immuno-electrochemical sensors for herbicides have been reported by Helali et al. (Helali et al. 2006).

In this chapter, we describe novel biosensors for evaluation of pollutant's toxicity based on the assessment of DNA alteration or damage caused by the pollutant. In particular, we will focus on the interactions of an herbicide atrazine, as the model compound, with DNA and on the methodology of quantifying the damage inflicted on DNA samples present in the solution or immobilized on a biosensor surface. The reports concerning atrazine genotoxicity indicate on one hand a mild DNA damage to mice leucocytes (Tannant et al. 2001) and on the other hand more significant damage in bullfrog tadpole erythrocytes (Clements et al. 1997) and human lymphocytes (Ribas et al. 1995). The biosensor techniques discussed here are based on the unique sensitivity of DNA hybridization to the damage and structural alterations of DNA strands caused by toxicants (Nowicka et al. 2010; Wang et al. 2009). These techniques can be applied to investigate toxicity of a wide range of pollutants, such as herbicides, pesticides, Cr(VI), and others.

MATERIALS AND METHODS

Oligonucleotides Used in Biosensor Construction

The type of oligonucleotides used in biosensor construction depends on the kind of binding to the substrate. While DNA fragments can be adsorbed directly on some substrates, e.g. gold, often to increase the binding strength, a SAM of functionalized surfactant is immobilized first on the substrate and then ssDNA or dsDNA chains are attached to the SAM via strong amide bonds, glutaraldehyde linkages, biotin-streptavidin binding, or other means. Alternatively, thiolated DNA can be attached directly to the substrate such as Au or Ag via thiolate bonds.

In the DNA biosensor construction, short oligonucleotides and calf thymus DNA are usually used as the recognition layer. The following list of oligonucleotides is provided below as an example of oligonucleotides that were used in hybridization biosensor construction fescribed in this Chapter.

Oligonucleotide sequences:

- NH_2- modified probe single-stranded DNA (NH_2-ssDNA$_{20b}$):
  ($5' \Rightarrow 3'$) $NH_2C_6H_{12}$-ATTCGACAGGGATAGTTCGA oligonucleotide (designed from the ssrA gene of *Listeria monocytogenes*, a common food pathogen)
- complementary target ssDNA:
  ($5' \Rightarrow 3'$) TCGAACTATCCCTGTCGAAT oligonucleotide

- complementary target ssDNA with one mismatch (C-A) at the probe:
  ($5' \Rightarrow 3'$) TCGAACTATCCCTATCGAAT oligonucleotide

For the oligonucleotide immobilization, a phosphate buffer saline solution (PBS) composed of 0.15 M NaCl, 2 mM KCl, and 0.02 M phosphate buffer with pH = 7.4, was used. For DNA hybridization, a solution of 1 M NaCl, 2 mM KCl, and 0.02 M phosphate buffer with pH = 7.4 was employed.

Procedures

Cleaning of gold disk electrodes. The electrodes were electrochemically pretreated first by cycling between 0 V and –1.8 V (hold 10 sec. at –1.8 V) in a 1 M NaOH with scan rate 50 mV/s, then cycling between 0 and 1.5 V (vs. Ag/AgCl) in 0.1 M H_2SO_4 solution until a stable voltammogram typical for a clean gold electrode was observed (Finklea et al. 1987).

Synthesis of gold nanoparticles. The Au nanoparticles were synthesized according to the published procedure (Stobiecka et al. 2010). Briefly, to obtain 5 nm AuNP, a solution of $HAuCl_4$ (10 mM, 2.56 mL) was mixed with a trisodium citrate solution (10 mM, 9.6 mL) and poured to distilled water (88 mL). The obtained solution was vigorously stirred and cold $NaBH_4$ solution (5 mM, 8.9 mL) was added dropwise. The solution slowly turned light gray and then ruby red. Stirring was maintained for 30 min. The obtained citrate-capped core-shell Au nanoparticles (AuNP) were stored at 4°C. Their size, determined by HR-TEM imaging and UV-Vis surface plasmon absorption was 5.0 nm. The concentrations of AuNP's are given in moles of particles per 1 L of solution.

DNA damage testing. A biosensor with immobilized DNA was soaked in a NB solution (100 µM) for 10 min, washed, and tested in PBS buffer (0.02 M), pH = 7.4 using a cyclic voltammetry (CV). The heights of the redox peaks of NB bound to the immobilized DNA are a measure of the NB uptake by a sensory film. The measurements of DNA damage were performed by immersing the biosensor in a Atz solution (100 µM) for 40 min, followed by washing with distilled water, soaking in NB solution (100 µM) for 10 min, washing, and CV testing in phosphate buffer (0.02 M, pH = 7.4). The change in the NB redox-peak heights was measured to determine the change in NB uptake caused by DNA interactions with Atz.

Voltammetric measurements. The DNA biosensors were immersed in a deoxygenated 0.02 M phosphate buffer (pH 7.4) and cyclic voltammograms were recorded in a potential range from 0 V to –0.6 V *vs.* Ag/AgCl reference.

Fluorimetric measurements. Stern-Volmer quenching and Scatchard plots experiments were performed in 0.02 M Tris-HCl buffer pH 7.5.

Ab-initio Calculations. Quantum mechanical (QM) calculations of electronic structures and molecular dynamics (MD) simulations for model systems to evaluate the interactions of atrazine with single base-pairs A-T and G-C were performed using advanced Hartree-Fock methods with 6-31G* basis set and pseudopotentials, semi-empirical PM3 method, and density functional theory (DFT) with B3LYP functional. The MD simulations and QM calculations were carried out using procedures embedded in Wavefunction (Irvine, CA, U.S.A.) Spartan 6. The electron density is expressed in atomic units, au^{-3}, where 1 au = 0.052916 nm and 1 au^{-3} = 6.7491×10^3 nm^{-3}.

BIOSENSORS FOR EVALUATING DNA ALTERATION AND DAMAGE BY HERBICIDES AND PESTICIDES

Different types of DNA-hybridization biosensors have been proposed for finding a single base-pair mismatch. Similar in principle are the biosensors designed for evaluating DNA alteration and damage since analytical signal in these sensors is also responding to problems with base-pair matching or conformation change induced by toxicants. Several DNA sensor designs that have been utilized in investigations of DNA alterations and damage by toxicants, such as herbicides and pesticides, are described and evaluated in the next sections. The main types of alterations and damage to the DNA encountered in structural studies are depicted in Fig. 13.1.

Horizontal Biosensors

The sensor presented in Fig. 13.2A (Design I) is based on electrostatic binding of DNA to positively charged poly-l-lysine (PLL). In this design, calf-thymus dsDNA (ctDNA) was used. The polyamine sublayer is immobilized on the Au electrode surface by adsorbing citrate anions. The experiments have been performed using solid film Au electrodes evaporated on a quartz crystal wafer piezoresonator and have enabled monitoring of apparent mass changes of the Au electrode associated with all stages including the citrate coating, PLL immobilization, and ctDNA attachment. Typical apparent-mass change transient for ctDNA-functionalized film formation on a QC/Au piezoelectrode is presented in Fig. 13.2B.

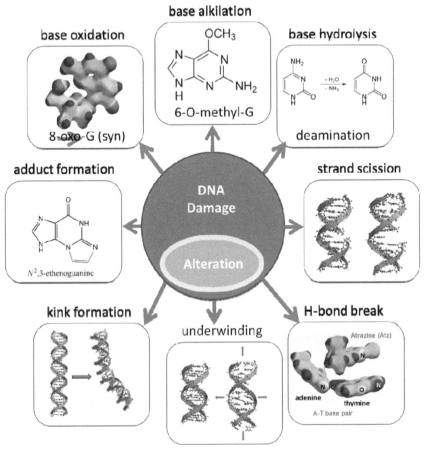

Figure 13.1. Types of the DNA damage and alterations. They are caused by oxidative stress and by toxicants. (Unpublished data).

High-density Vertical DNA Biosensors for Intercalation Measurements

The sensor in Fig. 13.3A (Design II) is based on designer oligonucleotide chains. A self-assembled monolayer of a mercaptopropionic acid (MPA), is used as the basal film. Owing to the high affinity of S-atoms of the thiol to the Au surface atoms, a strong thiolate bond is formed stabilizing the MPA SAM. The carboxylic functional group of MPA protruding to the solution

A

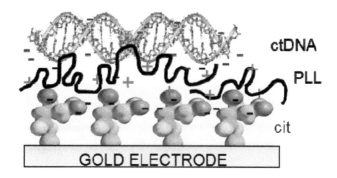

B

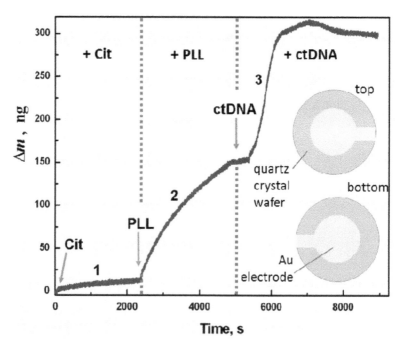

Figure 13.2. Horizontal biosensor QC/Au/Cit/PLL/ctDNA (design I) for the assessment of DNA damage by toxicants. **(A)** Schematic view; **(B)** nanogravimetric transient showing all stages of the biosensor preparation: (1) citrate (cit) adsorption in 10 mM Na$_3$Cit, (2) poly-L-lysine (PLL) immobilization from 30 ppm PLL solution, (3) calf thymus dsDNA (ctDNA) attachment from a ctDNA solution (45 μM base-pair); medium: 0.02 M PBS, pH = 7.45. The moments of solution injections are marked with arrows. (Adapted from (Hepel and Stobiecka 2010). With permission of Nova Science Publ.).

is then activated with N-(3-dimethylaminopropyl)-N'-ethylcarbodiimide (EDC). Next, the NH$_2$-modified probe ssDNA$_{20b}$ is immobilized on the SAM *via* the amide bond formation and monitored using EQCN, as illustrated in Fig. 13.3. After the oligonucleotide film had been immobilized on the sensor surface, it was hybridized with a complementary target oligonucleotide to form a vertical 20-bp dsDNA sensory film. The performance of this type of a sensor was examined using a NB redox dye as the probe intercalator. The linear dependence of the peak current of NB reduction on potential scan rate v confirms that this peak corresponds to the surface bound species and not to an NB dissolved in the solution. The charge analysis indicates that 1.65 NB molecules are incorporated per one dsDNA duplex (Wang et al. 2009).

The effect of interactions of herbicides with dsDNA immobilized on the sensor surface was investigated by soaking the sensor in a herbicide solution, followed by rinsing, and testing in PBS (Fig. 13.3C). Next, the sensor was soaked in NB solution (100 µM, 10 min), and rinsed with PBS. The tests performed in PBS show considerable increase of the NB peak current (curve 3) in comparison to the original peak current before the interaction with the herbicide.

The increase in the intercalation capacity of a dsDNA to NB dye is clearly associated with damaging interactions of Atz with the dsDNA duplex. The Atz interactions cause underwinding of dsDNA and increase the inter-base spacing in the base stacks leading to the increased uptake of NB probe intercalator. If Atz would have caused strand breaks or disassembly, the film capacity toward absorption of NB would have decreased and this is not observed.

The changes in the redox peak current of the intercalator probe NB can be quantified. Thus, the increase in the NB absorption capacity is given by:

$$\phi_{intercal} = 100\left[(I_{pc,Atz}-I_{b2})/(I_{pc,ini}-I_{b1})-1\right], \tag{1}$$

where $I_{pc,ini}$ and $I_{pc,Atz}$ are the cathodic peak currents for the reduction of NB absorbed in DNA on electrode, for a fresh DNA film (curve 2) and for a film with DNA damaged by the interactions with a herbicide (curve 3), respectively, and I_{b1} and I_{b2} are the background currents for these two peaks.

The value of $\phi_{intercal}$ calculated from the data of Fig. 13.3C, $\phi_{intercal} = 65\%$, indicates on a large DNA duplex expansion and underwinding caused by Atz. The tracing of NB leaching from a sensory film (not shown) indicates only a slow egress of NB taking place. It is also apparent that Atz repels NB from DNA quite effectively.

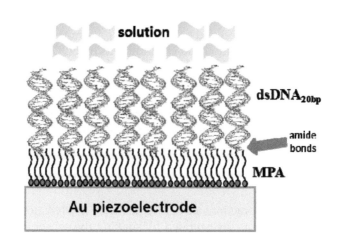

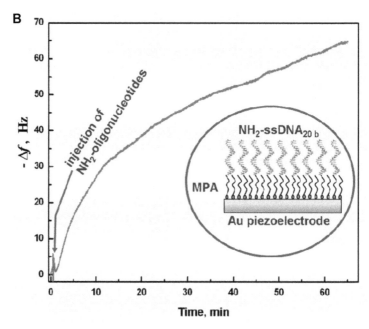

Figure 13.3. contd....

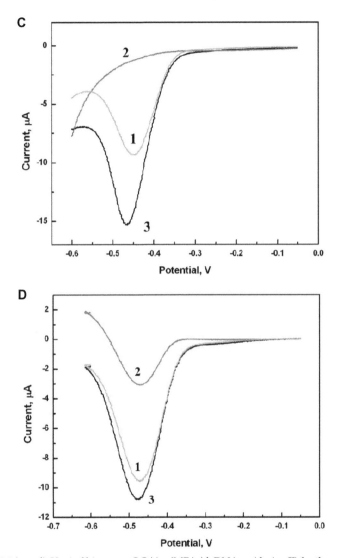

Figure 13.3 (contd). Vertical biosensor QC/Au/MPA/dsDNA$_{20bp}$ (design II) for the assessment of DNA damage by toxicants. **(A)** Schematic view; **(B)** apparent mass transient for a QC/Au/MPA/ssDNA piezosensor recorded after injection of a complementary 20-base oligonucleotide; inset: film structure after hybridization of dsDNA; solution: 0.02 M PBS, pH = 7.4; **(C-D)** linear potential scan voltammetry for a QC/Au/MPA/dsDNA$_{20bp}$ electrode without **(C)** and with a single C-A mismatch; **(D)** in the complementary oligonucleotide, recorded in 0.02 M PBS, pH = 7.45, after subsequent treatments: (1) after 10 min soaking in NB solution (100 μM); (2) after 40 min incubation in Atz solution (100 μM); (3) after 10 min soaking in NB solution (100 μM); scan rate v = 400 mV/s; *I-E* curves are background subtracted. (Adapted from (Hepel and Stobiecka 2010). With permission of Nova Science Publ.).

The high responsiveness of vertical dsDNA$_{20bp}$ sensors to DNA damage is accomplished owing to the high sensitivity to the base-pair mismatch of the hybridization process and favorable transduction of the information on the hybridization degree via surface-bound NB redox voltammetry. The high signal amplification in vertical dsDNA sensors is attributed to simultaneously acting two specific recognition mechanisms: the DNA hybridization and the propensity of NB to intercalate only to dsDNA. This amplification scheme results in a single base-mismatch sensitivity of these sensors, as illustrated in Fig. 13.3C,D.

Can DNA-modified Gold Nanoparticles Interact with Herbicides in Solution?

The utility of gold nanoparticles in identifying and monitoring subtle changes in dielectric function of the medium and ligand interactions with functionalized SAM-coatings of AuNP has recently been recognized. The functionalized AuNP's can serve as a sensitive and convenient platform for the determination of various analytes and to study interactions between biomolecules, drugs, toxicants, environmental pollutants, etc. The investigations of herbicide interactions with DNA embedded in a protective shell of AuNP described here demonstrate the feasibility of DNA damage evaluation using this technology.

The design of functionalized AuNP's for studies of atrazine interactions with DNA was based on the immobilization of positively charged PLL ligands on citrate-capped AuNP's. The following attachment of ctDNA on PLL-modified AuNP was electrostatically driven.

The interactions of NB and atrazine with ctDNA modified PLL-capped AuNP were investigated using resonance elastic light scattering spectroscopy (Fig. 13.4). The weakest scattering is observed for a PLL-capped AuNP solution (1.27 nM) in the presence of NB (10 µM), as illustrated in curve 3. Since the nanoparticles were not modified with ctDNA, only repulsive interactions of NB with AuNP are encountered in this case. The scattering increases when ctDNA-modified PLL-capped AuNP's are used instead of unmodified nanoparticles (curve 1). This increase is due to the increased size of nanoparticles which now accommodate ctDNA molecules immobilized on top of the PLL shell. Stronger scattering for these nanoparticles is expected since the scattering intensity I_{sc} increases with sixth power of the nanoparticle diameter a according to the Rayleigh equation which can be written in the form:

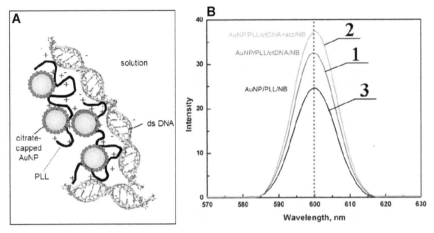

Figure 13.4. A ctDNA-modified PLL/Cit-capped AuNP$_{5nm}$ nanoparticle framework designed for the analysis of atrazine-induced alterations of DNA double-helix. **(A)** Schematic view; **(B)** resonance elastic light scattering spectra for: (1) AuNP/Cit/PLL/ctDNA + 10 μM NB solution without Atz, (2) AuNP/Cit/PLL/ctDNA + 10 μM NB solution with 33.3 μM Atz, and (3) AuNP/Cit/PLL + 10 μM NB solution. Conditions: 0.02 M PBS, pH = 7.45; C_{AuNP} = 1.27 nM, C_{ctDNA} 5 μM bp; all concentrations are final concentrations. (Adapted from (Hepel and Stobiecka 2010). With permission of Nova Science Publ.).

$$I_{sc} = I_0 N \frac{\left(1+\cos^2\theta\right)}{2R^2}\left(\frac{2\pi}{\lambda}\right)^4 \frac{\left(\left[n_2-n_1\right]^2-1\right)}{\left(\left[n_2-n_1\right]^2+2\right)}\left(\frac{a}{2}\right)^6 \tag{2}$$

where n_1 and n_2 are the refractive indices for the solution and particles, respectively, λ is the wavelength of incident light beam, θ is the scattering angle, N is the number of particles, and I_0 is the constant. When the AuNP/Cit/PLL/ctDNA nanoparticles were treated for 20 min with Atz (33.3 μM), even stronger light scattering was observed (curve 2). This is attributed to the expanded conformation of Atz-treated ctDNA in the shell of AuNP's. For λ = const and other experimental conditions (θ, R, I_0) unchanged, one obtains from eq. (2):

$$\frac{I_{sc,2}}{I_{sc,1}} = \frac{N_2 a_2^6}{N_1 a_1^6} = c_{rel} a_{rel}^6 \tag{3}$$

where indices 1,2 stand for the particles before and after Atz addition, respectively, $c_{rel} = N_2/N_1$ is the relative concentration of particles after addition of Atz (c_{rel} = 1 if no dehybridization occurs or 2 for a complete dehybridization of dsDNA (Stobiecka et al. 2010)), and $a_{rel} = a_2/a_1$ is the relative diameter of particles after addition of Atz. Due to the sixth–

power dependence of light scattering on particle size, the RELS based measurements of conformation changes of DNA caused by toxicant damage are extremely sensitive.

Nanostructured DNA Biosensors

The third type of biosensors (Design III), presented in Fig. 13.5A, utilizes monolayer-protected gold nanoparticles to increase the effective surface area of the sensor and enhance the analytical signal. A functionalized thiol aminoethanethiol (AET) was deposited first to form a SAM. Then, MPA-capped core-shell gold nanoparticles were activated with EDC and attached *via* covalent amide bonds to the AET-SAM. Finally, amine-derivatized oligonucleotide was immobilized on the EDC-activated nanoparticle shells *via* amide bonds and further hybridized with a matching oligonucleotide to obtain a dsDNA sensor.

Subsequent stages of the film development were monitored using the EQCN technique. The experiments performed with this sensor show a strong dependence of voltammetric redox currents on the addition of Atz (Fig. 13.5B). In comparison to the sensor Design II, the addition of AuNP has led to the analytical signal enhancement, as observed after interactions with Atz. The value of $\phi_{intercal}$ = 84 % was obtained.

Voltammetric Redox-labeled Hybridization Biosensors

The redox-labeled biosensors (Design IV, Fig. 13.6A) are based on the transmission of electric current from a redox center (e.g. ferrocene, Fc) covalently attached to dsDNA strands at the film/solution interface. Any damage to the dsDNA structure caused by toxicants, such as herbicides and pesticides, affects the current flow from the redox group to the electrode surface (the substrate) through the DNA.

In the biosensor employed in these studies a Fc-modified target ssDNA was hybridized to a biotinylated probe ssDNA immobilized through a biotin-streptavidin recognition binding to streptavidin and biotinylated thiol SAM. We have investigated the DNA damage and structural alterations, such as the double-helix underwinding, caused by several known herbicides: atrazine, 2,4-D, and glufosinate ammonium (GA) and representative compounds from the main groups of pesticides, including organophosphate, carbamate, and urea pesticides, *viz.* paraoxon-ethyl (PE), diflubenzuron (DFB), and carbofuran (CF) (Nowicka et al. 2010).

Freshly prepared biosensors were immersed in 100 μM solution of each herbicide/pesticide for a predetermined time, followed by rinsing and immersing in pure PBS buffer solution, pH = 7.4. The evaluation of

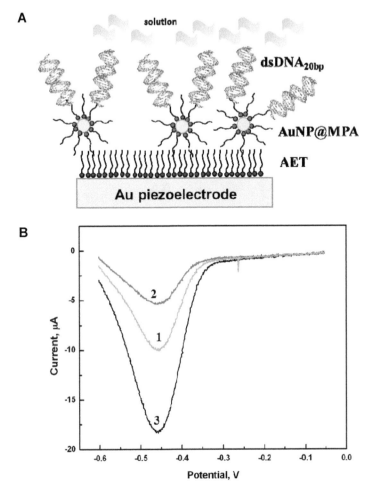

Figure 13.5. Nanostructured biosensor QC/Au/AET/AuNP@MPA-dsDNA$_{20bp}$ (design III) for the assessment of DNA damage by toxicants. **(A)** Schematic view; **(B)** linear potential scan voltammetry for the sensor in 0.02 M PBS, pH = 7.45, after subsequent treatments: (1) after 10 min soaking in NB solution (100 µM); (2) after 40 min soaking in Atz solution (100 µM); (3) after soaking in Atz and 10 min soaking in NB solution (100 µM); scan rate v = 400 mV/s; curves are background subtracted. (Unpublished data).

DNA damage was carried out by cyclic voltammetry in a PBS solution. The interaction of herbicides and pesticides with DNA which leads to structural changes in DNA double-helix, decreases the Fc oxidation current. The redox probe current does not decrease to zero indicating that toxicants do not damage DNA through the formation of strand breaks which could destroy the DNA hybridization and could result in complete current cessation but rather act as to cause an underwinding of the DNA, thereby decreasing the

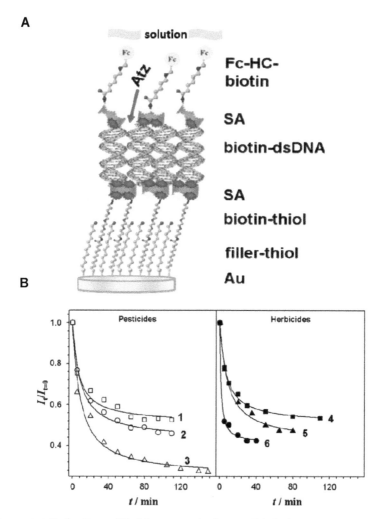

Figure 13.6. Redox Fc-modified biosensor $Au/bMUD/SA/b\text{-}dsDNA_{20bp}/SA/Fc$ (design IV) for the assessment of DNA damage by toxicants. **(A)** Schematic view; **(B)** effects of various pesticides and herbicides on the Fc redox current response: (1) carbofuran (CF), (2) diflubenzuron (DFB), (3) paraoxon-ethyl (PE), (4) 2,4-dichlorophenoxyacetic acid (2,4-D), (5) glufosinate ammonium (GA), (6) atrazine (Atz); $v = 50$ mV/s; 0.02 M PBS buffer (pH 7.4). (Adapted from (Nowicka et al. 2010). With permission of Elsevier).

film conductance. The interactions of herbicides and pesticides with DNA are solely responsible for the observed current diminution. It follows from the data collected (Nowicka et al. 2010) that the most detrimental influence on DNA hybridization among herbicides examined has atrazine for which the DNA damage δ_{DNA} defined as:

$$\delta_{DNA} = (I_{t=0} - I_t)/I_{t=0} \qquad (4)$$

approached 0.58. In the above equation, I is the peak current for Fc oxidation and t is the time of DNA exposure to the toxicant. For other herbicides tested, $\delta_{DNA} = 0.47$ for 2,4-D and $\delta_{DNA} = 0.52$ for GA. Among pesticides tested (Nowicka et al. 2010), CF caused the least damage ($\delta_{DNA} = 0.48$), followed by DFB ($\delta_{DNA} = 0.54$) and PE ($\delta_{DNA} = 0.72$) (Fig. 13.6) (Nowicka et al. 2010).

The experiments described above have shown that herbicides and pesticides cause considerable changes in DNA leading to the reduction of the charge transport between the electrode surface and the redox centers.

Stern-Volmer Quenching and Scatchard Plots

The evaluation of DNA damage by toxicants was also performed in solution (Fig. 13.7). The Nile Blue dye, after intercalating into the DNA duplex, becomes fluorescent-inactive and the quenching of fluorescence peak of the dye was observed after increasing concentration of DNA. After interaction of DNA with Atz, the increased quenching was found and the Stern-Volmer quenching constant changed from 0.51×10^6 L/mol to 0.59×10^6 L/mol. The change in the binding site size for a Nile Blue molecule bound to DNA from $n = 1.6$ to $n = 1.06$, following from the Scatchard plots, points to the expansion of the structure of DNA after the interaction with the herbicide, leading to the increased NB uptake. The binding constant changed from $K = 1.93 \times 10^6$ L/mol to $K = 0.89 \times 10^6$ L/mol for NB intercalation to DNA before and after DNA interaction with Atz, respectively.

Molecular Dynamics Simulations of Pesticide and Herbicide Interactions with DNA Double-helix

The MD simulation and QM calculation of electronic structure provide new insights to the understanding of intercalation processes in DNA. Using these techniques the interactions of herbicides and pesticides with model single- and double-stranded DNA can be investigated. For a given intercalator, it is possible to evaluate electrostatic interactions with phosphate-sugar backbone and hydrogen bonding in the major and minor groove of the double-helix. The latter interactions may further lead to the intercalation into the interbase space provided than no steric hindrance is encountered and the π-π orbital overlap in the stacking interactions between the intercalator and nitrogen bases is energetically favorable. The theoretical modeling may provide clear evidence of preferences of the molecular interactions in the minor or major groove and predict the type of conformational alterations in the DNA caused by the intercalation process.

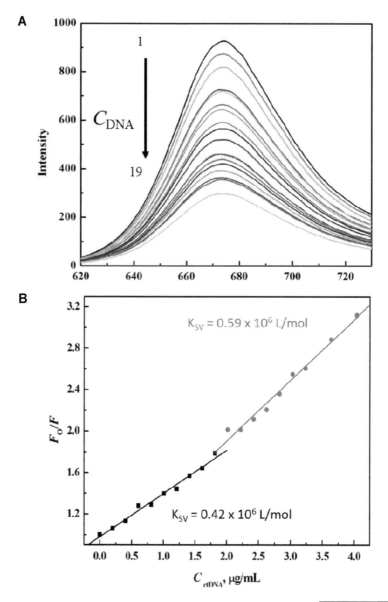

Figure 13.7. contd....

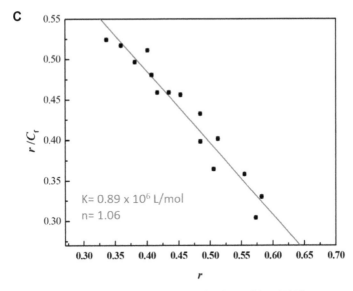

Figure 13.7 (contd.). Interactions of ctDNA with Nile Blue and Atz. **(A)** Fluorescence emission spectra; $C_{NB} = 2\,\mu M$, 20 mM Tris-HCl pH 7.5; C_{ctDNA} [μM/bp]: (1) 0, (2) 0.202, (3) 0.404, (4) 0.606, (5) 0.808, (6) 1.01, (7) 1.21, (8) 1.41, (9) 1.62, (10) 1.82, (11) 2.02, (12) 2.22, (13) 2.42, (14) 2.63, (15) 2.83, (16) 3.03, (17) 3.23, (18) 3.64, (19) 4.04; **(B)** Stern-Volmer quenching plots for NB fluorescence quenched by ctDNA after 40 min interaction with 100 μM Atz; and **(C)** Scatchard plot for binding of NB to ctDNA (same conditions); 20 mM Tris-HCl, pH 7.5. (Unpublished data).

The interactions of herbicides and pesticides with DNA have been analyzed in model systems using MD simulations and QM calculations. The general methodology applied in these investigations is based on examining the docking of the foreign molecule at the major and minor grooves of a model DNA double-helix, evaluating H-bonding and electrostatic interactions, followed by inserting the interacting molecule in a gap between two stacked bases in a model ssDNA or dsDNA.

The MD view of lateral interactions of Atz approaching the DNA double-helix from the major groove side at the AT base-pair is presented in Fig. 13.8. It is seen that this approach is stabilized by the formation of a hydrogen bond between the imine nitrogen of Atz and O4 oxygen of thymine of the AT base-pair. From the side-on configuration, Atz molecule may move to more stable stacking configuration with one of the bases. The analysis published recently (Wang et al. 2009), indicates that this interaction leads to tilting of the other base (here A) and results in changes of the DNA duplex conformation.

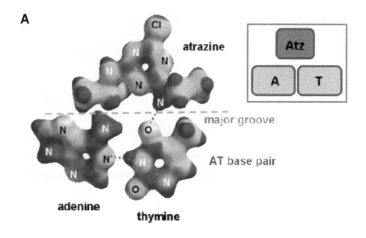

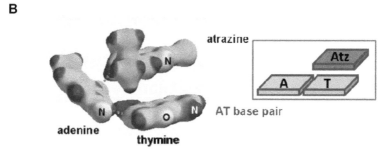

Figure 13.8. Initial stage of the interactions of atrazine with a AT base-pair. **(A)** In-plane interactions at the major groove, stabilized by hydrogen bonding with T; **(B)** intercalation of Atz, stabilized by stacking interactions with T and hydrogen bonding with A; atrazine-induced tilt of A with respect to T seen in (B); electron density surfaces ($d = 0.08$ a.u.) with mapped electrostatic potential (color coded: from negative-red to positive-blue). (Unpublished data).

Color image of this figure appears in the color plate section at the end of the book.

In Fig. 13.9, MD simulations of the interactions of Atz with a GC base-pair are presented. When Atz approaches the DNA double-helix from the major groove side, a stabilizing hydrogen bond is formed between the imine nitrogen of Atz and oxygen of guanine of the GC base-pair. During the movement of Atz molecule from the side-on to stacking conformation, the H-bond is keeping the group together.

In contrast to the weak groove binding, the intercalation of Atz leads to considerable interbase distance increase, resulting in the elongation of the dsDNA double-helix, and DNA duplex bending associated with small Atz size and inability of Atz to interact symmetrically with both bases in any given base-pair in dsDNA.

The MD investigations carried out using dsDNA$_{4bp}$ fragments (poly (A)·poly(T)) have shown that the interbase distance has substantially

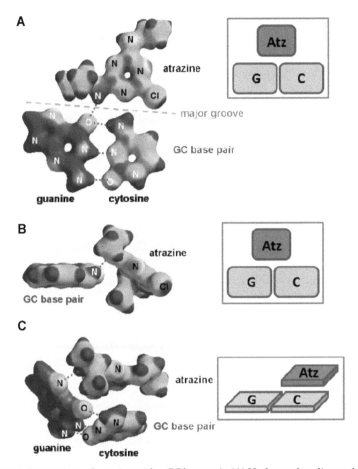

Figure 13.9. Interactions of atrazine with a GC base-pair. **(A)** Hydrogen bonding at the major groove; **(B)** approach towards intercalation; and **(C)** intercalation and atrazine-induced tilt of G with respect to C; electron density surfaces (d = 0.08 a.u.) with mapped electrostatic potential (color coded: from negative-red to positive-blue). (Unpublished data).

Color image of this figure appears in the color plate section at the end of the book.

increased resulting in longitudinal duplex expansion due to the intercalation of Atz (Wang et al. 2009).

Hence, the molecular dynamics simulations confirm the conformational transitions in dsDNA leading to the reduction in the number of base-pairs per turn and elongation of the duplex. This is consistent with experimental observations that dsDNA altered by the interactions with Atz has a higher capacity toward the sorption of NB. At the same time, the intercalation of Atz causes the dsDNA to bend due to the small size of Atz and its inability to participate in stacking interactions with both bases of a given base-pair.

APPLICATIONS TO OTHER AREAS OF HEALTH AND DISEASE

The DNA biosensors have recently gained importance, for instance in diagnostics, gene analysis, fast detection of biological warfare agents (Skladal et al. 2008), detection of genetically modified food (Stobiecka et al. 2007), and forensic applications. Various types of biosensors have been developed in the field of environmental monitoring (Pohanka and Skládal 2008) (Marty et al. 1998). The application of piezoimmunosensors have recently been reviewed (Pribyl et al. 2003). The DNA piezosensors have been applied to evaluate DNA/RNA aptamer interactions with proteins (Hianik 2007) and to determine DNA-protein binding constants (Hianik et al. 2004). The DNA sensors can serve as convenient devices for elucidation of anticancer drug interactions with DNA (Oliveira-Brett et al. 1998). The key part of these biosensors is a molecular probe that is built from oligonucleotides of well-defined sequence. That sequence must be complementary to the oligonucleotide sequence to be detected. The hybridization process is specific and only the complementary strands can hybridize. Many different transduction techniques have been applied to convert the information on the progress of DNA hybridization to an analytical signal. The unique sensitivity of the hybridization process to a single base mismatch makes these techniques suitable for testing pollutants and assessing damage caused to DNA (Nowicka et al. 2010). The studies of DNA damage by various agents have also been carried out by measuring DNA oxidation on mercury and carbon electrodes (Palecek 1996). The utility of magnetic bead-based label-free biosensors was recently discussed (Wang et al. 2001) and the effects of various intercalator probe molecules have been evaluated (Kelly et al. 1999) (Erdem et al. 2001).

The methods used for the transduction of DNA damage or sequence mismatch in hybridization process include voltammetric, optical, impedimetric, surface plasmon resonance, spectroscopic tagging of nucleotides under test, and EQCN techniques. The latter technique has been employed in highly sensitive biosensor applications (Halamek et al. 2001; Hianik et al. 2007; Pribyl et al. 2003; Pribyl et al. 2006; Stobiecka et al. 2007) including aptamers (Hianik et al. 2007) and immunosensors (Pribyl et al. 2006). The crystal resonance frequency decreases with increasing mass of the sensory film attached to the gold piezoelectrode (Hepel 1999). With EQCN, subnanogram levels of mass change can be detected. Higher sensitivity is achieved by using QCN crystals of higher frequencies (Hepel 1999).

CONCLUSION

We have presented several methods for monitoring the interactions of toxicants with dsDNA based on DNA biosensors. These methods can aid in assessing the damage and structural modifications caused to dsDNA by herbicides and pesticides and are suitable for screening of a wide range of herbicides and pesticides in different formulations and with different adjuvants. In particular, we have demonstrated that simple techniques with a variety of signal transduction schemes can be applied for detecting DNA alterations due to the herbicide interactions with DNA including intercalation to a DNA double-helix. On the basis of experimental results and with the help of MD simulations and QM calculations it becomes possible to evaluate the extent of dsDNA damage and possible weak spots in the double-helix structure. The weakest spots in DNA sequences are those with multiple TA base-pairs where unwinding is energetically favored because of the weakness in the strained hydrogen bonding there. Also, the base stacking interactions between T-T, T-A, and A-A bases are the weakest among other combinations of base stackings. Small molecules, like atrazine, fit neatly in the space between the stacked bases disrupting the local inter-helix hydrogen bonding and changing the DNA conformation because of the expansion of the inter-base distance in one side of a double-helix. This happens since the intercalator molecule is too small to match the entire base-pair in DNA.

It appears that the main action of many herbicides and pesticides is to change the DNA conformation, e.g. leading to underwinding, which makes DNA vulnerable to attacks by adjuvants of the herbicide preparation.

The determination of low concentrations of herbicides and pesticides in the environmental samples was not the objective of this study since such investigations require longer interaction time of pollutants with DNA. The focus was placed specifically on the development of means for rapid comparative testing which could then be applied for the development of assays for the assessment and screening of "manufacturer supplied" preparations. To facilitate accelerated testing, the use of high concentrations of herbicides and pesticides is advisable. Thus, the products showing little or no DNA damage at high doses have less chance to be harmful at lower concentrations than products that show a clear DNA damage.

The applications of dsDNA biosensors enable the DNA alterations and damage to be quantified. They can serve as new tools in systematic studies of toxicity and DNA damage caused by various pollutants. Among herbicides and pesticides tested by us, atrazine and paraoxon-ethyl cause the fastest and most severe damage to DNA. The methods used in these analyses and sensor preparations are relatively simple and very sensitive. With further developments and automatic fabrication, the cost and reliability of such sensors can be further improved. Moreover, the DNA biosensors can be miniaturized and in future produced as sensor chips or microsensor arrays.

KEY FACTS OF HERBICIDES AND PESTICIDES

- Herbicides and pesticides are bioactive compounds and their interactions with DNA may be harmful to human health.
- Environmental contamination by herbicides and pesticides is growing due to widespread application of intensive agricultural technologies.
- Monitoring of environmental pollution requires the development of new field-deployable, inexpensive and simple in use analytical platforms and the utilization of biosensors for pollution control is a promising solution.
- Sensors based on biorecognition principles and hybridization biosensors are promising candidates for accelerated tests of herbicide formulations for DNA alterations and damage.
- DNA-modified biosensors for assessment of DNA damage and alteration by herbicide and pesticide formulations offer simplicity of operation and scalability important for the development of sensor chips and microsensor arrays.

DEFINITIONS

- *Hybridization biosensor*: a sensor with ssDNA or dsDNA immobilized on the substrate or sensory film in which the degree of hybridization is modified by the interactions of DNA with analyte.
- *Transduction methods*: methods used to convert changes in physical and/or chemical properties of sensory films occurring during the interactions of analyte with film.
- *Analytical signal*: measurable signal, such as potential, conductance, a.c. frequency, absorbance, etc., generated in response to the interaction of an analyte with the sensory film and serving as a measure of analyte concentration.
- *Biorecognition*: specific binding of an analyte (guest) by a biomolecule (host) immobilized in a sensory film due to multiple interaction points, e.g. biotin-streptavidin, complementary ssDNA strands, antigene-antibody.
- *SAM*: self-assembled monolayer of organic surfactant molecules formed spontaneously owing to strong interactions between the

substrate (e.g. Au) and the surfactant molecule head (e.g. thiol group) and side-to-side attractions between tails (e.g. hydrocarbon chains) of surfactant molecules.

- *Core-shell gold nanoparticle*: a nanometer-sized gold particle (core) in the form of a sphere, rod, etc., coated with a protecting monolayer (shell).

SUMMARY POINTS

- DNA hybridization is highly sensitive to the presence of species interacting with DNA, especially to those able to intercalate into the dsDNA interbase area.
- Vertically bound ssDNA molecules provide higher density sensory films.
- Among the horizontal, vertical, and nanostructured DNA-biosensors, the latter are the most sensitive.
- The use of functionalized gold nanoparticles in sensory film design increases sensitivity, most likely due to the increased surface accessibility.
- Voltammetric and nanogravimetric transduction utilized in DNA-biosensors provides a single base-pair mismatch sensitivity.
- Comparison of DNA alterations by common herbicides and pesticides using ferrocene-labeled hybridization biosensors shows that most pronounced DNA alterations are observed for atrazine among herbicides and paraoxon-ethyl among pesticides.

ABBREVIATIONS

AET	:	aminoethanethiol
Atz	:	atrazine herbicide
AuNP	:	gold nanoparticles
bMUD	:	biotinylated 11-mercapto-1-undecanol
EDC	:	N-(3-dimethylaminopropyl)-N'-ethylcarbodiimide
EQCN	:	Electrochemical Quartz Crystal Nanogravimetry or: Electrochemical Quartz Crystal Nanobalance
MPA	:	mercaptopropionic acid
NB	:	Nile Blue redox dye
PBS	:	phosphate buffer saline solution
PLL	:	poly-l-lysine
SA	:	streptavidin

REFERENCES

Alavanja MCR, C Samanic, M Dosemeci, J Lubin, R Tarone, CF Lynch, C Knott, JA Hopin, J Barker, J Coble, DP Sandler and A Blair 2003. Use of Agricultural Pesticides and Prostate Cancer Risk in the Agricultural Health Study Cohort. Am J Epidemiol 157: 800–814.

Ashby J, H Tinwell, J Stevens, T Pastoor and GB Breckenridge. 2002. The Effects of Atrazine on the Sexual Maturation of Female Rats. Rugulatory Toxicol Pharmacol 35: 468–473.

Clements C, S Raph and M Petras 1997. Genotoxicity of Selected Herbicides in Rana Catesbeiana Tadpoles Using Alkaline Single-Cell Gel DNA Electrophoresis (Comet) Assay Environ Mol Mutagen 29: 277–288.

Del Carlo M, M Mascini, A Pepe, G Diletti and D Compagnone 2004. Screening of Food Samples for Carbamate and Organophosphate Pesticides Using an Electrochemical Bioassay. Food Chem 84: 651–656.

Erdem A, K Kerman, B Meric and M Ozsoz 2001. Methylene Blue as a Novel Electrochemical Hybridization Indicator. Electroanalysis 13: 219–223.

Fan W, T Yanase and H Morinaga 2007. Atrazine-Induced Aromatase Expression is Sf-1 Dependent: Implications for Endocrine Disruption in Wildlife and Reproductive Cancers in Humans. Environ. Health Perspect 115: 720–727.

Finklea HO, S Avery, M Lynch and T Furtsch 1987. Blocking Oriented Monolayers of Alkyl Mercaptans on Gold Electrodes. Langmuir 3: 409–413.

Freeman JL and AL Rayburn 2004. *In Vivo* Genotoxicity of Atrazine to Anuran Larvae. Mutation Res 160: 69–78.

Greenlae AR, TM Ellis and RL Berg 2004. Low-Dose Agrochemicals and Lawn-Care Pesticides Induce Developmental Toxicity in Murine Preimplantation Embryos. Environ Health Persp 112: 703–709.

Halamek J, M Hepel and P Skladal 2001. Investigation of Highly Sensitive Piezoelectric Immunosensors for 2,4-Dichlorophenoxyacetic Acid. Biosens Bioelectron 16: 253–260.

Helali S, C Matelet, A Abdelghani, MA Maaref and N Jaffrezic-Renault 2006. A Disposable Immunomagnetic Electrochemical Sensor Based on Functionalized Magnetic Beads on Gold Surface for the Detection of Atrazine. Electrochim. Acta 51: 5182–5186.

Hepel M. 1999. Electrode-Solution Interface Studied with Electrochemical Quartz Crystal Nanobalance pp 599–630. In: A. Wieckowski. [ed.] Interfacial Electrochemistry. Marcel Dekker, Inc, New York, USA.

Hepel M and M Stobiecka 2010. Interactions of Herbicide Atrazine with DNA. Nova Science Publishers, New York, USA.

Hianik T. 2007. DNA/Rna Aptamers—Novel Recognition Structures in Biosensing. Elsevier, Amsterdam, The Netherlands.

Hianik T, V Ostatna and Z Zajacova 2004. The Study of the Binding of Globular Proteins to DNA Using Mass Detection and Electrochemical Indicator Method J Electroanal Chem 564: 19–24.

Hianik T, V Ostatna, M Sonlajtnerova and I Grman 2007. Influence of Ionic Strength, Ph and Aptamer Configuration for Binding Affinity to Thrombin. Bioelectrochem 70: 127–133.

Hopenhayn-Rich C, ML Stump and SR Browning 2002. Regional Assessment of Atrazine Exposure and Incidence of Breast and Ovarian Cancers in Kentucky. Arch Environ Contam Toxicol 42: 127–136.

Kelly SO, EM Boon, JK Barton, NM Jackson and MG Hill 1999. Single-Base Mismatch Detection Based on Charge Transduction through DNA. Nucleic Acid Res 27: 4830–4837.

Marty J-L, B Leca and T Noguer 1998. Biosensors for the Detection of Pesticides. Analysis 26: M144–M149.

Mascini M. 2001. Affinity Electrochemical Biosensors for Pollution Control. Pure Appl Chem 73: 23–30.

Nowicka AM, A Kowalczyk, Z Stojek and M Hepel 2010. Nanogravimetric and Voltammetric DNA-Hybridization Biosensors for Studies of DNA Damage by Common Toxicants and Pollutants. Biophys Chem 146: 42–53.

Oliveira-Brett AM, TRA Macedo, R Raimundo, MH Marques and SHP Serrano 1998. Voltammetric Behaviour of Mitoxantrone at a DNA-Biosensor. Biosens Bioelectron 13: 861–867.

Palecek E. 1996. From Polarography of DNA to Microanalysis with Nucleic Acid-Modified Electrodes. Electroanalysis 8: 7–14.

Pohanka M and P Skládal 2008. Electrochemical Biosensors—Principles and Applications. J Appl Biomed 6: 57–64.

Pribyl J, M Hepel, J Halamek and P Skladal 2003. Development of Piezoelectric Immunosensors for Competitive and Direct Determination of Atrazine. Sensors Actuators B 91: 333–341.

Pribyl J, M Hepel and P Skladal 2006. Piezoelectric Immunosensors for Polychlorinated Biphenyls Operating in Aqueous and Organic Phases. Sensors Actuators B 113: 900–910.

Ribas G, G Ferenzilli, R Barale and R Marcos 1995. Herbicide-Induced DNA Damage in Human Lymphocytes Evaluated by Single-Cell Gel Electrophoresis (Scge) Assay Mutation Res 344: 41–54.

Schroeder JC, AF Olshan, R Baric, GA Dent, CR Weinberg, B Yount, JR Cerhan, CF Lynch, LM Schuman, PE Tolbert, N Rothman, KP Cantor and A Blair 2001. Agricultural Risk Factors for T(14;18) Subtypes of Non-Hodgkin's Lymphoma Epidemiol 12: 701–709.

Skladal P, J Pribyl and B Safar. Development and Testing of Portable Electrochemical Immunosensor System for Detection of Bioagents. pp 21–29. In: [eds.] 2008. Commercial and Pre-Commercial Cell Detection Technologies for Defence against Bioterror. IOS Press, Amsterdam The Netherlands.

Stobiecka M, JM Cieśla, B Janowska, B Tudek and H Radecka 2007. Piezoelectric Sensor for Determination of Genetically Modified Soybean Roundup Readyâ in Samples Not Amplified by Pcr. Sensors 7: 1462–1479.

Stobiecka M, J Deeb and M Hepel 2010. Ligand Exchange Effects in Gold Nanoparticle Assembly Induced by Oxidative Stress Biomarkers: Homocysteine and Cysteine Biophys Chem 146: 98–107.

Surralles J, J Catalan, A Creus, H Norppa, N Xamena and R Marcos 1995. Micronuclei Induced by Alachlor, Mitomycin-C and Vinblastine in Human Lymphocytes: Presence of Centromeres and Kinetochores and Influence of Staining Technique. Mutagenesis 10: 417–423.

Tannant AT, B Peng and AD Kligerman 2001. Genotoxicity Studies of Three Triazine Herbicides: *In Vivo* Studies Using the Alkaline Single Cell Gel (Scg) Assay. Mutation Res 493: 1–10.

Wang C, J Zhao, D Zhang and YZ 2009. Detection of DNA Damage Induced by Hydroquinone and Catechol Using an Electrochemical DNA Biosensor. Aust J Chem 62: 1181–1184.

Wang J, AN Kawde, A Erdem and M Salazar 2001. Magnetic Bead-Based Label-Free Electrochemical Detection of DNA Hybridization. Analyst 126: 2020–2024.

Zeljezic D, V Garaj-Vrhovac and P Perkovic 2006. Evaluation of DNA Damage Induced by Atrazine and Atrazine-Based Herbicide in Human Lymphocytes *in Vitro* Using a Comet and DNA Diffusion Assay. Toxicology *in Vitro* 20: 923–935.

14

Nitrate Determination by Amperometric Biosensors: Applications to Environmental Health

Erol Akyilmaz[1],* and Victor R. Preedy[2]

ABSTRACT

Nitrate is used in food processing, to enhance colour and to impart anti-microbial activity for the preservation of meat and meat products. However, nitrate can also be a contaminant that impacts on both environmental and human health. As a consequence, methods for accurate, fast and sensitive determination of nitrate are very important. Some spectrophotometric, chromatographic and electrochemical methods have been developed for nitrate determination. However, the requirements for low cost, a short time of analysis, sensitivity, simplicity, selectivity, stability and reliability can still not be met by a single analytical method, except by using biosensors. The main advantages offered by biosensors over conventional analytical techniques are the possibility of portability, miniaturization and working on-site, and the ability to measure pollutants in complex matrices with minimal sample preparation. Despite these advantages, the application of biosensors

[1]Ege University Faculty of Science Biochemistry Department 35100 Bornova/Izmir, Turkey; E-mail: erol.akyilmaz@ege.edu.tr
[2]Diabetes and Nutritional Sciences, School of Medicine, Kings College London, Franklin Wilkins Buildings, 150 Stamford Street, London SE1 9NU, UK; E-mail: victor.preedy@kcl.ac.uk
*Corresponding author

List of abbreviations after the text.

to the environmental field is still limited in comparison to medical or pharmaceutical applications. There are only a limited number of biosensoric studies on the determination of nitrate and nearly all of them are based on the enzyme nitrate reductase. However, in terms of applicability, nitrate reductase is fraught with practical issues when applied to biosensor assays. This is because nitrate reductase is not very active and there are problems associated with the direct transfer of electrons within the electrode systems used in such biosensors. Thus, microbial biosensors and inhibition type biosensors have been developed to address this. In this chapter, we focus on amperometric nitrate biosensors that have been applied to environmental health.

INTRODUCTION

Nitrate occurs as a contaminant in drinking water primarily from groundwater and wells, and in high concentrations it has harmful biological effects. There are many sources of nitrogen, both natural and anthropogenic (i.e. caused by human kind), that could potentially lead to the pollution of groundwater with nitrates. However, the anthropogenic sources can give rise to dangerous levels of nitrate. Waste materials are key anthropogenic sources of nitrate contamination of groundwater. For example, fertilized agricultural lands, industrial waste water, animal feedlots, some septic tanks and urban drainage are well known sources of nitrates. In terms of food production, the anti-microbial activities of nitrate and nitrite have been recognized for centuries and are still used for the preservation of meat produce (Deng-sheng et al. 2006; Wayne et al. 1991; King et al. 2000). Nitrate itself is not carcinogenic, but instead acts as a procarcinogen, meaning that it reacts with other chemicals to form carcinogenic compounds via a multiple step process. Nitrate is converted into nitrite after consumption and the nitrite reacts with natural or synthetic organic compounds known as secondary amines or amides in food or water to form new combinations of either nitrosamines or nitrosamides, many of which are carcinogens (Mirvish 1991). However, two of the well known conditions associated with inadvertent and excessive nitrate ingestion are the development of methemoglobinemia and goitre, as described in detail below.

Methemoglobinemia and Nitrate

Methemoglobinemia is an anaemia resulting from the oxidation of the ferrous iron in the hemoglobin to the ferric state, changing haemoglobin to methemoglobin. Methemoglobin cannot carry molecular oxygen so therefore it is produced at a higher rate than the body is able to convert back to haemoglobin. This can lead to cyanosis, tissue hypoxemia and in severe cases, death. The normal physiological concentration of methemoglobin in the blood is less than 2% of total haemoglobin. However symptoms of methemoglobinemia may not be apparent until methemoglobin levels are at or above 10% (Wright et al. 1999). Acquired methemoglobinemia can result from exposure to certain pharmaceutical preparations as well as chemical substances such as nitrates (though copper, sulphate, chlorite and other chemicals will have a similar effect) (Prachal and Gregg 2000; Manassaram et al. 2010). For example in a North African study on young children aged 1 to 7 yr living within a defined geographical region, two thirds of the population had access to water sources which contained abnormally high nitrate levels (Sadeq et al. 2008). One third of the study population had methemoglobinemia and this was associated with the nitrate level in the drinking water (Sadeq et al. 2008).

Iodine and Nitrate

The relationships between nitrate intake and effects on the thyroid have also merited concern. This is because nitrate competitively inhibits iodine uptake via the sodium iodine transporter (Gatseva and Argirova 2008). In addition to the effects of nitrate on the thyroid observed in animal studies and in livestock, epidemiological studies have revealed an antithyroid effect of nitrate in humans. If dietary iodine is available at an adequate range (corresponding to a daily iodine excretion of 150–300 µg/day), the effect of nitrate is weak, with a tendency to zero. However, the nitrate effect on thyroid function is strong if a nutritional iodine deficiency or marginal deficiency exists simultaneously (Höring 1992). The importance of nitrate in iodine nutrition is also emphasized by one example study which showed that the goitre rate (the visible facade of iodine deficiency) in pregnant women exposed to high levels of nitrate in drinking water was 35% compared to 9% in those exposed to normative levels of nitrate (Gatseva and Argirova 2009). In school children the goitre rate was 14% in those exposed to high nitrate levels and 5% in those with normative exposure (Gatseva and Argirova 2009).

Methods of Analysing Nitrates and Biosensors

Depending on the purpose, sensitive and selective methods are needed for both quantitative and qualitative determination of target analytes. Some analytical methods such as spectrometry, gas chromatography, chemiluminescence, and electrochemistry are available for the determination of nitrate especially in water samples. Table 14.1 shows some parameters of these techniques. However, the requirements for low cost, a short

Table 14.1. Methods for nitrate determination adapted to environmental sample analysis.

Technique	Matrix	Detection Limit (µM)	Detection Range (µM)	Reference
Spectrometric	Water	29	29–2100	(Madsen 1981)
Spectrometric Townshend	Tap water	0.05	50–600	(Devi and 1989)
Chemiluminescence	Atmospheric	0.016	0.016–16	(Yoshizumi and Aoki 1985)
Gas Chromatography	Waste water/plants	1.6	16–1600	(Englmaier 1983)
Electrochemical	Drinking/River water	2	5–60	(Noufi et al. 1990)
Electrochemical	Ground/Tap water	1.5	20–10000	(Zuther and Cammann 1994)
Electrochemical	Drinking water	2	25–10000	(Högg et al. 1994)
Electrochemical	Sewage/Water	10	10–200	(Davis et al. 2000)
Electrochemical	Water	2.8	2.8–80	(Solak et al. 2000)

time of analysis, sensitivity, simplicity, selectivity, stability and reliability can still not be met by a single analytical method. Biorecognition-based techniques, i.e. biosensors, are potential candidates to fulfil many of the above requirements, due to their high selectivity, sensitivity and speed of interaction between the biological component and the analyte itself. Nitrate determination methods can be classified into three main categories: *Direct*, through a reduced nitrogen species, and *indirect* (Sah1994; Glazier et al. 1998). Table 14.2 shows a variety of techniques that fall into these.

Table 14.2. Methods for Nitrate Determination (Glazier et al. 1998).

Direct
  nitration of phenols and colorimetry
  oxidation of organics and colorimetry
  ion-selective electrode detection
  direct UV-absorbance spectrophotometry
  gas chromatography after derivatization
  electrophoretic
  nitration of salicylic acid
  ion chromatography

Indirect
  atomic absorption spectrophotomety
  polarography

Reduced nitrogen species
  reduction to nitrite and
  spectrophotometry
  electrochemistry
  reduction to ammonia and
    colorimetry
    potentiometry
    conductimetry
  reduction to nitric oxide and
    chemiluminescence

"A biosensor is defined by the International Union of Pure and Applied Chemistry (IUPAC) as a self-contained integrated device that is capable of providing specific quantitative or semi-quantitative analytical information using a biological recognition element that is retained in direct contact with a suitable transducer". In the general scheme of a biosensor, the biological recognition element responds to the target compound and the transducer converts the biological response to a detectable signal, which can be measured electrochemically, optically, acoustically, mechanically, calorimetrically, or electronically, and then correlated with the analyte concentration (Byfield and Abuknesha 1994; Wilson 2005; Thevenot et al. 2001). The main advantages of biosensors over conventional analytical techniques are the possibility of portability, of miniaturization and working on-site, and the ability to measure pollutants in complex matrices with minimal sample preparation. They may also present advantageous analytical features, such as high specificity and sensitivity as a result of the biological materials and platforms used in their construction. Although many of the systems developed can not compete with conventional analytical methods in terms of accuracy and reproducibility, they can be used by industry to provide enough information for routine assay and screening of samples (Parellada et al. 1998).

Biosensors can be useful for the continuous monitoring of a contaminated area. In addition, biosensors offer the possibility of determining not only specific chemicals but also their biological effects, such as toxicity, cytotoxicity and genotoxicity or endocrine-disrupting effects. They can provide, finally, total and bioavailable or bioaccesible pollutant concentrations. Despite these advantages, the application of biosensors in the environmental field is still limited in comparison to medical or pharmaceutical applications. Nevertheless, the majority of the systems developed are prototypes that still need to be validated before being used extensively or before their commercialization. The need for low cost disposable systems for environmental monitoring has encouraged the development of new technologies and more suitable methodologies (Mozaz et al. 2006). Many different biosensor types have been developed for the determination of compounds with environmental relevance, like phenols (Nistor and Emneus 1999; Kotte et al. 1995; Hedenmo et al. 1997; Parellada et al. 1998) and pesticides (Skladal et al. 1996).

Nitrate Determination by Biosensors Based on Nitrate Reductase

Over the last 10 yr there have been increased reports on the use of amperometric biosensors for nitrate determination (Amine and Palleschi 2004) (Table 14.3). In simple terms, an amperometric biosensor is one that

Table 14.3. Some amperometric nitrate biosensors based on **nitrate reductase** or microbial activity (Amine and Palleschi 2004).

Detection	Biocatalyst	Detection Limit (µM)	Detection Range (µM)	Response time(s)	Reference
Amperometric	Nitrate reductase	5.0	5.0–100.0	10	Moretto et al. 1998
Amperometric	Nitrate reductase	3.0	3.0–18.0	<60	Glazier et al. 1998
Amperometric	Nitrate reductase	4.8	5.0–500	12	Kirstein et al. 1999
Amperometric	Bacteria	1.0	1.0–50.0	<30	Takayama 1998
Amperometric	Bacteria	3.6	4.0–86.0	30	Larsen et al. 2000
Amperometric	Bacteria	5.0	--	45	Meyer et al. 2002

provides a measurable current (proportional to the analyte) when a potential is applied across two electrodes. More indepth descriptions and principals are provided by D' Orazio (2003).

Biosensors developed for nitrate determination are generally based on nitrate reductase which is multiredox centred and catalyzes the conversion of nitrate to nitrite. In fact nitrate reductases are a whole family of molybdenum-containing enzymes with various structures depending on their derivation, i.e. eukaryotic, prokaryotic etc (for a detailed review see Tavares et al. 2006).

The usage of nitrate reductase in biosensory detection of nitrate is limited because of two main reasons. Firstly, nitrate reductases are generally not particularly active. In addition, conventional biomolecule immobilization procedures such as cross-linking, covalent binding, entrapment in gels or membranes induce a marked deactivation of the immobilized enzyme. As a consequence, the biosensors using this enzyme display weak sensitivities for nitrate. Secondly, the redox centres of nitrate reductase are deeply embedded in the protein structure, thus preventing a direct electron transfer to the electrode (Tavares et al. 2006). Small and mobile redox mediators such as methyl viologen are therefore used as electron shuttles between the immobilized enzyme and the electrode surface (Da Silva et al. 2004; Sohail and Adeloju 2008; Dinçkaya et al. 2010).

The first amperometric biosensor for nitrate determination was developed by Cosnier et al. (1994). The nitrate reductase was immobilized on a glassy carbon electrode coated with a polypyrrolic film containing viologen groups. The nitrate was assayed by potentiostating the electrode at -0.7 V vs. The biosensor showed a low detection limit of 0.4 µM. However, its activity dropped by 50% upon storage in the dry state at 5°C for 7 d (Cosnier et al. 1994).

Electrochemical biosensors for nitrate determination based on the immobilization of nitrate reductase within two different redox matrixes have been developed, one being a pure organic polymer and the other a clay-polypyrrole composite. Nitrate reductase is co-adsorbed with a poorly soluble pyrrole-viologen on an electrode surface and immobilized by physical entrapment during the subsequent electropolymerization of the adsorbed monomers (*Bioelectrode 1*). One bioelectrode design for the immobilization and electrical wiring of nitrate reductase is based on the entrapment of nitrate reductase in a laponite clay gel acting as an inorganic template for the electropolymerization of a new water-soluble pyrrole-viologen (**Bioelectrode 2**). The resulting performance for nitrate determination have been investigated by cyclic voltammetry. The sensitivity and detection limit of **Bioelectrode 2** for nitrate are 94.7 mA M^{-1} cm^{-2} and 0.5 μM, respectively (Da Silva et al. 2004).

Enzyme electrodes in several configurations including the enzyme nitrate reductase are applied in the bioelectrocatalyzed reduction of nitrate to nitrite. One enzyme electrode configuration includes the physical immobilization of nitrate reductase in a polythiophene-bipyridinium matrix formed by electropolymerization of 1-methyl-1'- (3-thiophene-3-yl) –pentyl - 4,4' - bipyridinium-bis- hexafluorophosphate (Willner et al.1992). The bipyridinium units of the polymer act as electron mediators between the electrode and the enzyme. The concentration of bioelectrogenerated nitrite depends on the concentration of nitrate in the electrolyte solution. A calibration curve representing the rate of nitrite formation as a function of nitrate concentration in the electrolysis cell allows quantitative analysis of nitrate in aqueous solutions.

Two other enzyme-electrode configurations include copolymerization of thiophene-bipyridinium and thiophene-3-acetic acid, onto a gold electrode and layered deposition of two polymers composed of thiophene-bipyridinium and thiophene-3-acetic acid onto the electrode (Willner et al. 1992). In another study on biosensors, the immobilization of nitrate reductase was performed by entrapment in a laponite clay gel and cross-linking by glutaraldehyde. In the presence of nitrate and methyl viologen, a catalytic current appeared at –0.60 V illustrating the enzymatic reduction of nitrate into nitrite via the reduced form of the freely diffusing methyl viologen (Cosnier et al.2008). Figure 14.1 shows the general detection principle of nitrate reductase based biosensors in the presence of methyl viologen (MV) as a mediator (Quan et al. 2005).

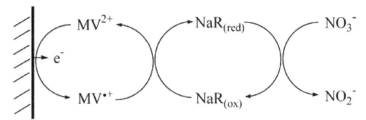

Figure 14.1. Detection principle of nitrate reductase based biosensor in the presence of methyl viologen as a mediator (Quan et al. 2005).

Viologens have been widely used as a mediatoric agent for amperometric nitrate biosensor based on nitrate reductase. Cosnier et al. (1997) have synthesized and electrochemically characterized a new series of amphiphilic viologens functionalized by one or two pyrrole groups. This functionalization has allowed the preparation of poly(pyrrole-viologen) films by oxidative electropolymerization. The electropolymerization of nitrate reductase + amphiphilic pyrrole viologen mixtures, previously adsorbed on electrode surfaces, provides the immobilization of the enzyme in N-substituted viologen polypyrrole films generated in aqueous solution. Furthermore, the electrogenerated redox polymers simultaneously entrap the enzyme and establish an electrical communication between nitrate reductase and the electrode surface. Since nitrate reductase catalyzes the reduction of nitrate to nitrite (mediated by the viologen redox couple), the resulting bioelectrodes potentiostated at −0.7 V act as a biosensing device for the amperometric detection of nitrate (Cosnier et al. 1997).

Two amperometric biosensors were prepared by immobilizing nitrate reductase derived from yeast on a glassy carbon electrode (GCE, $d = 3$ mm) or screen-printed carbon paste electrode (SPCE, $d = 3$ mm) using a polymer (poly(vinyl alcohol)) entrapment method by Quan et al . The sensors could directly determine the nitrate in an unpurged aqueous solution with the aid of an appropriate oxygen scavenger. The nitrate reduction reaction driven by the enzyme and an electron-transfer mediator, methyl viologen, at −0.85 V (GCE vs Ag/AgCl) or at −0.90 V (SPCE vs Ag/AgCl) exhibited no oxygen interference in a sulphite-added solution. Figure 14.2 shows the fabrication process of screen-printed carbon paste electrodes.

The sensitivity, linear response range, and detection limit of the sensors based on GCE were 7.3 nA/μM, 15–300 μM ($r^2 = 0.995$), and 4.1 μM (S/N = 3), respectively, and those of SPCE were 5.5 nA/μM, 15-250 μM ($r^2 = 0.996$), and 5.5 μM (S/N = 3), respectively. The disposable SPCE-based biosensor with a built-in well- or capillary-type sample cell provided high sensor-to-sensor reproducibility (RSD < 3.4% below 250 μM) and could be used for more than 1 mon in normal room-temperature storage conditions (Quan et al. 2005).

Badea et al. have proposed the use of a sensor system based on platinum electrodes modified with a cellulose acetate membrane or with poly(1,8-diaminonaphthalene) film for rapid amperometric detection of nitrates and nitrites in water by batch and flow injection analysis. They reported that the biosensor could be extended easily to the analysis of nitrites and nitrates in various types of samples such as food, soils, vegetables and fertilizers (Badea et al. 2001).

Moretto et al. (1998) have developed an electrochemical nitrate biosensor based on the ultrathin-film composite membrane concept. The composite membrane is prepared by electropolymerization of a thin anion-

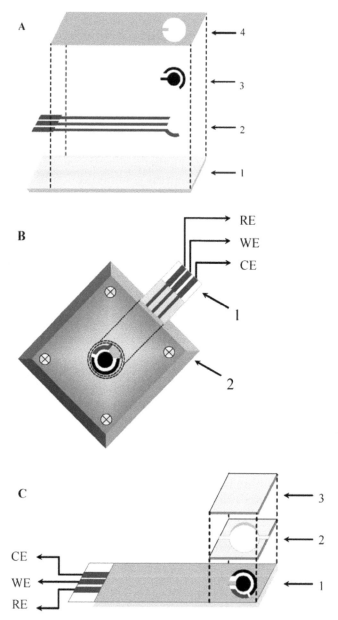

Figure 14.2. Screen-printed carbon paste electrode (Quan et al. 2005). **(A)** Fabrication process: 1, polyester substrate; 2, silver connectors and electrode site; 3, carbon paste electrodes; and 4, insulator. **(B)** SPCE with sample well: 1, SPCE; 2, 100-μL sample well block. and **(C)** SPCE with capillary sample cell: 1, SPCE; 2, capillary fluidic path formed on double-sided tape; and 3, cover plate. Key : SPCE=Screen-Printed Carbon Paste Electrode. RE: Reference electrode, WE: Working electrode and CE: Counter electrode.

permselective coating of 1-methyl-3-(pyrrol-1-ylmethyl)pyridinium across the surface of a microporous support membrane. Figure 14.3 shows a schematic diagram of the ultrathin-film composite membrane based nitrate sensor (Moretto et al. 1998). This film separates the analyte solution from an internal sensing solution which contains the enzyme nitrate reductase and an electrocatalyst (methyl viologen). The ultrathin anion-permselective film prevents loss of these components from the internal sensing solution, yet allows the analyte, NO_3^-, to enter the internal solution. The nitrate is reduced enzymatically, yielding the oxidized form of the enzyme, which is reduced again by the electrogenerated methyl viologen radical cation (Moretto et al. 1998). The nitrate concentration in the analyte solution is proportional to the catalytic methyl viologen reduction current. This sensor shows good sensitivity to nitrate, with a detection limit of 5.4 µM and a dynamic range which extends up to 100 µM NO_3^- (Moretto et al. 1998).

Ferreyra and Solis (2004) have developed an amperometric nitrate biosensor based on the enzyme-catalyzed reduction of nitrate with nitrate reductase and phenosafranin in solution as the enzyme regenerator.

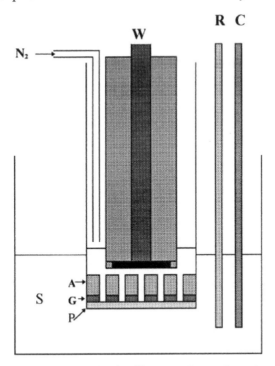

Figure 14.3. Schematic diagram of the ultrathin-film composite membrane-based nitrate sensor (Moretto et al. 1998). **(A)** anopore membrane, **(G)** gold film, **(P)** poly 1-methyl-3-(pyrrol-1-ylmethyl)pyridinium tetrafluoborate (poly-MPP), **(W)** working electrode, **(R)** reference electrode, **(C)** counter electrode, **(S)** sample solution.

This works at lower potentials compared to the more common methyl viologen mediator (Ferreyra and Solis 2004). When nitrate reductase, phenosafranin and substrate were together, cyclic voltammograms showed the enzyme-catalyzed reduction of nitrate, with relatively slow kinetics. Analytical parameters for the enzyme-modified electrode in the presence of phenosafranin for the determination of nitrate content in water were assessed, including a recovery assay for nitrate added to a river water sample. The excellent results obtained from the recovery assay of nitrate in river water indicated the absence of significant interferences in this matrix (Ferreyra and Solis 2004).

Glazier et al. (1998) have proposed an amperometric biosensor based on nitrate reductase used for the determination of nitrate in fertilizer and water samples. The biosensor consisted of nitrate reductase held by a dialysis membrane onto a Nafion-coated glassy carbon electrode. Methyl viologen was allowed to absorb into the Nafion layer, which acted as a reservoir for the electron mediator. The assays conducted with this electrode compared well with colorimetric and potentiometric assays of the same samples (Glazier et al. 1998).

Kirstein et al. (1999) co-adsorbed nitrate reductase enzyme (1.7.99.4) along with a selected mediator onto a graphite electrode. The screening of 19 mediators showed that azure A gave the highest current density, the short response time, a low detection limit, and long lifetime. The response time of the sensors was 12 s, the nitrate detection limit was 4.8 μmol l^{-1} and the half lives of the sensors were of the order of 1 to 2 d (Kirstein et al. 1999).

Nitrate Determination with Microbial Biosensors

Because of the instability and problems in the electron carrying features of nitrate reductase Kobos et al. (1979) used intact bacterial cells as a biocatalyst in the construction of biosensors for nitrate. Bacteria were spread onto a gas permeable membrane of an ammonia electrode and were held in place by means of a dialysis membrane. Nitrate was reduced to ammonia by a two-step process involving the enzymes nitrate and nitrite reductase's contained in the bacterial cells. The response of the sensor was linear over the range of 0.01–0.8 mM nitrate. The response time was around 6 min. The bacterial electrode retained its full activity over 2 wk provided that it was stored in growth medium when not in use (Kobos et al. 1979).

A microscale bacterial sensor for online determination of nitrate in marine sediments was developed by Larsen et al. (1997). The principle of detection of nitrate was based on bacterial reduction of NO_3 to N_2O and subsequent electrochemical detection of the produced N_2O. The sensor showed a low detection limit of 1 μM and short response time, less than 60 s. A major disadvantage of the sensor is a limited lifetime (2-4 d) primarily due to decreasing catalytic activity of the silver cathode in the N_2O electrode (Larsen et al. 1997).

Takayama (1998) developed a whole-cell microbial biosensor based on *Paracoccus denitrificans* as the biocatalyst for the determination of nitrate. The denitrifying bacteria were immobilized under a dialysis membrane on the surface of a graphite electrode. Durohydroquinone was used as an electron transfer mediator between the electrode and respiratory chain of the bacterial cell. The response of the sensor decreased gradually with continuous use, to 60% of the original response after 5 d. The biosensor was susceptible to interference from nitrites but when hexacyanoferrate (II), $Fe(CN)_6^{4-}$, was used as a mediator, the electrode exhibited nitrite selectivity (Takayama, 1998).

Larsen et al. (2000) used a microscale biosensor for *in-situ* monitoring of nitrate in an activated sludge plant. The biosensor was based on the diffusion of nitrate through a tip membrane into a dense mass of bacteria which converted the ions into nitrous oxide for subsequent electrochemical detection. The only interfering substance was nitrite. Due to the short response time and low detection limit, the proposed nitrate biosensor offered superior resolution in terms of both time and concentration as compared to other techniques for *in-situ* monitoring of wastewater. The lifetime of the applied biosensor was limited to a maximum of 5 d due to the detachment of the ion permeable membrane that separated the bacteria incorporated into the sensor from the wastewater (Larsen et al. 2000).

KEY FACTS OF NITRATE AND GOITRE

- Beside its use as fertilizers in agriculture and also as an anti-microbial agent in the food industry, nitrate is the end oxidative product of nitric oxide and other nitrogen species and hence constitutes a well-known environmental contaminant.
- Nitrate contamination is largely encountered in ground and stream water.
- Nitrate affects the efficiency of the sodium iodide symporter which allows sodium and iodide to enter thyroid cells.
- Reduced iodide uptake means that there is insufficient iodine for hormones thyroxine (also termed T4) and triiodothyronine (also termed T3), each molecule of which contains three and four molecules of iodine respectively.
- Thyroxine is more metabolically active than triiodothyronine but reduced circulating levels of both will stimulate the pituitary to produce thyroid stimulating hormone.

- In turn, thyroid stimulating hormone will cause the thyroid glands to enlarge in an endeavour to produce more thyroxine and triiodothyronine.
- The enlarged thyroid gland is called a goitre (also spelt as goiter).
- The goitre is the visible façade of iodine deficiency diseases.
- Reduced iodine availability and thus iodine deficiency diseases can be reflected in a variety of medical problems depending on the severity of the deficiency and if it affects pregnant women, children or adults.
- In pregnancy iodine deficiency diseases and conditions include abortions, stillbirths, congenital abnormalities, increased prenatal and infant mortality.
- In children iodine deficiency diseases and conditions will cause mental deficiency, psychomotor defects, impaired school performance.
- In adults iodine deficiency diseases and conditions will cause mental fatigue and lethargy.

DEFINITIONS

- *Amperometry*: It is a subclass of voltammetry in which the electrode is held at a constant potential for various lengths of time.
- *Cyclic voltammetry*: Cyclic voltammetry is an electrolytic method that uses electrodes and an unstirred solution so that the measured current is limited by analyte diffusion at the electrode surface. The electrode potential is ramped linearly to a more negative potential, and then ramped in reverse back to the starting voltage. The forward scan produces a current peak for any analytes that can be reduced through the range of the potential scan.
- *Genotoxicity*: It describes a deleterious action on a cell's genetic material affecting its integrity. Genotoxic substances are known to be potentially mutagenic or carcinogenic, specifically those capable of causing genetic mutation and contributing to the development of tumours.
- *Mediator:* An organic or inorganic substance that is capable of smooth electron transfer between an electrode and a biocatalyst.
- *Methemoglobinemia:* A disorder characterized by the presence of a higher than normal level of methemoglobin (metHb) in the blood. Methemoglobin is a form of the oxygen-carrying protein haemoglobin, in which the iron in the heme group is in the Fe^{3+} (ferric) state, not the Fe^{2+} (ferrous) of normal haemoglobin.
- *Methylviologen:* A mediator that has been used to shuttle electrons to nitrate and nitrite reductases.

- *Nitrate Reductase*: An oxidoreductase enzyme which contains molybdenum and reduce nitrate (NO^{3-}) to nitrite (NO^{2-}).

SUMMARY POINTS

- Nitrate is an important analyte in such diverse materials as fertilizers, foods, livestock feeds, wastewater, and drinking water. It is a widespread contaminant of groundwater and surface waters and is a potential human health threat, especially to infants, causing the condition known as methemoglobinemia.
- Nitrate is able to interfere with iodine metabolism causing goitre.
- Most amperometric biosensors developed for nitrate determination are based on the enzyme nitrate reductase but there are a few microbial biosensors.
- A large number of dyes and other redox species have been examined for their ability to act as mediators for nitrate reductase in amperometric electrodes.
- Use of biosensors nitrate has been determined for environmental monitoring especially for spring or groundwater samples.
- There are a few biosensor studies for nitrate determination in food samples.
- There is a need for researchers to develop new materials and techniques for more specific, sensitive, reliable, fast and routine analysis of nitrate.

ABBREVIATIONS

GCE	:	Glassy Carbon Electrode
IUPAC	:	International Union of Pure and Applied Chemistry
MV	:	Methyl Viologen
NaR	:	Nitrate Reductase
SCE	:	Standard Calomel Electrode
SPCPE	:	Screen-Printed Carbon Paste Electrode

REFERENCES

Amine A and G Palleschi. 2004. Phosphate, Nitrate, and Sulfate Biosensors. Anal Lett 37: 1–19.

Badea M, A Amine, G Palleschi, D Moscone, G Volpe and A Curulli. 2001. New electrochemical sensors for detection of nitrites and nitrates. J Electroanal Chem 509: 66–72 .

Byfield MP and RA Abuknesha. 1994. Biochemical aspects of biosensors. Biosens Bioelectron 9: 373–400.

Cosnier S, C Innocent and Y Jouanneau. 1994. Amperometric determination of nitrate via nitrate reductase immobilized and electrically wired at the electrode. Anal Chem 66: 3198–3201.

Cosnier S, B Galland and C Innocent. 1997. New electropolymerizable amphiphilic viologens for the immobilization and electrical wiring of a nitrate reductase . J Electroanal Chem 433: 113–119.

Cosnier S, S Da Silva, D Shan and K Gorgy. 2008. Electrochemical nitrate biosensor based on poly(pyrrole–viologen) film–nitrate reductase –clay composite. Bioelectrochemistry 74: 47–51.

Da Silva S, D Shan and S Cosnier. 2004. Improvement of biosensor performances for nitrate determination using a new hydrophilic poly(pyrrole-viologen) film, Sens Act B: Chemical 103: 397–402.

Davis J, MJ Moorcroft, SJ Wilkins, RG Compton and MF Cardosi. 2000. Electrochemical detection of nitrate at a copper modified electrode under the influence of ultrasound. Electroanalysis 12: 1363–1367.

Deng-sheng X, L Yin, Z Chun-mei,Y Sheng-hui and W Song-lin. 2006. Antimicrobial effect of acidified nitrate and nitrite on six common oral pathogens *in vitro*. Chin Med J 119: 1904–1909.

Devi S and A Townshend. 1989. Determination of nitrate by flow-injection analysis with an on-line anion-exchange column Anal Chim Acta 225: 331–338.

D' Orazio P. 2003. Biosensors in clinical chemistry. Clinica Chimica Acta 334: 41–69.

Dinçkaya E, E Akyilmaz, MK Sezgintürk and FN Ertaş . 2010. Sensitive Nitrate Determination in Water and Meat Samples by Amperometric Biosensor. Prep Biochem Biotechnol 40: 119–128.

Englmaier P. 1983. Nitrate analysis by gas-liquid chromatography using the nitration of 2,4-dimethylphenol in sulphuric acid J Chromatogr A 270: 243–251.

Ferreyra NF and VM Solís. 2004. An amperometric nitrate reductase –phenosafranin electrode: kinetic aspects and analytical applications. Bioelectrochemistry 64: 61–70.

Gatseva, PD and MD Argirova. 2008. High-nitrate levels in drinking water may be a risk factor for thyroid dysfunction in children and pregnant women living in rural Bulgarian areas. Int. J. Hyg. Environ. Health 211: 555–559.

Gatseva PD and MD Argirova. Iodine nutrition in Bulgaria: results from the national strategy for elimination of iodine deficiency disorders in Bulgaria Epidemiologic studies on risk population groups. pp 1169–1177. In: VR Preedy, G Burrow and RR Watson [eds.] 2009. Comprehensive Handbook of Iodine Elsevier San Diego USA.

Glazier SA, ER Campbell and WH Campbell. 1998. Construction and Characterization of Nitrate Reductase-Based Amperometric Electrode and Nitrate Assay of Fertilizers and Drinking Water. Anal Chem 70: 1511–1515.

Hedenmo M, A Narvaez, E Dominguez and I, Katakis. 1997. Improved mediated tyrosinase amperometric enzyme electrodes. J Electroanal Chem 425: 1–11.

Högg G, G Steiner and K Cammann. 1994. Development of a sensor card with integrated reference for the detection of nitrate Sens Act B (18-19): 376–379.

Höring H. 1992. Der Einfluss von Umweltchemicalien auf die Schilddrüse. [The influence of environmental chemicals on the thyroid.] Bundesgesundheitsblatt 35: 194–197.

King MD, EM Dick and WR Simpson. 2000. A new method for the atmospheric detection of the nitrate radical (NO_3). Atmos Environ 34: 685–688.

Kirstein D, L Kirstein, F Scheller, H Borcherding, J Ronnenberg, S Diekmann and P Steinrucke. 1999. Amperometric nitrate biosensors on the basis of *Pseudomonas stutzeri* nitrate reductase. J Electroanal Chem 474: 43–51.

Kobos RK , DJ Rice and DS Flournoy. 1979. Bacterial membrane electrode for the determination of nitrate. Anal Chem 51: 1122–1125.

Kotte H, B Gruendig, KD Vorlop, B Strehlitz and U Stottmeister. 1995. Methylphenazonium-Modified Enzyme Sensor Based on Polymer Thick Films for Subnanomolar Detection of Phenols. Anal Chem 67: 65–70.

Larsen LH, T Kjaer. and NP Revsbech. 1997. A microscale NO-biosensor for environmental applications. Anal Chem 69: 3527–3531.

Larsen LH, LR Damgaraad, T Kjaer, T Stenstrom, A Lynggaard-Jensen and NP Revsbech. 2000. Fast responding biosensor for on-line determination of nitrate/nitrite in activated sludge. Wat Res 34: 2463–2468.

Madsen BC. 1981. Utilization of flow injection with hydrazine reduction and photometric detection for the determination of nitrate in rain-water. Anal Chim Acta 124: 437–441.

Manassaram DM, LC Backer, R Messing, LE Fleming, B, Luke. and CP Monteilh. 2010. Nitrates in drinking water and methemoglobin levels in pregnancy: a longitudinal study, Environ Health 9: 60–72.

Meyer RL, T Kjaer and NP Revsbech. 2002. Nitrification and denitrification near a soil-manure interface studied with a nitrate-nitrate biosensor. Soil Sci Soc Am J 66: 498–506.

Mirvish SS. The significance for human health of nitrate, nitrite, and n-nitroso compounds. In: Bogardi et al. [eds.] 1991. Nitrate Contamination: Exposure, Consequence, and Control. NATO ASI Series, Vol. G30. Springer Verlag. Berlin.

Moretto LM , P Ugo, M Zanata, P Guerriero and CR Martin. 1998. Nitrate biosensor based on the ultrathin-fllm composite membrane concept. Anal Chem 70: 2163–2166.

Mozaz SR, MJ Lopez de Alda. and D Barceló. 2006. Biosensors as useful tools for environmental analysis and monitoring, Anal Bioanal Chem 386: 1025–1041.

Nistor C and J Emnéus. 1999. Bioanalytical tools for monitoring polar pollutants. Waste Manag 19: 147–170.

Noufi M, C Yarnitzky and M Arie. 1990. Determination of nitrate with a flow-injection system combining square-wave polarographic detection with on-line deaeration. Anal Chim Acta 234: 475–478.

Parellada, J, A Narvaez, MA Lopez, E Dominguez, JJ Fernandez, V Pavlov and I Katakis. 1998. Amperometric immunosensors and enzyme electrodes for environmental applications. Anal Chim Acta 362: 47–57.

Prachal JT and XT Gregg. Red cell enzymopathies. pp 561–575 In: R Hoffman, EJ Banz and SJ Shattil [eds.] 2000. Hematology Basic Principles and Practice. Churchill Livingstone, New York, USA.

Quan D, JH Shim, JD Kim, HS Park, GS Cha and H Nam. 2005. Electrochemical determination of nitrate with nitrate reductase -immobilized electrodes under ambient air. Anal Chem 77: 4467–4473.

Sadeq M, CL Moe, B Attarassi, I Cherkaoui, R ElAouad and L Idrissi. 2008. Drinking water nitrate and prevalence of methemoglobinemia among infants and children aged 1–7 years in Moroccan areas. Int J Hygiene Environ Health 211: 546–554.

Sah RN. 1994. Nitrate-nitrogen determination: a critical review. Commun. Soil Sci Plant Anal 25: 2841–2869.

Skladal P, M Fiala and J Krejci. 1996. Detection of pesticides in the environment using biosensors based on cholinesterases. Int J Env Anal Chem 65: 139–148.

Sohail M and SB Adeloju. 2004. Electroimmobilization of nitrate reductase and nicotinamide adenine dinucleotide into polypyrrole films for potentiometric detection of nitrate, Sens. Act B: Chemical 133: 333–339.

Solak AO, P Gulser, E Gokmese and F Gokmese. 2000. A new differential pulse voltammetric method for the determination of nitrate at a copper plated glassy carbon electrode. Turk. Mikrochim. Acta 134: 77–82.

Takayama K. 1998. Biocatalyst electrode modified with whole-cells of for the determination of nitrate. Bioelectroch Bioener 45: 67–72.

Tavares P, AS Pereira, JJ Moura and I Moura. 2006. Metalloenzymes of the denitrification pathway. J Inorg Biochem 100: 2087–2100.

Thévenot DR, K Toth, RA Durst and GS Wilson. 2001. Electrochemical biosensors: recommended definitions and classification. Biosens Bioelectron 16: 121–131.

Wayne RP, I Barnes, P Biggs, JP Burrows, CE Canosa-Mas, J Hjorth, G Le Bras, GK Moortgat, D Perner, G Poulet, G Restelli and H Sidebottom. 1991. The nitrate radical: Physics, chemistry, and the atmosphere. Original Research Article Atmospheric Environment. Part A. General Topics 25: 1–203.

Willner I, E Katz, N Lapidot and P Bauerle. 1992. Bioelectrocatalysed reduction of nitrate utilizing polythiophene bipyridinium enzyme electrodes. Bioelectrochem Bioener 29: 29–45.

Wilson JS. 2005. Sensor Technology Handbook, Elsevier, Amsterdam/Boston.

Wright RO, WJ Lewander and AD Woolf. 1999. Methemoglobinemia: etiology, pharmacology, and clinical management. Ann Emerg Med 34: 646–656.

Yoshizumi K and K Aoki. 1985. Determination of nitrate by a flow system with a chemiluminescent NO$_x$ analyzer. Anal Chem 57: 737–740.

Zuther F and K Cammann. 1994. A selective and long-term stable nitrate sensor. Sens Act B (18-19): 356–358.

15

Antibody-based Biosensors for Small Environmental Pollutants: Focusing on PAHs

Candace R. Spier,[1,a] Michael A. Unger[1,b] and
Stephen L. Kaattari[1,c,*]

ABSTRACT

Polycyclic aromatic hydrocarbons (PAHs) are classified as priority pollutants because of their high toxicity and suspected carcinogenicity. Traditional methods for monitoring PAHs in the environment are time-consuming, labor-intensive, and expensive. Biosensors are presently being developed for the environmental monitoring of these molecules. One means of accomplishing this is to couple anti-PAH antibody molecules with electronic transducers to provide for the rapid quantification of PAHs. Although not immunogenic on their own, antibodies to PAHs can be produced by coupling advances in synthetic chemistry and immunology. Once specific antibodies have been created, a variety of transduction methods (i.e. optical, piezoelectric, and electrochemical) can be employed to quantify this signal. Full descriptions of these PAH immunosensors are provided,

[1]Virginia Institute of Marine Science, College of William & Mary, PO Box 1346, Gloucester Point, Virginia 23062-1346 USA.
[a]E-mail: cspier@vims.edu
[b]E-mail: munger@vims.edu
[c]E-mail: kaattari@vims.edu
*Corresponding author

List of abbreviations after the text.

concentrating on the speed, sample volume, and detection methods employed. Advantages and drawbacks of these technologies are compared to traditional PAH analytical methods.

INTRODUCTION

Environmental Concern Associated with PAHs

PAHs are natural constituents of petrochemical products and can be produced during the incomplete combustion of organic materials. PAHs enter the aquatic environment via natural oil seeps, oil spills, atmospheric deposition, industrial effluents, runoff in storm events, and other means (ATSDR 1995) where they can cause potentially harmful effects to humans and wildlife (US EPA 2010). Some PAHs are designated as priority pollutants by the US EPA and the EU due to high toxicity and suspected carcinogenicity (i.e. BaP) and have regulated limits as low as 0.2 µg/l in drinking water (ATSDR 1995; US EPA 2009).

Traditional Laboratory-based Analytical Methods for PAH Quantification

Most standard analytical methods for PAHs employ chemical extraction, followed by GC-MS or HPLC. Federal agencies such as the US EPA routinely use these methods in their guidelines for environmental sample analysis (e.g. Method 8270). However, as these methods are time-consuming and labor-intensive, we currently possess no economical options for frequent monitoring.

Biosensors as Alternative Analytical Methods for PAH Quantification

Biosensors are being developed as alternatives or supplements to classical analytical chemistry (see review by Rodriguez-Mozaz et al. 2006). This goal involves making the technology rapid, user-friendly, portable, sensitive, accurate, reliable, and inexpensive (Van Emon and L. Gerlach 1998). Biosensors also cost less than traditional analytical techniques, require fewer reagents, and they possess faster turnaround times enabling higher sample throughput. On-site technologies require minimal use of power, less dangerous reagents, and produce less toxic waste. Advances in manufactured materials and miniaturization are facilitating portability and on-site operation of biosensors (Rodriguez-Mozaz et al. 2006).

Biosensors are defined as analytical devices that integrate a biologically-derived receptor with a transducing device. A variety of biorecognition elements such as; enzymes, whole cell receptors, DNA, and antibodies have been used. These are then linked with a transducer, such as an electrochemical, optical, piezoelectrical, or thermal device, which converts the biorecognition event into a quantifiable signal (Nakamura and Karube 2003). Fluorophores or enzymes are often used to augment or generate detectable signals. The transduced signal is then recorded as a digital output, typically via a computer using dedicated software (Fig. 15.1).

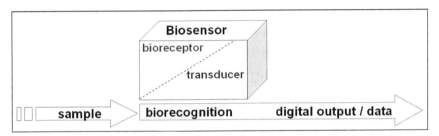

Figure 15.1. Schematic of a generalized biosensor. The bioreceptor and the transducer are linked such that the transducer quantifies the biorecognition event occurring between the sample and the bioreceptor. The arrows indicate user interaction, in which a sample may require some manipulation prior to introduction to the biosensor.

Ideally, biosensors should operate automatically with a user simply introducing the sample and the sensor producing a digital result. The sample may undergo user manipulation prior to introduction (as shown by an incomplete arrow in Fig. 15.1), then the biosensor is responsible for detecting the analytes of interest and translating this biorecognition event into a digital output. The resulting data is then compared to a standard curve to estimate the concentration of analyte in the sample.

Unfortunately, the development of environmental biosensors is still in its infancy with automation and portability often not being accomplished. Given the proper modifications in electrical and mechanical engineering, the described tools show great potential for autonomous operation, and are therefore included in this chapter. We focus on biosensors that use antibodies for the detection of PAHs, specifically 1) the production of the antibodies and 2) the functionality of the entire sensor system. Numerous other immunoassays exist for PAHs, such as enzyme-linked immunosorbent assays, but these have been reviewed previously (Fahnrich et al. 2002).

Development of Antibodies to Small Molecules

The success of immunosensors is reliant on the antibody molecule itself. Antibodies are commonly employed biorecognition elements because they possess exquisite molecular specificity and affinity for their target. Serum antibodies are the products of discrete antibody secreting B cells, each recognizing the analyte (antigen) with a distinctive binding site (Pressman and Grossberg 1968). Antibodies in this form have been employed for clinical diagnosis since the 1930's. However, with the advent of immortal B cell lines (hybridomas), a single mAb can be produced in virtually limitless quantities (Harlow and Lane 1988). Besides being economical, this also eliminates the great heterogeneity of specificities that burdens serum antibody (polyclonal) analysis (Fig. 15.2). Fortunately, for small molecule analysis, the actual structure recognized by an antibody is approximately the size of the analytes we wish to target. However, in order to induce an antibody response, this small structure must be part of a larger immunogenic molecule (carrier), while preserving the characteristic structural features of the target analyte. The physical linking of the analyte to the carrier can provide additional sites that may elicit antibodies. The use of mAb technology permits the facile selection of only those antibodies that specifically bind the analyte. Thus the goal of developing highly specific, economical biosensor reagents are uniquely met with mAbs (Fig. 15.2).

In the production of analyte-immunogens, a wide variety of conjugation techniques are available to conjugate small analytes, such as PAHs to proteins (Table 15.1 and Hermanson 1996). Figure 15.3 provides an example of a target analyte, functionalized analyte, and analyte-protein conjugate.

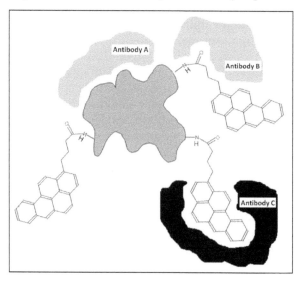

Figure 15.2. Antibody recognition diversity of the immune response. For our purposes, the desired type of antibody is "Antibody C" that solely recognizes the analyte. "Antibody A" is shown as recognizing the carrier molecule, while "Antibody B" binds the linking arm of the conjugate (Adapted from Vanderlaan et al. 1988). The amorphous shapes represent protein structures.

Table 15.1. Summary of antibodies to PAHs.

Antibody	Species	Type	Derivatized PAH analyte	Reference
BAP-13	mouse	monoclonal	benzo(a)pyrenyl-1-butyric acid	Suchanek et al. 2001, Scharnweber et al. 2001
–[1]	sheep	polyclonal	np[2]	Gift from Abuknesha used in Goryacheva et al. 2007
–[1]	rabbit	polyclonal	benzo(a)pyrene-6-isocyanate	Vo-Dinh et al. 1987
7B2.3	mouse	monoclonal	dibenzothiophene-4-acrylic acid	Spier et al. 2009
–[1]	rabbit	polyclonal	4-(1-pyrenyl)butyric acid	Meisenecker et al. 1993
ANT 16B4	mouse	monoclonal	hemisuccinate of 9-hydroxymethylanthracene	Quelven et al. 1999
4D5 and 10C10	mouse	monoclonal	benzo(a)pyrene-6-isocyanate	Gomes and Santella 1990
22F12	mouse	monoclonal	γ-(7-benzo[a]pyrenyl)-butyric acid	Matschulat et al. 2005[3]
–[1]	rabbit	polyclonal	9,10-Dihydrobenzo[a]pyren-7(8H)-one-7-(O-carboxymethyl)-oxime	Roda et al. 1991
anti-BaP-BSA	mouse	monoclonal	4-oxo-4-(benzo[a]pyrene)butyrate	Miura et al. 2003
anti-fluorene	rabbit	polyclonal	fluorenyl-methylazide. oxy-carbonyl-o-succinimide	Professor F. Le Goffic of ENSCP, Paris France
anti-pyrene	rabbit	polyclonal	pyrene derivative	Used in Pérez and Barceló 2000

Commercially available anti-PAH antibodies

Source	Antibody	Species	Type	Reference
Novus Biologicals	BAP-13	mouse	monoclonal	see above
Abcam	–	rat	polyclonal	"
	BAP-13	mouse	monoclonal	"
ExBio	BAP-13	mouse	monoclonal	"
Abraxis		mouse	np	"
Santa Cruz Biotechnology, Inc.	4D5 & 10C10	mouse	monoclonal	"
SDIX	α-Phen-33	mouse	monoclonal	Used in Moore et al. 2004, Fähnrich et al. 2003
	–	rabbit	polyclonal	

1. '–' no name designation
2. np = not provided
3. a total of 14 mAbs were examined in this paper

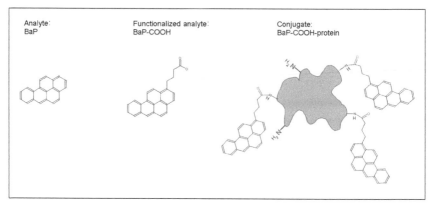

Figure 15.3. Examples of a target PAH analyte, a functionalized analyte, and an analyte-protein conjugate. The amorphous shape of the conjugate represents a protein molecule.

Anti-PAH Antibodies Presently in Scientific Literature

Antibodies are known to cross-react with molecules of similar structure (Nording and Haglund 2003), which presents a great challenge with PAH analysis. Therefore, mAbs are selected with the greatest specificity for the PAH of interest. This requires screening the mAbs against a broad panel of PAHs and other molecules (such as PCBs, NPs, etc.). Several anti-PAHs antibodies have been developed; albeit the rigor of screening of each antibody varied greatly (Table 15.1).

REVIEW OF PAH IMMUNOSENSORS

The most common assay format for the detection of small molecules is a competitive inhibition assay wherein a signal-generating antibody is blocked from binding to an antigen-coated surface by the presence of free analyte within a sample (Fig. 15.4). The antibody and sample are mixed and introduced to the antigen-coated surface. The resulting signal is inversely proportional to the amount of analyte in the sample. Transducers of this binding event vary and are organized according to type (electrochemical, optical, or piezoelectric). For clarification, a table describing the various types of transducers and the principles behind the measured values is provided (Table 15.2).

Electrochemical

Electrochemical devices measure electrical signals produced from a chemical reaction. (Electrochemical cells either produce electrical energy

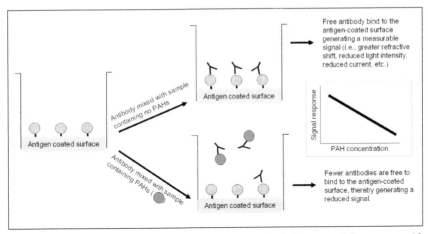

Figure 15.4. A typical competitive inhibition assay. An antibody (shown as 'Y') competes with free PAHs in a sample for binding to the surface. The resulting signal is inversely proportional to the PAH concentration.

Table 15.2. The types of transducers used in PAH immunosensors.

Transducer / Detection		Measurement
Electrochemical	Capacitance	Storage of electric energy (energy storage potential)
	Amperometric	Change in current while applying a fixed potential
Piezoelectric	Quartz crystal microbalance	Change in shape producing a measurable frequency
Optical	Fluorescence	Change in light intensity
	Infrared	Change in light intensity
	Surface plasmon resonance Reflectometric/interference	Change in refraction/reflection of light

from a chemical reaction, or produce a chemical reaction by introducing electrical energy.) When biomolecules are involved, the biorecognition event is used to generate the chemical reaction. For immunosensors, typically the antigen is immobilized onto the surface of an electrode, which ensures that the biorecognition event is occurring near the sensing (or working) electrode surface (see review by Grieshaber et al. 2008). However, as the binding of antibody to the antigen-coated electrode does not produce a substantial current, the antibody is conjugated to an enzyme which is capable of producing the needed chemical reaction. These enzyme-labeled antibodies are then introduced to the solution and will bind to the antigen-coated electrodes. In a second step, unbound antibodies are removed and a substrate is added to the antibody-bound enzymes to generate an electroactive product (usually hydrogen peroxide or oxygen), altering the electrical properties of the buffer. These electrical properties generate a measurable change in the current (amperometric) or the energy storage potential of a solution (capacitance).

Capacitance

Electrolytic capacitance measures the ability of the cell to hold an electrical charge. The capacitance is dependent on the dielectric properties of the electrodes. For sensors employing a biological molecule, the biological molecule serves as an insulating layer, such that the distance between the electrodes is influenced by the thickness of the biological layer bound on the electrode surface. For biorecognition, either the antibody or the antigen is immobilized onto the surface of the electrode. As the corresponding antibody binds the immobilized antigen or vice versa, a decrease in capacitance is measured proportional to thickness of the dielectric layer, which is due to the amount of biological molecule bound.

A capacitance immunosensor for detecting PAH-conjugates has been described by Liu and colleagues (1998). In this system, the PAH-BSA antigen or the anti-PAH mAb is immobilized onto the surface of gold plated electrodes. For both formats, 10 µl of sample solution were transferred to the electrode surface and allowed to incubate for 15 min before linear sweep voltammetry was employed detecting the cell's capacitance. With the antigen-immobilized format, specificity for the anti-PAH mAb was demonstrated as compared with the nonspecific binding of a mouse IgG molecule. With the mAb-immobilized format, antigen samples containing pyrene-BSA and BaP-BSA both bound to the mAb while a solution of BSA did not. This was the extent of the investigation. Nonetheless, it demonstrated this immunosensor's potential for PAH detection as it could employ competitive inhibition where the analyte conjugates could be blocked by PAHs.

Amperometric

In amperometric devices, the current associated with the reduction or oxidation of the enzymatic product is monitored while applying a fixed potential between the cell's electrodes. The current is proportional to the concentration of the electroactive product in solution, and therefore to the concentration of antibody binding to the electrode's surface. Consequently, the current is inversely proportional to the amount of PAH in solution, which inhibits antibody binding to the surface, dampening the current change. Importantly, amperometric transducers are less affected by sample turbidity, quenching, or interference from absorbing and fluorescing compounds which can confound optical biosensors.

Liu and colleagues (2000) have described an amperometric immunosensor for the detection of BaP. A pyrene derivative was immobilized onto a gold-plated electrode. BaP was detected through its competitive inhibition of anti-PAH antibodies binding to the pyrene-coated electrode. Presumably, the BaP and antibody solution were allowed to incubate with the analyte-modified electrode for 40 min, as this was noted as the amount of time provided for the antibody solution alone to incubate with the electrode. Then, the electrode was immersed in a solution containing the redox probe $Fe(CN)_6^{3-/4-}$, such that bound antibodies would increase the hydrophobic layer therefore decreasing the redox current. Cyclic voltammetry measurements were conducted and the authors report quantification of BaP from 25.2 to 1,260 µg/l.

Wei and colleagues (2009) described a similar electrochemical immunosensor for BaP employing ruthenium(II)polypyridine derivative as a redox label. Electrodes were coated with anti-PAH antibodies. Competitive inhibition was performed by depositing a 5 µl solution containing a PAH analyte (BaP, acenaphthene, fluorene, fluoranthene, pyrene, and 1-pyrenebutyric acid) and the redox-labeled-PAH onto the electrode and was incubated for 180 min. Cyclic voltammetry measurements were conducted and the LOD for BaP was 2.4 µg/l while the other PAHs were examined for cross-reactivity.

Recently, Wang and colleagues (2011) described another amperometric immunosensor for BaP. They increased the number of analyte (derivatized pyrene)-BSA antigens immobilized onto the electrode surface by employing dendritic-nanosilica to increase the surface area of the working electrode. This subsequently increased the number of antibodies taking part in the reaction. The electrode was incubated with a 5 µl mixture of anti-PAH mAb and BaP for 120 min for a competitive inhibition assay. An enzyme-labeled secondary antibody was then introduced to the electrode and incubated for 60 min, immersed in a hydroquinone solution, followed by exposure to a current. This sensor detected BaP with a linear range of 2.5 to 2,520 µg/l, and a LOD of 2.0 µg/l.

Guilbault's laboratory (Fähnrich et al. 2003) has explored the use of screen-printed electrodes for the electrochemical detection of PAHs. They concluded that the most sensitive and fastest assay format was yielded by immersing the phenanthrene-BSA coated electrode in a solution of mAb and an enzyme-labeled secondary antibody, with or without the PAH sample, and controls for 90 min. The electrode was rinsed and a +300mV potential was applied. After the current reached a flat baseline, a substrate was added, which produced a change in the current. This method detected phenanthrene from 0.5 to 45 µg/l with a LOD of 0.8 µg/l. Specificity analysis was conducted, resulting in varying degrees of cross-reactivities with the 16 EPA priority pollutant PAHs, most notably, the 3- and 4-ring PAHs. Lastly, spiked water samples (river and tap) were evaluated and the system showed slightly decreased sensitivity.

In a subsequent report, Guibault's laboratory (Moore et al. 2004) refined this method for analysis of a single drop (10 µl) of a spiked environmental sample (sea, river, tap, and mineral water). This method utilizes the displacement of pre-adsorbed mAbs onto the antigen-coated electrode surface by the introduction of phenanthrene in the sample. The electrode was immersed with the sample solution for 30 min, then washed and incubated with an enzyme-labeled secondary antibody for 30 min. Following another wash step, the substrate was added and linear sweep voltammetry was employed to detect the current produced at the working electrode. The linear range of detection for phenanthrene was 2 to 100 µg/l.

Piezoelectric

Quartz crystals have the unique property of resonating at a given frequency upon application of an electric field. A detectable shift in frequency is produced through a perturbation to the crystal, such as the adsorption of mass onto its surface, as is the case with most antibody-based piezoelectric biosensors. The greater the change in mass, the greater the change in frequency. As antibody molecules possess 600 times the mass of PAH molecules (i.e. 150 kDa compared to 250 Da), piezoelectric techniques employ an antigen-coated platform. PAH detection is conducted via competitive inhibition. Without PAHs present in a sample, antibodies can bind unimpeded to the surface, decreasing the frequency. However, with increasing PAH concentrations, there is a decrease in the frequency modulation by the antibody.

A piezoelectric immunosensor for PAHs was developed by Liu and colleagues (1999). This method employs the immobilization of a BaP-BSA antigen onto a quartz crystal. An anti-PAH mAb was then allowed to bind the sensor surface for 30 min before the introduction of a 100 µl BaP sample. At this step, the flow of the BaP sample was stopped for 30 min to allow for a competitive displacement immunoreaction to occur. The sample was then washed away and the frequency recorded. Because the sensor surface is pre-exposed to mAb molecules, an increase in PAH concentration reaching the sensor surface causes more antibodies to be displaced with an increase in resonance frequency. This technique provided BaP quantification from 1,260 to 2,500 µg/l. Cross-reactivity was demonstrated with pyrene and naphthalene.

Boudjay and colleagues (2010) describe another piezoelectric immunosensor for the detection of BaP. Instead of immobilizing the analyte-conjugate (a larger molecule), a layer of the derivatized analyte (a smaller molecule), pyrene butyric acid, was used to coat the surface of the crystals. Antibody and analyte solutions (500 µl) were incubated for 60 min prior to introduction to the sensor surface. The mAb and BaP solutions were pumped over the sensor surface at a flow rate of 25 µl/min and the frequency shift of the antibody adsorption onto the crystal surface was recorded. This method resulted in quantification from 756 to 2,520 μg/l.

Optical

Optically-based transducers possess the broadest range of mechanisms for transduction. In the case of PAHs, light can be emitted from the analyte itself or can be generated from a fluorescent label affixed to the antibody, the antigen, or the analyte. Light can be detected as it is reflected or refracted from the sensor's surface in response to antibody-PAH binding.

Surface Plasmon Resonance

SPR-based biosensors operate by detecting a change in the refractive index caused by adsorption of a molecule onto the sensor surface. The larger the molecule, the larger the change in the refractive index. Since antibodies are approximately 600 times larger than PAH, derivatized PAHs are covalently attached to the sensor surface to capture the larger antibody and effect a substantive change in the refractive index. SPR-based PAH immunosensors utilize a competitive inhibition format where the concentration of PAHs in the sample is inversely proportional to the change in refractive index. If there are no PAHs in the sample, all of the antibodies will bind to the sensor surface generating a greater shift response. Two SPR-based immunosensors have been described for the detection of BaP.

Gobi and Miura (Gobi et al. 2003; Miura et al. 2003; Gobi and Miura 2004) have described both a macro- and a micro-flow-cell, SPR-based immunosensors for the detection of BaP. Both methods used the same anti-BaP mAb developed against a BaP-BSA immunogen. This same immunogen, BaP-BSA, was also used as the antigen in all of their studies. These methods require an antibody and sample incubation step of 5 min, with the incident angle reaching a plateau in approximately 15 min resulting in a total time of 20 min for sample analysis. They reported a BaP quantification range of 0.01 to 300 μg/l and no-cross reactivity with a second analyte, HBP. This is the largest range of quantification (10^5) reported for any PAH immunosensor to date, however results were as much as ± 2 times the actual BaP concentration. The micro-flow-cell method requires 300 μl volume of a sample.

A second SPR-based method was described by Dostalek and colleagues (2007). Their immunosensor used a 30 min analyte and antibody incubation step. The sample and antibody mixture was permitted to flow over the sensing channel for 10 min, followed by a wash step for 5 to 7 min. In all, it took 45 min from the incubation of the sample with the antibody to obtaining a signal response. The LOD for BaP was 0.05 μg/l with quantification up to 0.18 μg/l. Selectivity of this method was examined in two ways; by testing the cross-reactivity of the antibodies (anti-ATR, anti-2,4-D, anti-4-NP, and anti-BaP) for recognition of the coated antigens (ATR-BSA, 2,4-D-BSA, BaP-BSA, and 4-NP-OVA) and by incubating the anti-BaP antibody with a mixture of four analytes (BaP, ATR, 2,4-D, and 4-NP) at one specified concentration in buffer. Although the anti-BaP antibody significantly cross-reacted with the 4-NP-OVA antigen (64%), the anti-BaP antibody reportedly demonstrated no discrepancy when incubated with the BaP analyte alone or in the mixture.

Fluorescence

Fluorescence detection is considered a more sensitive technique than other spectrophotometric method. Fluorescence is produced when sufficient energy is absorbed to excite a valence electron from its ground state to an excited state and a photon is emitted after its relaxation to the ground state. In the immunosensors that follow, this fluorescence is produced from either the excitation of the PAH molecule itself or by a fluorescent tag conjugated to one of the reagents.

Natural PAH fluorescence. The first anti-PAH method employing fluorescence detection was conducted by Vo-Dihn (1987). Fluorescence, an inherent property of BaP, was emitted by the BaP molecule upon exposure to a laser. Rabbit anti-BaP pAbs were covalently attached to the tip of a fiber-optic cable. This served as the probe, which was placed in a sample solution and incubated for 10 min. The probe was then rinsed with PBS and exposed to a helium cadmium laser in order to elicit fluorescence. In all, sample analysis was reportedly completed in approximately 12 min. The signal was proportional to the number of BaP molecules bound to the tip of the probe. The sample size needed was 5 μl, and in this volume 1fmol could be detected, which is equivalent to 0.25 μg/l. However, this technology suffers from high non-specific binding of PAH to the cable itself.

Polarized fluorescence. Yadavalli and Pishko (2004) employed an anti-PAH mAb labeled with a fluorophore. A solution containing analyte and antibody was passed through a microfluidic channel where it was exposed to polarized excitation light. The antibody bound to the analyte resulted in a larger, slower rotating entity that was reflected by the increase in fluorescent anisotropy. It is worth noting that fluorescence polarization is the only PAH immunosensor method, described to date, that does not use any form of immobilized antigen nor antibody. The authors noted that typical assays were conducted on <10 μl samples, however it was unclear how much was used in their experiments. They noted a LOD of 10 to 40 nM, which, if this was detected in a 10 μl sample, would result in a LOD of approximately 2,000 μg/l. The authors analyzed other aromatics (benzene, toluene, and anthracene), but could not conclusively report their detection. No mention of antibody incubation time or speed of analysis was provided.

Goryacheva and colleagues (2007) employed fluorescence polarization, where contrary to the previous study, they fluorescently labeled an analyte tracer and not the antibody. Similar to the previous study, there was no need to separate the bound and unbound antibodies, because as the fluorescently-labeled tracer is bound, there would be an increase in fluorescence polarization. In this investigation, the method was based on competitive inhibition with PAHs in the sample competing with the labeled tracers for antibody binding. Two anti-PAH antibodies, a mAb and sheep pAb, were employed. Different PAH tracers were explored (2, 4, and 5 rings), as well as characterization of cross-reactivity of the mAb/antisera with 16 PAHs. The pAb detected all analytes, while the mAb preferentially bound the 4- to 5-ring PAHs. No incubation time was required, as both mAb and pAb came to equilibrium immediately upon mixing the reagents, but total analysis time was not provided. No matrix effects were evident when environmental river samples were spiked with known amounts of BaP.

Fluorescence intensity. Spier and colleagues (2011) report a completely automated PAH immunosensor based on detecting the intensity of a fluorescently-labeled mAb. Antibody characterization demonstrated that the mAb recognized 3- to 5-ring PAHs (Spier et al. 2009). Analyte-protein conjugates were immobilized on polymethylmethacrylate beads loaded into a flow-cell positioned in front of a laser. As the mAb came to equilibrium quickly, no additional incubation with the sample was required. A 400 μl solution of sample and antibody were passed over the bead-packed flow-cell, where the fluorescently-labeled antibodies would either bind to the antigen-coated beads, or were inhibited from binding by PAHs in solution. For a sample containing no PAHs, the tagged mAbs provide the greatest fluorescence intensity. Thus, the fluorescent signal was inversely proportional to the concentration of PAH. The linear range of PAH quantification was 0.3 to 30 μg/l. Samples were analyzed in 3 min, with an additional 7 min for cleaning and preparing the sensor for the next sample. This study evaluated natural water samples contaminated with a complex mixture of PAHs. Validation analysis with GC-MS demonstrated good correlation and accurate quantification of 3- to 5-ring PAHs.

Reflectometric interference UV/VIS spectroscopy

One of the first automated immunosensors for BaP detection was described by Lange and colleagues (2002). The aim of their study was characterization of anti-BaP pAb rather than to produce a rapid, portable instrument. They demonstrated that the pAb (Vo-Dinh et al. 1987) were specific for BaP and not for chrysene and pyrene. This method employed a flow-injection system, an analyte-coated transducer surface, and a UV/VIS spectrometer. The principle of this method was that as antibodies bound to the analyte-coated transducer surface, the reflectance pattern of the continuous white light aimed at the transducer surface changed and was measured with good resolution. The antibody reaction was based on competitive inhibition where the maximum signal was produced when no BaP was present and all of the antibodies bound the sensor coated surface. Increasing BaP concentrations caused fewer antibodies to bind to the surface and therefore less reflected white light was observed. This method had a working range of 3 to 70 µg/l. As speed was not a concern of this study, pre-incubation of antibody with sample was conducted, but no times were provided, however a signal response was generated in less than 250 s.

Infrared

Boudjay and colleagues employed IR absorption technology for the development of a BaP immunosensor (2009). An IR signal was produced when the antibody binds a BaP molecule—a new IR band formed, indicative of the presence and stretching of the aromatic C-H band of BaP upon interaction with the antibody. Therefore, the concentration of BaP in a sample is directly proportional to the integrated area of this band. This method relied on the direct immobilization of the mAb onto the sensor surface and required no additional labels. A 10 mL solution of BaP was exposed to the antibody-coated surface for 60 min, washed and then the IR spectrum recorded. This label-free, direct IR method demonstrated a LOD of 1,260 µg/l with a measuring range up to 2,520 µg/l. In a subsequent publication, Boudjay and colleagues (2010) reported the employment of this IR method in an indirect competitive format, where the BaP-surrogate antigen was coated onto the immobilized surface, yielding the same LOD of 1,260 µg/l.

CONCLUSIONS

The majority of PAH immunosensors possess detection limits in the µg/l range (Table 15.3) making them suitable for measuring environmentally relevant concentrations. Rapid analysis is a desirable trait for biosensors and most of these can complete PAH analysis in minutes. For sensors requiring more than 30 min, sample incubation was required to allow the antibody to come to equilibrium with any available analytes. While

Table 15.3. Features and specifications of PAH immunosensors.

Transducer / Detection			Target PAH analyte(s)	LOD / range (µg/l)	Speed (mins)	Vol (µl)	Matrix	Reference
Electrochemical	Capacitance	Linear sweep voltammetry	BaP-BSA	na	15	10	buffer	Liu et al. 1998
		Cyclic voltammetry	BaP	25.2-1,260	40	np	buffer	Liu et al. 2000
			BaP, acenaphthene, fluorene, fluoranthene, pyrene, 1-pyrenebutyric acid	2.4	180	5	buffer	Wei et al. 2009
	Amperometric		BaP	2.0, 2.5-2,520	180	5	buffer	Wang et al. 2011
		Screen printed electrodes	Phen and 16 PAHs	0.8, 0.5-45	90	150	spiked river, tap, sea, and mineral	Fähnrich et al. 2003
		Linear sweep voltammetry	phenanthrene	2-100	60	10	buffer	Moore et al. 2004
Piezoelectric	QCM		BaP, pyrene, naphthalene	1,260-2,520	30	100	buffer	Liu et al. 1999, Liu et al. 2000
			BaP	756-2,520	60	500	buffer	Boujday et al. 2010
	SPR	macro-flow cell and micro-flow cell	BaP	0.01-300	20	300	buffer	Miura et al. 2003, Gobi et al. 2003, Gobi and Miura 2004
		Multichannel	BaP	0.05-0.18	45	300	buffer	Dostalek et al. 2007
Optical	Fluorescence	Fiber optic	BaP	0.25	12	5	buffer	Vo-Dinh et al. 1987
			Pyrene	~2,000	np	10	buffer	Yadavalli and Pishko 2004
		Polarization	BaP + 16 PAHs	0.9	np	50	spiked river water	Goryacheva et al. 2007
		Intensity	3- to 5-ring PAHs	0.3-30	3	400	naturally contaminated river and run-off	Spier et al. 2011
	Reflectometric interference spectroscopy	White light	BaP	3-70	np	np	buffer	Lange et al. 2002
	IR	infrared	BaP	1,260-2,520	60	150	buffer	Boujday et al. 2010, Boujday et al. 2009

Target analyte = PAHs detected in the assay.

Speed = Defined as the amount of time (estimated to the best extent possible) it takes from sample introduction to electronic result. In most cases, protein immobilization is required, but is performed in advance. In the case where the sample requires incubation with the antibody reagent, external to the sensor, this time is included in the speed value.

Buffer = Solutions vary, such as some contain 10% ethanol. For those details, please refer to the reference provided.

np = not provided

sensitivity and speed are desirable traits for biosensor methodologies, biosensors must transcend the controlled situation of the laboratory to make field assessments routine and reliable. Namely, matrix effects must be evaluated with complex environmental samples and PAH mixtures, as well as validation of results with traditional analytical methods. Portability, automation, and user-friendliness are issues of paramount importance.

Comparisons of PAH Immunosensors and Classical Analytical Methods

The main goal for developing PAH immunosensors is to perform rapid, on-site PAH analysis. This is to replace or assist traditional methods, which require samples to be analyzed by laboratory-bound techniques such as GC-MS or HPLC. To accomplish this, biosensor methods must employ samples that can be assessed in the field without extensive manipulations, such as extraction or concentration. The biosensor methods require small volumes, ranging from 5 to 500 μl per sample. In comparison, for low-level detection, classical analytical methods require up to 1 L per sample and the analytes of interest must first be extracted into an organic solvent and then concentrated. For analysis of a single sample, the analytes are then separated and individually identified. Depending on the number of identifiable analytes, this process can take hours to days to complete. Although biosensor methods do not provide resolution of individual analytes, sample analysis can be completed in minutes. In return, a much greater number of samples can be assessed with less time, effort and expense.

As reported in this chapter, most of these biosensors have only been tested with single analyte samples. However, PAHs in the environment commonly exist as complex mixtures. Moreover, these compounds possess similar molecular structures and electron density. As demonstrated in several anti-PAH antibody characterization studies, these antibodies are highly cross-reactive with other PAHs (Nording and Haglund 2003). Therefore, if the final goal is to employ the biosensor for precise analyte identification during environmental analysis, samples containing relevant complex mixtures of PAHs must be resolved and confirmed with traditional analytical methods (Rodriguez-Mozaz et al. 2006).

Antibody Incubation Times

The analysis time for the biosensors presented here is reported as the amount of time required from the moment the sample is introduced to the point at which a quantitative assessment is transmitted. This includes incubating the sample with the antibody, typically conducted for 30 to 60 min. However, most of the methods described do not utilize antibodies for which kinetic analyses have been completed. If an antibody comes to equilibrium with a sample in a very short amount of time, this incubation step can be reduced or eliminated. This is the case for the antibodies employed in the Goryacheva and colleagues (2007); Spier and colleagues (2011) studies. If kinetic analysis were to be conducted for the other antibodies employed, analysis times may be shortened in the future.

Comparison of Label-free and Labeled Reagents

Both, label-free (Fig. 15.5) and labeled methods have their advantages. With less reagents employed in the label-free approaches, less steps can be

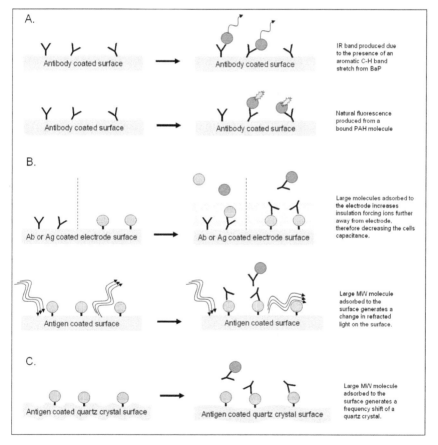

Figure 15.5. Label-free detection methods. **(A)** infrared spectroscopy or natural fluorescence emitted from a PAH molecule, **(B)** electrical storage (capacitance), surface plasmon resonance, or light reflectance/interference; and **(C)** piezoelectric.

involved. When labels are employed, they require covalent attachment to the antibody or antigen. This conjugation process can inadvertently place labels within the antibody binding site, thereby potentially decreasing the antibody's activity. Unlabeled reagents often rely on the intrinsic properties of the reagents for detection. For instance, the natural fluorescence of PAH molecules allow label-free detection (Fig. 15.5A). However, protein molecules and other natural organic substances may also fluoresce at a similar wavelength. Fluorescent labels can provide a narrow targeted wavelength and optical filters can be incorporated to increase the specificity of the signal detection. Another intrinsic property that is often exploited is the reagent's mass (Fig. 15.5B and C). Binding large proteins can change the dielectric properties of the solution, alter the refraction of light, or generate a frequency shift—all of which are directly measured.

Reusability of Biosensors

Reversible binding of antibody-antigen systems (i.e. sensor reusability) can be exploited to increase the useable life of some biosensors. To regenerate surfaces (removing previously reacted antibodies or antigens) a variety of chaotropic and proteolytic reagents have been used, including; urea (Liu et al. 1999), $MgCl_2$ (Liu et al. 2000), pepsin (Gobi et al. 2003; Miura et al. 2003), and NaOH with acetonitrile (Dostalek et al. 2007). An immunosensor has been reused successfully as many as 50 times, while others have experienced high retention of the previous reactants. This is very likely a function of the antibody's affinity or non-specific binding with the sensor surface. Likewise, the choice of chaotropic reagent will vary depending on the types of materials incorporated into the lines and seals of the sensor instrument. While this is certainly an avenue that can be explored, it is unclear if it is critical for successful field deployment. Regeneration would involve more reagents, but may also eliminate the need for numerous sensing surfaces and could reduce the reagents required in preparatory stages (i.e. antigen or antibody coating).

FUTURE PERSPECTIVES

While the original goal of developing biosensors was for the rapid quantification of environmental contaminants, other applications may be of value. Antibody platforms could be used as a preparatory method for isolating and concentrating a specific analyte within a sample (Fishman et al. 1998; Pérez and Barceló 2000). Multi-analyte detection has been briefly explored (Miura et al. 2003; Dostalek et al. 2007) and may yield extraordinary uses for biosensors. Namely, biosensors could be employed to survey an area with unknown contaminants. Traditional methods require prior knowledge about a pollutant in order to determine the proper extraction method. Information can be missed if a particular analyte is not retrieved or is lower than the LOD of the method. Once detected on-site by a sensitive biosensor, a larger sample can be collected for compound specific analysis by traditional methods saving time and money. Another possible use of biosensors is to monitor the dose during a toxicological study. Especially important when volatilization, sorption, metabolism, etc. of the analyte is of concern and the toxicant dose may be changing on a time scale not conducive for traditional analysis.

APPLICATIONS TO OTHER AREAS OF HEALTH AND DISEASE

PAH detection is not only occurring in aquatic environmental samples, but also has been completed extensively in food sources (EU 2002) by GC-MS methods. Human exposure to PAHs is also conducted via examining urine samples for PAH metabolites (Jongeneelen et al. 1985), and antibody-based assays have already been developed for one PAH metabolite (Marco et al. 1993). Similarly, human exposure to PAHs typically results in the formation of PAH-DNA adducts, which can serve as another biochemical molecule to detect via PAH biosensor (Vo-Dinh et al. 1987). Likewise, the methods presented in this chapter are only specific for PAHs due to the incorporated antibodies. By changing to other antibodies, these biosensors could be adapted to quantify any molecule for which an antibody can be developed.

KEY FACTS OF PAH IMMUNOSENSORS (ANTIBODY-BASED BIOSENSORS)

- Immunosensors use an antibody which specifically binds a PAH (the antigen) and the transducer reports this reaction in a digital format.
- Signal transduction format will influence speed, accuracy and sensitivity of the sensor.
- Although highly specific, antibodies commonly exhibit cross-reactivity with related compounds, thus comprehensive antibody characterization must be conducted.
- Immunosensors report a single concentration value for each sample, even if the sample contains multiple PAH analytes.
- PAH immunosensors are sensitive with LOD at environmentally relevant levels (µg/l).

- Immunosensors can be used for rapid field monitoring of environmental PAHs.
- New, inventive antibody specificities can greatly increase the applications of this technology.

SUMMARY POINTS

- PAHs are of great environmental and human health concerns, requiring new methods for faster (real-time) analysis.
- PAHs rarely exist in the environment as a single analyte. Biosensors should be evaluated in environmental contexts.
- Immunogens have been successfully designed to produce anti-PAH antibodies.
- The key to a biosensor is the specificity and affinity of the antibody molecule used and the sensitivity of the transducer.
- Numerous PAH immunosensors have been developed in the last decade.
- These immunosensors have been demonstrated as fast, sensitive methods for PAH quantification.
- What is needed for future immunosensor development?

  - portability and on-site demonstration
  - analysis of contaminated environmental samples
  - characterization of antibody binding kinetics
  - development of more extensive and specific anti-PAH detection
  - increased automation
  - increased robustness to address matrix interference
  - validation of quantification

ACKNOWLEDGEMENTS

Support was provided by a grant from NOAA-CICEET (Cooperative Institute for Coastal and Estuarine Environmental Technology) University of New Hampshire. Spier was also funded by U.S. National Science Foundation GK-12 (Division of Graduate of Education 0840804). VIMS Contribution #3230.

ABBREVIATIONS

2,4-D	:	2,4-dichlorophenoxyacetic acid
4NP	:	4-nonylphenol
ATR	:	atrazine
BaP	:	benzo[*a*]pyrene

BSA	:	bovine serum albumin
EU	:	European Union
GC-MS	:	gas chromatography mass spectrometry
HBP	:	hydroxybiphenyl
HPLC	:	high performance liquid chromatography
IgG	:	immunoglobulin G
IR	:	infrared
KLH	:	keyhole limpet hemocyanin
LOD	:	limit of detection
mAb	:	monoclonal antibody
MW	:	molecular weight
OVA	:	ovalbumin
PAH	:	polycyclic aromatic hydrocarbon
pAb	:	polyclonal antibody
SPR	:	surface plasmon resonance
UV/VIS	:	ultraviolet/visible
US EPA	:	Unites States Environmental Protection Agency

REFERENCES

ATSDR. 1995. Agency for Toxic Substances and Disease Registry. Toxicological profile for polycyclic aromatic hydrocarbons.

Boujday S, C Gu, M Girardot, M Salmain and C Pradier. 2009. Surface IR applied to rapid and direct immunosensing of environmental pollutants. Talanta 78: 165–170.

Boujday S, S Nasri, M Salmain and CM Pradier. 2010. Surface IR immunosensors for label-free detection of benzo[*a*]pyrene. Biosens Bioelectron 26: 1750–1754.

Dostalek J, J Pribyl, J Homola and P Skladal. 2007. Multichannel SPR biosensor for detection of endocrine-disrupting compounds. Anal Bioanal Chem 389: 1841–1847.

EU. 2002. Polycyclic aromatic hydrocarbons—occurrence in foods, dietary exposure and health effects. SCF/CS/CNTM/PAH/29 ADD1.

Fahnrich KA, M Pravda and GG Guilbault. 2002. Immunochemical detection of polycyclic aromatic hydrocarbons. Anal Lett 35: 1269–1300.

Fähnrich KA, M Pravda and GG Guilbault. 2003. Disposable amperometric immunosensor for the detection of polycyclic aromatic hydrocarbons using screen-printed electrodes. Biosens Bioelectron 18: 73–82.

Fishman HA, DR Greenwald and RN Zare. 1998. Biosensors in chemical separations. Annu Rev Biophys Biomol Struct 27: 165–198.

Gobi KV and N Miura. 2004. Highly sensitive and interference-free simultaneous detection of two polycyclic aromatic hydrocarbons at parts-per-trillion levels using a surface plasmon resonance immunosensor. Sensor Actuat B Chem 103: 265–271.

Gobi KV, M Sasaki, Y Shoyama and N Miura. 2003. Highly sensitive detection of polycyclic aromatic hydrocarbons and association constants of the interaction between PAHs and antibodies using surface plasmon resonance immunosensor. Sensor Actuat B Chem 89: 137–143.

Gomes M and RM Santella. 1990. Immunologic methods for the detection of benzo[*a*]pyrene metabolites in urine. Chem Res Toxicol 3: 307–310.

Goryacheva IY, SA Eremin, EA Shutaleva, M Suchanek, R Niessner and D Knopp. 2007. Development of a fluorescence polarization immunoassay for polycyclic aromatic hydrocarbons. Anal Lett 40: 1445–1460.

Grieshaber D, R MacKenzie, J Vörös and E Reimhult. 2008. Electrochemical biosensors–sensor principles and architectures. Sensors 8: 1400–1458.

Harlow E and D Lane. Monoclonal antibodies. pp 139–243. In: Antibodies: A Laboratory Manual.1988. Cold Springs Harbor Laboratory Press, Cold Springs Harbor, NY, USA.

Hermanson GT. 1996. Bioconjugate techniques. Academic Press, San Diego, CA, USA.

Jongeneelen FJ, RBM Anzion, CM Leijdekkers, RP Bos and PT Henderson. 1985. 1-hydroxypyrene in human urine after exposure to coal tar and a coal tar derived product. Int Arch Occup Environ Health 57: 47–55.

Lange K, G Griffin, T Vo-Dinh and G Gauglitz. 2002. Characterization of antibodies against benzo[a]pyrene with thermodynamic and kinetic constants. Talanta 56: 1153–1161.

Liu M, GA Rechnitz, K Li and QX Li. 1998. Capacitive immunosensing of polycyclic aromatic hydrocarbon and protein conjugates. Anal Lett 31: 2025–2038.

Liu M, QX Li and GA Rechnitza. 1999. Flow injection immunosensing of polycyclic aromatic hydrocarbons with a quartz crystal microbalance. Anal Chim Acta 387: 29–38.

Liu M, QX Li and GA Rechnitz. 2000. Gold electrode modification with thiolated hapten for the design of amperometric and piezoelectric immunosensors. Electroanalysis. 12: 21–26.

Marco M, BD Hammock and MJ Kurth. 1993. Hapten design and development of an ELISA for the detection of the mercapturic acid conjugates of naphthalene. J Org Chem 58: 7548–7556.

Matschulat D, A Deng, R Niessner and D Knopp. 2005. Development of a highly sensitive monoclonal antibody-based ELISA for detection of benzo[a]pyrene in potable water. Analyst 130: 1078–1086.

Meisenecker K, D Knopp and R Niessner. 1993. Development of an enzyme-linked immunosorbent assay for pyrene. Anal Method Instrum 1: 114–118.

Miura N, M Sasaki, KV Gobi, C Kataoka and Y Shoyama. 2003. Highly sensitive and selective surface plasmon resonance sensor for detection of sub-ppb levels of benzo[a]pyrene by indirect competitive immunoreaction method. Biosens Bioelectron 18: 953–959.

Moore EJ, MP Kreuzer, M Pravda and GG Guilbault. 2004. Development of a rapid single-drop analysis biosensor for screening of phenanthrene in water samples. Electroanal 16: 1653–1659.

Nakamura H and I Karube. 2003. Current research activity in biosensors. Anal Bioanal Chem 377: 446–468.

Nording M and P Haglund. 2003. Evaluation of the structure/cross-reactivity relationship of polycyclic aromatic compounds using an enzyme-linked immunosorbent assay kit. Anal Chim Acta 487: 43–50.

Pérez S and D Barceló. 2000. Evaluation of anti-pyrene and anti-fluorene immunosorbent clean-up for PAHs from sludge and sediment reference materials followed by liquid chromatography and diode array detection. Analyst 125: 1273–1279.

Pressman D and AL Grossberg. 1968. The structural basis of antibody specificity. W. A. Benjamin, NY, USA.

Quelven E, S Tjollyn, L Rocher, G Mille and J Fourneron. 1999. Development of a monoclonal antibody against polycyclic aromatic hydrocarbons. Polycycl. Aromat Comp 13: 93–103.

Roda A, MA Bacigalupo, A Ius and A Minutello. 1991. Development and applications of an ultrasensitive quantitative enzyme immuno-assay for benzo[a]pyrene in environmental samples. Environ Technol 12: 1027–1035.

Rodriguez-Mozaz S, MJ Lopez de Alda and D Barcelo. 2006. Biosensors as useful tools for environmental analysis and monitoring. Anal Bioanal Chem 386: 1025–1041.

Scharnweber T, M Fisher, M Suchanek, D Knopp and R Niessner. 2001. Monoclonal antibody to polycyclic aromatic hydrocarbons based on a new benzo[a]pyrene immunogen. Fresenius J Anal Chem 371: 578–585.

Spier CR, ES Bromage, TM Harris, MA Unger and SL Kaattari. 2009. The development and evaluation of monoclonal antibodies for the detection of polycyclic aromatic hydrocarbons. Anal Biochem 387: 287–293.

Spier CR, GG Vadas, SL Kaattari and MA Unger. 2011. Near-real-time, on-site, quantitative analysis of PAHs in the aqueous environment using an antibody-based biosensor. Environ Toxicol Chem 30: 1557–1563.

Suchanek M, T Scharnweber, M Fisher, D Knopp and R Niessner. 2001. Monoclonal antibodies specific to polynuclear aromatic hydrocarbons. Folia Biol (Praha) 47: 106–107.

US EPA. 2009. National primary drinking water regulations. EPA 816-F-09-0004.

US EPA. 2010. What is nonpoint source pollution? EPA 841-F-96-004A.

Vanderlaan M, BE Watkins and L Stanker. 1988. ES&T critical review: Environmental monitoring by immunoassay. Environ Sci Technol 22: 247–254.

Van Emon JM and CL Gerlach. 1998. Environmental monitoring and human exposure assessment using immunochemical techniques. J Microbiol Meth 32: 121–131.

Vo-Dinh T, BJ Tromberg, GD Griffin, KR Ambrose, MJ Sepaniak and EM Gardenhire. 1987. Antibody-based fiberoptics biosensor for the carcinogen benzo[*a*]pyrene. Appl Spectrosc 41: 735–738.

Wang C, M Lin, Y Liu and H Lei. 2011. A dendritic nanosilica-functionalized electrochemical immunosensor with sensitive enhancement for the rapid screening of benzo[*a*]pyrene. Electrochim Acta 56: 1988–1994.

Wei MY, SD Wen, XQ Yang and LH Guo. 2009. Development of redox-labeled electrochemical immunoassay for polycyclic aromatic hydrocarbons with controlled surface modification and catalytic voltammetric detection Biosens Bioelectron 24: 2909–2914.

Yadavalli VK and MV Pishko. 2004. Biosensing in microfluidic channels using fluorescence polarization. Anal Chim Acta 507: 123–128.

Exoelectrogens-based Electrochemical Biosensor for Environmental Monitoring

Liu Deng,[1,a] Li Shang[2] and Shaojun Dong[1,b,*]

ABSTRACT

There are increasing reports on the exoelectrogens that certain bacteria, which form biofilms on conductive materials, can achieve a direct electron transfer with the electrode surface, without the aid of mediators. Several mechanisms for electron transfer to electrodes have been proposed, including direct electron transfer via outer-surface c-type cytochromes, long-range electron transfer via microbial nanowires and electron flow through self-excreted soluble electron shuttles. The mechanisms that are most important depend on the type of microorganisms and the thickness of the anode biofilm. The capability of exoelectrogens to connect their metabolisms directly in an external electrical power supply is very exciting and extensively researched. Great progress has been made on exploring the possibilities of exoelectrogens applications. Coating exoelectrogens on the electrode surfaces have recently become

[1]State Key Laboratory of Electroanalytical Chemistry,Changchun Institute of Applied Chemistry, Chinese Academy of Sciences, Changchun, Jilin, 130022, P. R. China.
[a]E-mail: dengliu@ciac.jl.cn
[b]E-mail: dongsj@ciac.jl.cn.
[2]Institute of Applied Physics and Center for Functional Nanostructures (CFN), Karlsruhe Institute of Technology (KIT), Wolfgang-Gaede-Strasse 1, 76131 Karlsruhe, Germany; E-mail: lishang208@gmail.com
*Corresponding author

List of abbreviations after the text.

popular in biosensor design for environmental monitoring. Microbial biosensors can be used in the continuous monitoring of a contaminated area, which are applicable in the determination of the "biological oxygen demand" (BOD) as well as the toxicity of certain compounds. In this chapter, we provide an overview of recent advances in materials and methods that are used to construct exoelectrogens-based biosensor for environmental monitoring.

INTRODUCTION

Increasing scientific and social concern in environmental pollution control, especially on water quality control, has recently led to a growing number of initiative and legislative actions being extensively enacted in this area. To monitoring the water quality with rapidity, simplicity, sensitivity and easy maintenance is very important. Analytical instrumental methods, including HPLC, GC-MS and LC-MS, require complex pretreatment and analysis protocols. Biomonitoring methods using vertebrates or fish have drawbacks of long detection time, as well as the ethic and moral issue. The need of portable and inexpensive systems for environmental monitoring has stimulated the development of new methods. In recent years, the use of whole microorganism cell as the biological sensing element has demonstrated great potential as an alternative to the conventional analytical methods. Due to the excellent performance of microbial biosensors in environmental monitoring, numerous microbial biosensors by measuring light, fluorescence, color or electrochemical signals have been developed. A BOD determination with photodiode was developed based on reading the luminescence intensity of luminous bacterium *Photobacterium phosphoreum* in wastewater. The fluorescence intensity of microbial communities excited by UV at 430 nm was in proportion to the assimilable organic contaminants amount, based on which the toxicity value can be measured by the difference of fluorescence intensity. Despite the good sensitivity and stability of these methods based on photometric or fluorescence techniques, some compounds could interfere with the performance during the test procedure. The electrochemical measurements are of particular interests for *in situ* measurements, due to their convenience to manipulate, anti-interference ability and short response time.

Most BOD and toxicity electrochemical biosensors rely on the measurement of the bacterial respiration rate with a transducer, commonly employing the Clark-type oxygen electrode or redox mediator as the terminal electron acceptor in microbial respiration. These methods have a number of limitations, such as questionable accuracy, irreproducibility and labor intensive, and consequently they are not suitable for process control and real-time environmental monitoring on-site. In the early 2000s, the discovery of exoelectrogens, which can transfer electrons outside the cell to conductive materials while complete oxidize organic compounds, opens a new perspective (Logan 2008). Various names for these bacteria were used, such as "exoelectrogens", "electrogenic microorganisms" or "electrochemically active bacteria". Here they will be referred as exoelectrogens. Exoelectrogens are the subject of a new area of interdisciplines involving electrochemistry, microbiology and chemical engineering.

Increasing efforts have been made to characterize exoelectrogens. Various bacteria with exoelectrogenic activity have been discovered, *Geobacteraceae, Gammaproteobacteria, Betaproteobacteria, Rhizobiales, Glostridia, Shewanella*, etc. (Logan 2008). Moreover, techniques for optical, electrochemical and genetic engineering have been developed for these bacteria. Exoelectrogens have significant impact in many fields, including biotechnology, sustainable energy development and bioremediation (Erable et al. 2010). For example, they have great potential in the fields of wastewater treatment, renewable energy recovery and the production of hydrogen or methane (Logan 2008). The direct electron transfer (DET) of exoelectrogens toward the electrode is quite attractive in the electrochemical biosensor, which is sensitive to changes in the metabolic status of the cellular biocatalyst and simplifies the construction procedure. The amount of electron transfer to the electrode is found in proportion of the exoelectrogens metabolic activity. Thus the metabolic activity of exoelectrogens can be directly measured by the generated electricity. At the same time, the metabolic activity of the microorganism is related to the organic substrate in the growing medium. In addition, toxic substances have an inhibitory effect on the metabolism of microorganisms and in case of exoelectrogens, an inhibitory effect on the transfer rate of electrons to the electrode. So that exoelectrogens-based electrochemical biosensors are applicable for environmental monitoring. The major advantage of these novel biosensors over the conventional biosensors is that they donot need complex transducers to read the response and transform it to the output signal. Since the value can be directly measured by the electricity generated, the exoelectrogens-based biosensor can be easily used as efficient online biosensor. The purpose of this chapter is to summarize the understanding of extracellular electron transfer mechanism in exoelectrogens. Significant advancements on increasing the performance of the exoelectrogens-based electrochemical biosensor have been made recently by modifying biosensor architecture and materials.

WORKING PRINCIPLE OF EXOELECTROGENS-BASED BIOSENSORS

The extracellular electron transfer of exoelectrogens at the electrode surface has not been precisely defined. At a first stage, the bacteria have to colonize on the electrode surface and manufacture the specific structure or the enzymes for the electron transfer outside the cell. Then the bacteria use the citric acid cycle (CAC) for oxidation of substrates, producing three different electron carriers (NADH, FADH, and GTP) (Logan 2008). The concentrations of reduced species in the cell are central to the regulation of many different processes. When the NADH pool is full, for example, the CAC cycle has to suspend as it will become a thermodynamics unfavorable reaction. The cell respiration chain will stop until the electrons can flow to another electron acceptor. As we will discuss below, the electrode could act as the electron acceptor in the case of exoelectrogens, thus the electricity would be generated in the presence of cathode and resistor. Electrical current is a direct linear measure for the metabolic activity of exoelectrogens. This can be easily measured and used to indicate a direct metabolic rate for specific respiratory processes. Therefore, information such as, the concentration of the monitored organic substance and the presence of some toxic species, can be made available online.

THE ELECTRON TRANSFER MECHANISM OF EXOELECTROGENS

To understand the electron transfer to the electrode during the respiration chain is crucial for the exoelectrogens-based biosensor. Initial understanding of electron transfer by bacteria to electrodes come from studies of dissimilatory metal-reducing bacteria such as *Geobacter* and *Shewanella* species (Aelterman et al. 2008; Bretschger et al. 2007; Gorby et al. 2006; Marsili et al. 2008; Hartshorne et al. 2009). Several mechanisms have been proposed for direct electron transfer in exoelectrogens. The exoelectrogens transfer electrons via either in direct contact between the cell or the electrode surface or the self-produced mediator. Three main mechanisms are distinguished for the electron transfer from exoelectrogens to the electrode as described below.

The Direct Contact by Outer-membrane Cytochromes

A number of proteins in the cytoplasmic membrane, periplasm, and outer membrane that are involved in extracellular electron transfer have been identified by mutagensis and biochemical studies. Biochemical and genetic characterizations demonstrate that the exogenous electron transfer involves outer-membrane cytochromes (Fig. 16.1). For example, *Shewanella oneidensis* had 42 c-type cytochromes, among which were 14 c-type cytochromes identified with four or more heme-binding sites (Meyer et al. 2004). *Geobacter sulfurreducens* were subsequently sequenced, and it was found to contain

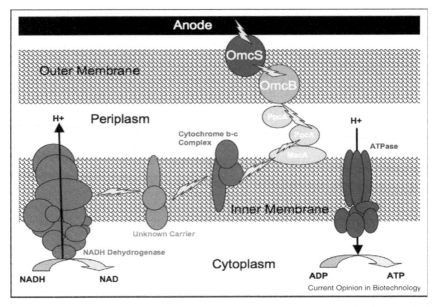

Figure 16.1. Model for *Geobacter sulfurreducens* electron transfer to the anode of a microbial fuel cell from NADH derived from organic matter oxidation. Energy conservation results from proton pumping associated with inner membrane electron transport. Subsequent electron transfer steps function merely to dispose of electrons. Electron transfer to the anode is depicted via an outer surface cytochrome as has been proposed for cells in direct contact with the anode surface. Adapted from Lovely 2008.

73 c-type cytochromes that contained two or more heme groups (Methé et al. 2003). These cytochromes (c-Cyts) located in various positions at the inner and outer membranes, which acted as a "molecular wire" and played an important role in the electron transfer process. The role of c-type cytochromes playing in the extracellular electron transfer was investigated (Hartshorne et al. 2009). The results demonstrated that the OmcA, OmcB and MtrC were exposed on the cell surface and were thought to participate directly in the electron transfer to the electrode surface. In addition, MtrcAB and OmcZ were shown to obtain electrons from the host electron transport chain and pass through the membrane, as well as MtrcA and MtrcB were helpful in this process (Bouhenni et al. 2010).

The Use of Self-excreted Mediators

Some bacteria can produce soluble electron shuttles for the extracellular electron transfer. Rabaey and coworkers demonstrated that phenazine production by a strain of *Pseudomonas aeruginosa* stimulated electron transfer for several bacterial strains (Rabaey et al. 2005). Pyocyanins produced by

P. aeruginosa could function as a signal for the regulation of quorum sensing-controlled genes, which is related to a possible role of electron transfer for cell-cell communication (Pham et al. 2008). The electricity could be generated with *E. coli* K12 HB101 in an air-cathode MFC when this bacterium was "electrochemically-evolved" in fuel cell environments through natural selection (Qiao et al. 2008a). As shown in Fig. 16.2, it was assumed that the self-excreted mediators in the electrochemically-evolved *E. coli* were the hydroquinone, which was released from the quinone pool located at the cytoplasm membrane. Studies with *Shewanella oneidensis* suggested that soluble electron shuttles were the mediators for the electron transfer to the electrode (Marsili et al. 2008). The endogenous redox mediator can serve as a reversible terminal electron acceptor to transfer electrons from the bacterial cell to the electrode surface.

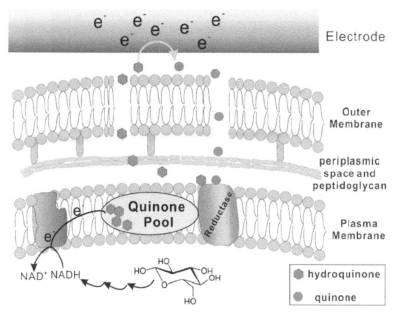

Figure 16.2. A hypothetical mechanism for extracellular electron transport of evolved *E. coli* cells. Adapted from Qiao et al. 2008b.

The Microbial Nanowires

The discovery of nanowires (Fig. 16.3) introduces a whole new dimension to the study of extracellular electron transfer. These conductive, pilus-like structures, identified so far in *Geobacter sulfurreducens* PCA, *Shewanella oneidensis* MR-1, a phototrophic cyanobacterium *Synechocystis* PCC6803, and the thermophilic fermenter *Pelotomaculum thermopropionicum*, appear

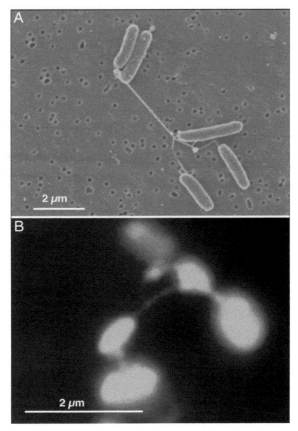

Figure 16.3. (A) SEM image of MR-I. **(B)** Epifluorescence micrograph of MR-1 stained with the fluorescent nonspecific protein-binding stain NanoOrange in liquid medium. Adapted from Gorby et al. 2006.

to be directly involved in extracellular electron transfer (Logan et al. 2006). After disrupting pili gene of *G. sulfurreducens,* the bacterium's reduce ability was eliminated (Reguera et al. 2005). Deletion of the genes associated with two *c*-type cytochromes (MtrC and OmcA) in *S. oneidensis* resulted in poorly conductive nanowires, loss of electrochemical activity and the ability to reduce insoluble electron acceptors (Bretschger et al. 2007). Evidence to support the possibility of cell respiration using these uninsulated nanowires was initially based on measurements of electrical conductance in the z plane (across the wire diameter). The nanowires produced by *Shewanella oneidensis* MR-1 exhibit nonlinear electrical transport properties along their length. These nanowire structures allowed the direct reduction of a distant electron acceptor (Gorby et al. 2006). The fermentative bacterium *Pelotomaculum thermopropionicum* was observed to be linked to the methanogen *Methanothermobacter thermautotrophicus* by an electrically conductive appendage, which provided the first direct evidence for interspecies electron transfer.

APPLICATION OF EXOELECTROGENS IN BIOSENSORS

Electrode Material

Using the optimized electrode material with high surface area and good electrocatalytic property is one of the most efficient approaches to improve the performance of electrochemical biosensor. The ideal microbial biosensor platform should provide a favorable microenvironment to maintain the microbe activity and minimize kinetic barriers of the substrate and product. To date, carbon materials (carbon-based electrodes in paper, cloth, and foam forms) have been a good choice for the microbial biosensor construction owing to their good biocompatibility in the microbial mixture and high specific area (Chang et al. 2005; Stein et al. 2010; Moon et al. 2004). Nevertheless, they have poor conductivity for the electron transfer between microbes and the electrode surface. Some highly conductive metallic materials, such as Au and Cu usually result in the denaturalization of redox-active proteins, such as cytochromes. The surface chemical modification is the common strategy to improve the electrode surface property. Recent advances in nanotechnology have enabled the development of new platforms aimed at more sensitive and faster toxic substrate detection.

A unique nanostructured polyaniline (PANI)/mesoporous TiO_2 composite with uniform nanopore distribution and high specific surface area was synthesized (Qiao et al. 2008b). The optimized composite, with 30 wt % PANI, exhibited the best bio- and electrocatalytic performance. This modified electrode offered several potential advantages over commercial carbon materials as electrode material for exoelectrogens-based devices. A carbon nanotube (CNT)/PANI composite has also been evaluated as an electrode material (Qiao et al. 2007). The addition of CNTs to PANI increased the specific surface-area of the electrode and enhanced the charge transfer capability so as to cause considerable improvement of the electrochemical activity. In comparison to the commercial carbon material, this nanocomposite exhibited a superior and specific electrocatalytic effect for the exoelectrogens-based devices. Recently, our group reported a new, simple microbial biosensor platform based on carbon fiber mat from the silk fiber (Deng et al. 2010a). The S-derived carbon fiber mat possesses amino, pyridine and carbonyl functional groups. As shown in Fig. 16.4, these natural-existed functionalities allowed the Au@Pt hybrid nanomaterials to self-assembly on the carbon fiber surface and provided a biocompatible microenvironment for bacteria. These Au@Pt urchilike NPs on the electrode

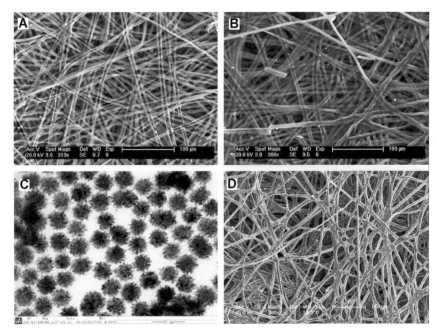

Figure 16.4. SEM micrographs for silk fibers **(A)**, S-derived carbon fibers **(B)**, TEM micrograph **(C)**, for Au@Pt urchilike NPs, the Au@Pt urchilike NPs deposited S-derived carbon fibers **(D)**. Adapted from Deng et al. 2010a.

surface provided the necessary conduction pathways and allowed efficient electron tunneling, which accelerated electrical communication between microbes and the electrode surface. The quinone groups presented on carbon fiber surfaces could mimic the quinone-containing constituents of natural microbial electron acceptors, which facilitated the direct electrocatalysis process of microbes. The combination of nanomaterials and carbon material fiber, which facilitated directional electron transfer, provides a wider scope for the design of advanced microbial biosensor.

The Architecture of the Biosensor

Practical applications of exoelectrogens-based electrochemical biosensor will require an architecture that not only produces high electricity, but also is economical for mass production. The first exoelectrogens-based electrochemical biosensor was constructed based on a two-chamber microbial fuel cell (MFC)-type electrochemical device (Kang et al. 2003). As shown in Fig. 16.5, in a two-chamber MFC, two electrodes (anode and cathode) each are placed in two chambers joined by a proton exchange membrane (PEM). The electricity generated from the microbial fuel cell was

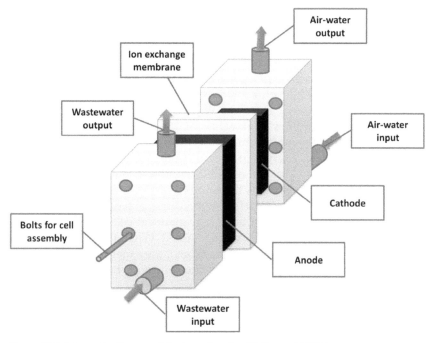

Figure 16.5. Schematic diagram of the two-chamber MFC type BOD biosensor.

directly proportional to the strength of the wastewater. This observation suggests the possibility to use it as a BOD sensor. Various kinds of two-chamber MFC-type biosensor have been developed (Kumlanghan et al. 2007; Chang et al. 2005; Chang et al. 2004; Kim et al. 2007). However, it would be difficult to apply this two-compartment architecture in integrated systems for continuous online testing. The main disadvantage of a two-chamber MFC is that the cathode must be aerated to provide oxygen. In addition, reducing their cost is essential, e.g. PEMs such as Nafion are quite expensive. To make a more compact and simple system with reduced cost of operation, Lorenzo et al. tested a single-chamber MFC (SCMFC), with an air cathode as a BOD biosensor (Lorenzo et al. 2009). An air-cathode MFC provided potential advantages over the two-chamber system, because neither aeration, recycling, nor chemical regeneration of catholyte was required. The biosensor performance was evaluated in terms of measurement range, response time, reproducibility and operational stability. When artificial wastewater was used as fuel, the biosensor output had a linear relationship with the BOD concentration of up to 350 mg BOD cm^{-3} and stability over 7 mon of operation. Under optimum condition, the biosensor exhibited a good correlation between COD concentration and current output towards the real wastewater, demonstrating the applicability of this system to real treatment effluents.

Although the exoelectrogens-based MFC can be applied in various types of biosensors including the BOD sensor and toxicity monitoring sensor, the unstable cell potential of the electrodes in MFC and long enrichment time (2 to 4 wk) with careful maintenance limited their application. A three-electrode electrochemical cell with a poised electrode for exoelectrogens inoculation was found to be useful for developing novel bioelectrochemical devices, as it can easily enrich exoelectrogens without the complicated MFC format including oxygen supplies (Seok-Min et al. 2007). Most microorganism cell walls and membranes contain numerous proteins, lipid molecules, teichoic acids, and lipopolysaccharides that contribute to the characteristic charge. The cell is negatively charged at pH 7, because the number of carboxyl and phosphate groups exceeds the number of amino groups. So they can be attached on the electrode surface at a positive potential. Exoelectrogens were successfully enriched in an electrochemical cell using a positively poised working electrode (+0.7 vs. Ag/AgCl). When the enrichment was completed, the bioelectrochemical reaction of the exoelectrogens-enriched electrode exhibited a good linear relationship with the concentration of BOD.

EXOELECTROGENS BASED BIOSENSORS FOR ENVIRONMENTAL MONITORING

BOD Sensor

One important application of the exoelectrogens based microbial biosensor is for the pollutant analysis and *in situ* process-monitoring and control. The proportional correlation between the electricity generated by exoelectrogens and the strength of the wastewater enable them to function as BOD sensors. A number of works have shown a good linear relationship between the generated electricity and the strength of the wastewater in a quite wide BOD concentration range (Moon et al. 2004; Hyunsoo et al. 2005; Chang et al. 2004; Chang et al. 2005). At the low BOD range, the current values increase with the BOD value linearly. However, a high BOD concentration requires a long response time. During this stage, the anodic reaction was limited by substrate concentration. This monitoring mode can be applied to real-time BOD determinations for either surface water, secondary effluents or diluted high BOD wastewater samples. The exoelectrogens-based biosensor with a single-chamber was used as BOD sensor (Kumlanghan et al. 2007). The sensor response was linear against the concentration of glucose for up to 25 g L^{-1}. The detection limit was found as 0.025 g L^{-1}. The major advantage of this sensor was the short measuring time required (between 3 and 5 min). This type of sensor was also used to estimate the quantities of biodegradable organic matter present in the wastewater (Lorenzo et al. 2009). The exoelectrogens-based BOD sensors are advantageous over other types of BOD sensors because they exhibited excellent operational stability and good reproducibility and accuracy. The exoelectrogens enriched BOD biosensor showed long-term operational stability without extra maintenance for over 5 yr (Kim et al. 2003), far longer in the service life span than other types of BOD sensors reported in the literature.

High concentration of nitrate is contained in effluents from sewage works and certain industrial wastewater. A considerable amount of oxygen diffuses from the cathode compartment into the anode compartment. In the presence of higher redox potential (nitrate and oxygen), the output electricity of MFC would be reduced (Kim et al. 2003). To obtain an accurate BOD value, these electron acceptors should be removed from the samples. The inert gases such as nitrogen were used to eliminate the oxygen in samples (Chang et al. 2004; Kim et al. 2003). At the same time, the respiratory inhibitors were applied to eliminate the influence of these electron acceptors. The current generation from exoelectrogens was inhibited by inhibitors of NADH dehydrogenase, coenzyme Q and quinol-cytochrome *b* oxidoreductase, but not by terminal oxidase inhibitors (Chang et al. 2005). The azide and cyanide are known to inhibit the activity of terminal enzyme of the respiration chain. The use of the respiratory inhibitors was therefore recommended for the accurate BOD measurement of environmental samples containing nitrate and/or oxygen with an MFC-type BOD sensor (Fig. 16.6). These results could be useful in operating MFC-type BOD sensors.

Detection Toxic Substance

The amplitude of the exoelectrogens-based biosensor is proportional to the level of metabolic activity of the biocatalyst. The toxic substance will influence the activity of the biocatalyst, so we can detect the toxicity according to the exoelectrogens-based biosensor signal change. Kim et al. reported a biomonitoring system using exoelectrogens-based MFC as the biocatalyst for detecting the inflow of toxic substances (Kim et al. 2007). When toxic substances (an organophosphorus compound, Pb, Hg, and PCBs) were added to the microbial fuel cell, rapid decreases in the current were observed. The inhibition ratios caused by inflow of these toxic substances (1 mg L^{-1}) were 61%, 46%, 28% and 38%, respectively (Fig. 16.7). These systems were able to detect the toxin. However, the main drawback associated with this microbial biosensor is the slow electron transfer between the microbial cell wall and the electrode surface. In order to facilitate the electron transfer rate, great efforts have been devoted to effectively overcome the kinetic barriers. Our group constructed a toxicity biosensor based on Au@Pt NPs modified S-derived carbon fiber (Deng et

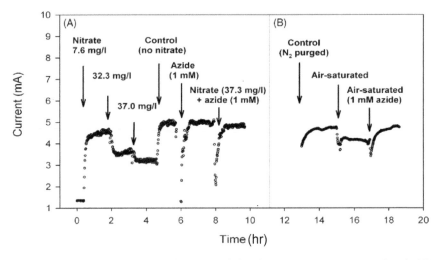

Figure 16.6. The current generation from MFCs fed with AW containing nitrate and azide **(A)** and with aerated AW containing azide **(B)**. AW with the BOD concentration of 113.5 mg/l was fed to MFCs at a feeding rate of 0.35 ml/min. Different nitrate concentrations and gas treatments (N_2 and air) are indicated in the figure. Adapted from Chang et al. 2005.

al. 2010a). The nanomaterials-modified electrode facilitates the directional electron transfer, which were valuable to the advancement of the biosensor. The electrochemical-evolved *E.coli* was enriched on the electrode surface via three-electrode system. This biosensor was also tested for the detection of organophosphate pesticides, fenamiphos. The relative inhibition of *E. coli* activity was linear with the log of fenamiphos concentration at the concentration range of 0.5–36.6 mg/L with the lowest observable effect concentration (LOEC) of 0.09 mg/L.

In most cases, the presence of a toxic compound will be an exceptional event. Ideally there should be only a change in current, when there is a toxic compound present in the water fed to exoelectrogens-based biosensor. This means that it is crucial to assure that under non-toxic conditions the current remains stable to prevent false positive alarms. This is especially important as the non-toxic conditions are the rule and toxic events are the exception. Several researches have investigated the influence of substrate concentration, pH and temperature on the overpotential and the electricity of an exoelectrogens based biosensor (Moon et al. 2004; Stein et al. 2010; Jadhav and Ghangrekar 2009). Stein et al. studied the influence of pH and bicarbonate concentration on electricity (Stein et al. 2010). They found that the pH had 0.43% change in the current density per mV and bicarbonate 0.75%/mV. Therefore, the strict control of pH and bicarbonate concentration in a small range is necessary. The temperature dependence during the enrichment procedure was also investigated (Patil et al. 2010). It was found that elevated

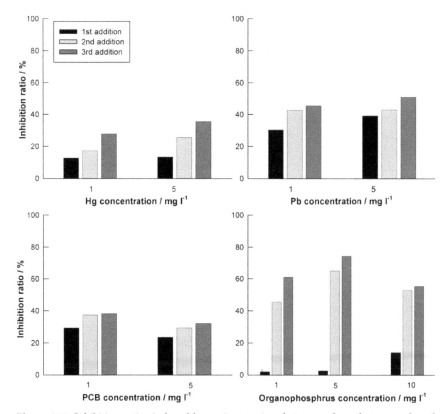

Figure 16.7. Inhibition ratios induced by various toxic substances of exoelectrogens based biosensor. Adapted from Kim et al. 2007.

temperatures during initial biofilm growth not only accelerated the biofilm formation process, but also affected the bioelectrocatalytic performance of these biofilms. The biofilm grown at a certain temperature would have a much better performance, so that different temperature would influence the electrochemically activity of the formed biofilm. All biofilms possess the similar operational temperature range between lower (0°C) and upper (50°C) temperature-defining the working limits of the exoelectrogens-based electrochemical biosensor.

OUTLOOK

Exoelectrogens can be obtained principally from natural sites such as soils, seawater and freshwater sediments, or from samples collected from a wide range of different microbial rich environments (sewage sludge, activated sludge, or industrial and domestic effluents). The exploitation of exoelectrogens' metabolism to catalyze or to control electrochemical reactions, which naturally occur in the environment, could hopefully lead to major technology breakthroughs. A multidisciplinary approach and intensive researches are in progress exploring the possibilities of exoelectrogens applications and improvement in the process design and configuration to suit the desired applications. Implementation of such bioelectrochemical systems is not straightforward because certain microbiological, technological, and economic challenges need to be resolved.

In the near future, development of exoelectrogens-based electrochemical biosensor for the real environmental applications will be devoted to address the issues including the efficiency and cost of materials, physical architecture and chemical limitations. We also need a better understanding of bacterial electron transfer to a surface at a molecular level, in order to provide an electrode material with excellent conductivity and biocompatibility for the microbial biosensor fabrication. For example, recent researches have experimentally demonstrated the feasibility of increasing the electron transfer of exoelectrogens using nanomaterials (Peng et al. 2010; Wu et al. 2010). The OM c-type cytochromes wire efficiency of the *S. oneidensis* MR-1 could be improved greatly by Fe_3O_4/Au nanocomposites (Deng et al. 2010b). The electron propagation from OM c-type cytochromes could go along the Fe_3O_4/Au nanocomposites assembled nanostructure instead of the traditional electron transfer between adjacent bacteria. In addition, the presence of Fe_3O_4/Au nanocomposites may provide a favorite biocompatible environment for the proteinaceous electron transfer components necessary for extracellular electron transfer. The bioparticles Pd_0 can act as highly active catalysts for electrode reactions of *D. desulfuricans* (Wu et al. 2010). Wu et al. presented the evidence that Pd nanoparticles bound to the microbes may participate in the electron transfer process (Fig. 16.8). The incorporation of biocompatible conductive nanomaterials into bacterial films opens up a new avenue to accelerate the evolution of technological applications. Another technical challenge is to identify and select microbial communities, which involves increased substrate degradation rates, efficient electron transfer mechanisms, or increased electrical conductance of the biofilm matrix. Genome-scale metabolic modeling coupled with genetic engineering may yield strains that can enhance current production. The exoelectrogens-based biosensors will facilitate rapid development by genetically engineering bacteria to achieve stable and high electricity generation. If appropriate strategies can be devised, it may be possible to recover microorganisms capable of higher rates of electron transfer between microorganisms and electrodes than currently available strains. Controlling the anode potential on the first day after establishment of biofilm limits the microbial competition on the anode and encourages exoelectrogens growth

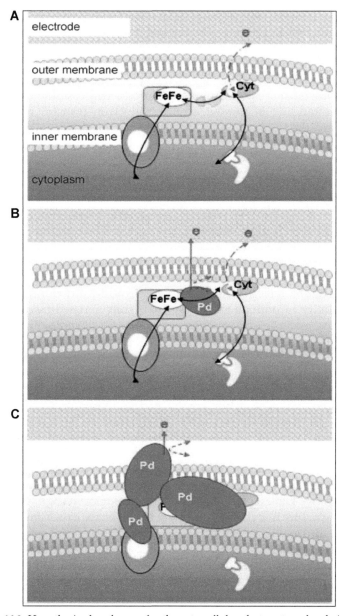

Figure 16.8. Hypothesized pathways for the extracellular electron-transfer chain between the cell and the electrode: **(A)** in the absence of Pd_0 (via periplasmic cytochromes and hydrogenases); **(B)** in the presence of Pd_0 at a low loading; **(C)** in the presence of Pd_0 at a high loading. (FeFe) is an iron-only hydrogenase, Cyt is a c-type cytochrome, and areas shaded gray are the Pd_0 nanoparticles. Adapted from Wu et al. 2010.

Color image of this figure appears in the color plate section at the end of the book.

(Erable et al. 2009). This strategy helps to maximize power harvesting from the microbial electrode without losing the substrate for competitive metabolisms (methanogenesis, for example). More studies with "real" wastewaters but not "synthetic" wastewaters are also required to evaluate the real potential of the technology. At the same time, biosensor architecture needs improvements before a marketable product is possible. Furthermore, the understanding of the extracellular electron transfer mechanism will expand the application of exoelectrogens-electrode interactions to environmental monitoring.

APPLICATION TO OTHER AREAS OF HEALTH AND DISEASE

The aim of developing exoelectrogens-based biosensors is to be of use as environmental quality monitoring tools, such as detecting the toxicity or endocrine activity, and they can also detect a compound or a group of compounds based on the specific biorecognition of a molecule. Exoelectrogens-based biosensors can provide information on the bioavailability of pollutants directly from soils and sediment samples, which often undergo an extraction step to draw out contaminants into a liquid solution relevant for assessing the potential damage caused by a substance.

For human health and disease, exoelectrogens-based biosensors provide a way to measure and assess the potential risk that chemical compounds could pose on humans and on the environment when humans are exposed. As we discussed above, information on the concentration of the monitored organic substance and the presence of some toxic species, can be made available online by exoelectrogens-based biosensors. The direct electron transfer of exoelectrogens between microorganism and electrodes provide exoelectrogens-biosensors with a superior selectivity and sensitivity. Due to the specificity of the biological recognition and the amplifying nature imparted by substrate turnover, the exoelectrogens have advantages in the sensing process. The high capacity of resisting interference for exoelectrogens-biosensors make them particularly suitable for use in serum, blood, urine, as well as other complex body fluid. Other potential applications of exoelectrogens-based biosensors include food analysis, biopharmaceuticals and clinical medicine.

SUMMARY

- A microbial biosensor is an analytical device that couples microorganisms with a transducer to enable rapid, accurate and sensitive detection of target analytes in fields as diverse as medicine, environmental monitoring, defense, food processing and safety. Microbial biosensors use the respiratory and metabolic functions of the microorganisms to detect a substance that is either a substrate or an inhibitor of these processes.

- The direct electron transfer (DET) of exoelectrogens towards the electrode is quite attractive in electrochemical biosensors. The amount of electron transfer to the electrode is found in proportion of the exoelectrogens metabolic activity, which can be directly measured by the electricity generated. At the same time, we know that toxic substances have an inhibitory effect on the metabolism of microorganisms and in case of exoelectrogens, an inhibitory effect on the transfer rate of electrons to the electrode. Correspondingly, exoelectrogens-based electrochemical biosensors are applicable for environmental monitoring.

- To understand the electron transfer to the electrode during the respiration chain is crucial for exoelectrogens-based biosensors. Three main mechanisms are distinguished for electron transfer from exoelectrogens to the electrode: direct electron transfer via outer-surface c-type cytochromes, long-range electron transfer via microbial nanowires, and electron flow through a self-excreted soluble electron shuttles. The mechanisms that are most important depend on the microorganisms and the thickness of the electrode biofilm.

- The performance, sensitivity and reliability of the exoelectrogens-based biosensor are dictated by exoelectrogens immobilization. A full understanding of the surface morphology of the exoelectrogens-electrochemical biosensor is necessary to guarantee the correct interpretation of the experimental results.

- The incorporation of biocompatible conductive nanomaterials into bacterial films opens up a new avenue to accelerate the development of the exoelectrogens-based biosensor. The electron propagation can go along the nanomaterial-assembled structures instead of the traditional electron transfer between adjacent bacteria. In addition, the presence of nanomaterials may provide a favorite biocompatible environment for the proteinaceous electron transfer components of extracellular electron transfer.

- Exoelectrogens-electrodes have recently become popular in biosensor design for environmental monitoring. Microbial biosensors can be used in the continuous monitoring of a contaminated area. They are applicable in the determination of the "biological oxygen demand" (BOD) as well as the toxicity of certain compounds.

- Other potential applications of exoelectrogens based biosensors include food analysis, biopharmaceuticals, clinical medicine and medical diagnosis.

KEY FACTS

Electrochemical monitoring is of special interest for *in situ* measurements, the advantages of electrochemical measurements include:

- Speed
- Can be used online and *in situ*
- Highly sensitive and can detect relatively low activities
- Reproducible
- Easy to manipulate and multiply
- Easy for miniaturization

  Because of these properties, electrochemical biosensors are especially useful for monitoring environmental pollution, as they can be used in the site that has to be monitored as well as monitoring accurately in water, turbid water or soil.
- A microbial fuel cell (MFC), which converts biochemical energy into electrical energy through the catalytic activities of microorganisms, has attractive advantages in that electricity generation arises from the utilization of organic materials
- Exoelectrogens, which use the electrode as the sole electron acceptor, can completely oxidize organic compounds. Several mechanisms for electron transfer to electrodes have been proposed, including: direct electron transfer via outer-surface c-type cytochromes, long-range electron via microbial nanowires, electron flow through a soluble electron shuttles.
- The direct electron transfer (DET) of exoelectrogens towards the electrode is quite attractive in electrochemical biosensor, which is sensitive to the changes in the metabolic status of the cellular biocatalyst and simplifies the construction procedure. The amount of electron transfer to the electrode is found in proportion of the exoelectrogens metabolic activity.
- The ideal microbial biosensor platform should provide a favorable microenvironment to maintain the microbe activity and minimize kinetics barriers of substrate and product. Nanomaterials are expected to be good candidates for electrode improvements.

DEFINITIONS

- *Biosensors*: a device incorporating a biological sensing element connected to a transducer.
- *Exoelectrogens*: a kind of microorganism, that can use the electrode as the sole electron acceptor while oxidizing the organic materials.
- *Microbial fuel cell*: A microbial fuel cell (MFC) or biological fuel cell is a bio-electrochemical system that drives a current by mimicking bacterial interactions found in nature.
- *Nanomaterials*: materials with morphological features on the nanoscale, especially those that have special properties stemming from their nanoscale dimensions.
- *Electrode*: an electrical conductor used to make contact with a nonmetallic part of a circuit (e.g. a semiconductor, an electrolyte or a vacuum).
- *BOD*: is a chemical procedure for determining the amount of dissolved oxygen needed by aerobic biological organisms in a body of water to break down organic material present in a given water sample at certain temperature over a specific time period.
- *COD*: Nearly all organic compounds can be fully oxidized to carbon dioxide with a strong oxidizing agent under acidic conditions. In environmental chemistry, the COD test is commonly used to indirectly measure the amount of organic compounds in water.
- *In situ*: in place.

ACKNOWLEDGEMENT

This research was supported by the National Natural Science Foundation of China (Nos. 21075116, 20935003 and 20820103037) and 973 Project (2009CB930100, 2011CB911002 and 2010CB933600).

ABBREVIATIONS

AW	:	artificial wasterwater
BOD	:	Biochemical oxygen demand
CAC	:	Citric acid cycle
CNT	:	Carbon nanotube
COD	:	Chemical oxygen demand
DET	:	Direct electron transfer
GC-MS	:	Gas chromatography-mass spectrometry

HPLC	:	High-performance liquid chromatography
LC-MS	:	Liquid chromatography-mass spectrometry
MFC	:	Microbial fuel cell
OM	:	outer membrane
PANI	:	Polyaniline
PEM	:	Proton exchange membrane
SCMFC	:	Single-chamber microbial fuel cell

REFERENCES

Aelterman PS Freguia, J Keller, W Verstraete and K Rabaey. 2008. The anode potential regulates bacterial activity in microbial fuel cells. Appl Microbiol Biotechnol 78: 409–418.

Bouhenni, RA, JV Gary, JC Biffinger, S Shirodkar, K Brockman, R Ray, P Wu, BJ Johnson, EM Biddle, MJ Marshall, LA Fitzgerald, BJ Little, JK Fredrickson, AS Beliaev, BR Ringeisen, and DA Saffarinia. 2010. The Role of Shewanella oneidensis MR-1 Outer Surface Structures in Extracellular Electron Transfer. Electroanalysis 22: 856–864.

Bretschger O, A Obraztsova, CA Sturm, IS Chang, YA Gorby, SB Reed, DE Culley, CL Reardon, S Barua, MF Romine, JZ Zhou, AS Beliaev, R Bouhenni, D Saffarini, F Mansfeld, BH Kim, JK Fredrickson and KH Nealson. 2007. Current production and metal oxide reduction by *Shewanella oneidensis* MR-1 wild type and mutants. Appl Environ Microbiol 73: 7003–7012.

Chang IS, JK Jang, GC Gil, M Kim, HJ Kim, BW Cho and BH Kim. 2004. Continuous determination of biochemical oxygen demand using microbial fuel cell type biosensor. Biosens Bioelectron 19: 607–613.

Chang IS, H Moon, JK Jang and BH Kim. 2005. Improvement of a microbial fuel cell performance as a BOD sensor using respiratory inhibitors. Biosens Bioelectron 20: 1856–1859.

Deng L, SJ Guo, M Zhou, L Liu, C Liu and SJ Dong. 2010a. A silk derived carbon fiber mat modified with Au@Pt urchilike nanoparticles: A new platform as electrochemical microbial biosensor. Biosens. Bioelectron 25: 2189–2193.

Deng L, SJ Guo, ZJ Liu, M Zhou, D Li, L Liu, GP Li, EK Wang and SJ Dong. 2010b. To boost c-type cytochrome wire efficiency of electrogenic bacteria with Fe_3O_4/Au nanocomposites. Chem Commun 46: 7172–7174.

Erable B, MA Roncato, W Achouak and A Bergel. 2009. Sampling natural biofilms: a new route to build efficient microbial anodes. Environ Sci Technol 43: 3194–3199.

Erable B, M Narcis, MM Duteanua, CD Ghangrekara and K Scott. 2010. Application of electro-active biofilms. Biofouling 26: 57–71.

Gorby YA, S Yanina, JS McLean, KM Rosso, D Moyles, A Dohnalkova, TJ Beveridge, IS Chang, BH Kim, KS Kim, DE Culley, SB Reed, MF Romine, DA Saffarini, EA Hill, L Shi, DA Elias, DW Kennedy, G Pinchuk, K Watanabe, S Ishii, B E Logan, KH Nealson and JK Fredrickson. 2006. Electrically conductive bacterial nanowires produced by *Shewanella oneidensis* strain MR-1 and other microorganisms. Proc Natl Acad Sci USA 103: 11358–11363.

Hartshorne RS, CL Reardon, D Ross, J Nuester, TA Clarke, AJ Gates, PC Mills, JK Fredrickson, JM Zachara, L Shi, AS Beliaev, MJ Marshall, M Tien, S Brantley, JN Butt and DJ Richardson. 2009. Characterization of an electron conduit between bacteria and the extracellular environment. Proc Natl Acad Sci USA 106: 22169–22174.

Hyunsoo M, IS Chang, JK Jang, KS Kim, J Lee, RW Lovitt and BH Kim. 2005. Online monitoring of low biochemical oxygen demand through continuous operation of a mediator-less microbial fuel cell. J Microbiol Biotechnol 15: 192–196.

Jadhav GS and MM Ghangrekar. 2009. Performance of microbial fuel cell subjected to variation in pH, temperature, external load and substrate concentration. Bioresource Technol 100: 717–723.

Kang KH, JK Jang, TH Pham, H Moon, IS Chang and BH Kim. 2003. A microbial fuel cell with improved cathode reaction as a low biochemical oxygen demand sensor. Biotechnol Lett 25: 1357–1361.

Kim BH, IS Chang, GC Gil, HS Park and HJ Kim. 2003. Novel BOD (biological oxygen demand) sensor using mediator-less microbial fuel cell. Biotechnol Let 25: 541–545.

Kim M, MS Hyun, GM Gaddb and HJ Kim. 2007. A novel biomonitoring system using microbial fuel cells. J Environ Monit 9: 1323–1328.

Kumlanghan A, J Liu, P Thavarungkul, P Kanatharana and B Mattiasson. 2007. Microbial fuel cell-based biosensor for fast analysis of biodegradable organic matter. Biosens Bioelectron 22: 2939–2944.

Logan BE. 2008. Microbial Fuel Cells. John Wiley & Sons, Hoboken, New Jersey, USA.

Lorenzo MD, TP Curtis, IM Head and K Scott. 2009. A single-chamber microbial fuel cell as a biosensor for wastewaters. Water Research 43: 3145–3154.

Lovely DR. 2008. The microbe electric: conversion of organic matter to electricity. Cur Opin in Biotechnol 19: 564–571.

Marsili E, DB Baron, I Shikhare, D Coursolle, JA Gralnick and DR Bond. 2008. *Shewanella* secretes flavins that mediate extracellular electron transfer. Proc Natl Acad Sci USA 105: 3968–3973.

Methé BA, KE Nelson, JA Eisen, IT Paulsen, W Nelson, JF Heidelberg, D Wu, M Wu, N Ward, MJ Beanan, RJ Dodson, R Madupu, LM Brinkac, SC Daugherty, RT DeBoy, AS Durkin, M Gwinn, JF Kolonay, SA Sullivan, DH Haft, J Selengut, TM Davidsen, N Zafar, O White, B Tran, C Romero, HA Forberger, J Weidman, H Khouri, TV Feldblyum, TR Utterback, SE Van Aken, DR Lovley and CM Fraser. 2003. Genome of Geobacter sulfurreducens: Metal Reduction in Subsurface Environments. Science 302: 1967–1969.

Meyer TE, AT Tsapin, I Vandenberghe, L de Smet, D Frishman, KH Nealson, MA Cusanovich and JJ van Beeumen. 2004. Identification of 42 Possible Cytochrome C Genes in the Shewanella oneidensis Genome and Characterization of Six Soluble Cytochromes. OMICS 8: 57–77.

Moon H, IS Chang, KH Kang, JK Jang and BH Kim. 2004. Improving the dynamic response of a mediator-less microbial fuel cell as a biochemical oxygen demand (BOD) sensor. Biotechnol Let 26: 1717–1721.

Patil S, F Harnisch and U Schröer. 2010. Toxicity Response of Electroactive Microbial Biofilms-A Decisive Feature for Potential Biosensor and Power Source Applications. Chem Phys Chem 11: 2834–2837.

Peng L, SJ You and JY Wang. 2010. Carbon nanotubes as electrode modifier promoting direct electron transfer from *Shewanella oneidensis*. Biosens Bioelectron 25: 1248–1251.

Pham TH, N Boon, P Aelterman, P Clauwaert, L D Schamphelaire, L Vanhaecke, K D Maeyer, M Höfte, W Verstraete and K Rabaey. 2008. Metabolites produced by *Pseudomonas* sp. enable a Gram positive bacterium to achieve extracellular electron transfer. Appl Microbiol Biotechnol 77: 1119–1129.

Qiao Y, CM Li, SJ Bao and QL Bao. 2007. Carbon nanotube/polyaniline composite as anode material for microbial fuel cells. J Power Sources 170: 79–84.

Qiao Y, CM Li, SJ Bao, Z Lu and Y Hong. 2008a. Direct electrochemistry and electrocatalytic mechanism of evolved *Escherichia coli* cells in microbial fuel cells. Chem Commun 1290–1292.

Qiao Y, SJ Bao, CM Li, XQ Cui, ZS Lu and J Guo. 2008b. Nanostructured polyaniline/titanium dioxide composite anode for microbial fuel cells. ACS Nano 2: 113–119.

Rabaey K, N Boon, M Hofte and W Verstraete. 2005. Microbial phenazine production enhances electron transfer in biofuel cells. Environ Sci Technol 39: 3401–3408.

Reguera G, KD McCarthy, T Mehta, JS Nicoll, MT Tuominen and DR Lovley. 2005. Extracellular electron transfer via microbial nanowires. Nature 435: 1098–1101.

Seok-Min Y, CH Choi, M Kim, MS Hyun, SH Shin, DH Yi and HJ Kim. 2007. Enrichment of electrochemically active bacteria using a three-electrode electrochemical cell. J Microbial Biotechnol 17: 110–115.

Stein NE, HVM Hamelers and CNJ Buisman. 2010. Stabilizing the baseline current of a microbial fuel cell-based biosensor through overpotential control under non-toxic conditions. Bioelectrochemistry 78: 87–91.

Wu XE, F Zhao, N Rahunen, JR Varcoe, C Avignone-Rossa, AE Thumser and RCT Slade. 2010. A Role for Microbial Palladium Nanoparticles in Extracellular Electron Transfer Angew Chem Int Ed 49: 1–5.

17

Detecting Waterborne Parasites Using Piezoelectric Cantilever Biosensors

Sen Xu[1,a] and Raj Mutharasan[1,b,*]

ABSTRACT

Molecular and immuno-based methods for detecting waterborne parasites in current use are both time-consuming and laborious. A great need exists for developing rapid and inexpensive methods for water quality and environmental monitoring. In this chapter, we review the application of piezoelectric-excited millimeter-sized cantilever (PEMC) sensor for the detection of waterborne parasites. Cantilever physics and working principle, sensor fabrication and surface functionalization, flow experimental design, as well as examples of detecting C. parvum and G. lamblia are presented. PEMC sensor is a mass-sensitive biosensor whose resonant frequency decreases when the mass of the sensor increases due to target binding. PEMC sensor functionalized with a specific antibody was exposed to target parasites in a flow apparatus. When the parasite binds to the surface-immobilized antibody, mass of the sensor increases and causes decrease of sensor resonant frequency. Real-time monitoring of resonant frequency changes was used for low concentration parasite detection.

[1]Department of Chemical and Biological Engineering, Drexel University, Philadelphia, PA 19104, USA.
[a]Email: xuzhonghou@gmail.com
[b]Email: mutharasan@drexel.edu
*Corresponding author

List of abbreviations after the text.

Limit of detection (LOD) for *C. parvum* using PEMC sensor was 5 oocysts/mL in 25% milk medium. In the dynamic range of 50 to 10,000 oocysts/mL the sensor response is characterized by a semi-log relationship between resonant frequency response and *C. parvum* oocysts concentration. In 25% milk background, both binding kinetics was slower and total sensor response was lower (~45%) than in water-like medium. LOD for *G. lamblia* was 10 cysts/mL in both buffer and complex matrixes (tap water and river water). Feasibility of analyzing at a low concentration of 1 cyst/mL in a one liter sample is also described.

INTRODUCTION

Parasites in finished water, if left undetected, can detrimentally impact both public safety and economy. *Cryptosporidium parvum* is a common parasitic protozoan responsible for numerous waterborne and foodborne outbreaks of diarrheal disease, causing a majority of gastrointestinal parasitic infections globally (Fayer et al. 2000). In the United States alone, an estimated 300,000 cases of cryptosporidiosis occur each year causing 66 deaths (Mead et al. 1999). Median infective dose is estimated as 87 oocysts for the Iowa calf isolates (Okhuysen et al. 1999). Oocysts have been found in surface water samples with concentrations ranging from 0.1 to 10,000 oocysts per 100 L (Lisle and Rose 1995; Mons et al. 2009). *Giardia lamblia* is a flagellated enteric protozoan parasite and a causative agent of human giardiasis (Adam 2001). Infection occurs by ingestion of *G. lamblia* cysts present in contaminated water and food, or by fecal-oral route. *Cryptosporidium* and *Giardia* have been classified as important human pathogens in the WHO Neglected Disease Initiative (Savioli et al. 2006). Since ingestion of as few as several (oo)cysts may cause cryptosporidiosis or giardiasis, there is a great need for a sensitive method for detecting and monitoring *C. parvum* and *G. lamblia* in source and finished water samples at ultralow concentrations.

For detecting waterborne pathogens, the popular methods are those based on culture and colony counting methods, polymerase chain reaction (PCR), and immuno-methods such as enzyme-linked immunosorbent assay (ELISA). All these approaches exhibit high selectivity and reliability. Culturing and plating method is the oldest bacterial detection technique; nowadays it is still a standard detection method because of its accuracy and high sensitivity. The largest drawback of this method is the long assay time (several days to weeks) and is not applicable to parasites. PCR can provide the ultimate sensitivity since it relies on gene amplification, but it also has the drawback of complexity, contamination, high cost, lack of portability and the need for trained personnel. ELISA, on the other hand, exhibits poor sensitivity (10^4–10^5/mL) compared with the culture and PCR method. The environmental protection agency (EPA) has established Method 1623 for simultaneous detection of *Cryptosporidium* spp. and *Giardia* spp. (USEPA 2005). The method involves multiple steps of filtration, immunomagnetic separation, and a detection method based on immunofluorescence, with confirmation through vital dye staining (4,6-diamidino-phenylindole, DAPI) and differential interference contrast microscopy. This method is tedious and requires trained personnel for conducting the assay. As a consequence, timely management decisions are not made. Therefore, there is a need for a rapid method for detecting waterborne parasites.

In the past 20 years, biosensors have attracted considerable interest because they can provide an equally reliable assay in a short time. The application areas range from clinical analysis to environmental and industrial processes monitoring. According to the *International Union of Pure and Applied Chemistry (IUPAC)*, a biosensor consists of three components: the biorecognition molecule that recognizes and binds the target of interest with high selectivity, the transducer that converts the binding reaction into a measurable signal, and the output system which amplifies and displays the signal in a useful form (Thevenot et al. 1999). Compared with conventional analytical and culture methods, biosensors offer significant advantages of being rapid and field portable. These features have attracted a great deal of interest in developing biosensors for waterborne parasites and environmental monitoring. Several reports have appeared in the literature for detecting *C. parvum* using surface plasmon resonance (SPR) (Kang et al. 2008) and quartz crystal microbalance with dissipation monitoring (QCM-D) (Poitras et al. 2009) in both buffer and water samples; and *G. lamblia* and *C. parvum* detection using a filter-based microfluidic device (Zhu et al. 2004). Other techniques, using electrochemical impedance spectroscopy (EIS) and gold nanoparticles have also been attempted for parasites detection.

In this chapter, we emphasize on the cantilever biosensor technology developed in our laboratory, namely piezoelectric-excited millimeter-sized cantilever (PEMC) sensor, for waterborne parasites detection. Cantilever sensors can be operated in two different modes: bending mode (static mode), in which deflection of the sensor is measured when target of interest binds to the sensor surface; resonant mode (dynamic mode), in which decrease in resonant frequency is measured as target of interest binds. PEMC sensor operates in the resonant mode. Similar to SPR and QCM, PEMC has the advantage of label-free, near real-time detection in a flow arrangement. Mass change sensitivity at levels of femtogram (10^{-15} g) was obtained in flow conditions (Maraldo et al. 2007b). The PEMC sensor has been used for the detection of a variety of pathogens and toxins due to its high mass-change sensitivity. In this chapter, we summarize our results obtained in detecting *C. parvum* and *G. lamblia* at very low concentration. Cantilever physics and experimental arrangement used are included, and typical detection examples are illustrated for demonstrating the working principle and effectiveness of the sensor platform.

CANTILEVER PHYSICS AND WORKING PRINCIPLE

PEMC sensor is a macro-cantilever, which comprises of a lead zirconate titanate (PZT) and a non-piezoelectric silica glass layer of a few millimeters in length and 1 mm in width (Fig. 17.1). The PZT layer of the cantilever acts both as an actuating and a sensing element. The detection of biological entities requires the immobilization of a recognition molecule, such as an antibody or a receptor molecule, on the sensor surface. When the target of interest binds to sensing surface, the effective mass of the cantilever increases and causes the decrease of resonant frequency. Sensing is achieved by tracking changes in resonant frequency.

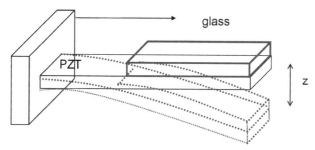

PZT: Lead zirconate titanate ($Pb[Zr_xTi_{1-x}]O_3$, $0<x<1$)

Figure 17.1. Schematic representation of a PEMC sensor.

When an electric field is applied across the thickness of PZT, it extends along its length and leads to bending of the bound glass. Periodically altering the field makes the PZT-glass composite cantilever vibrate. The natural frequency of a cantilever is a function of flexural modulus and mass density of the composite cantilever. When the excitation frequency of the cantilever matches its natural mechanical frequency, it resonates and undergoes higher than normal bending stress. The electro-mechanically active PZT exhibits a sharp change in electrical impedance that is conveniently measured using an impedance analyzer.

In general, the natural frequency of a beam with a flexural rigidity of EI, where E is the modulus of elasticity and I is the inertia moment, can be obtained by solving the general equation representing transverse mechanical vibration, and is given by Eq. (1):

$$EI(\frac{\partial^4 y}{\partial x^4}) + (\rho wt + m_a)(\frac{\partial^2 y}{\partial x^2}) + (c_o + c_v)(\frac{\partial y}{\partial \tau}) = 0 \qquad (1)$$

In the above equation, y is the displacement along the thickness of the cantilever, x is length along the cantilever, τ is time and ρ is density. The term c_o is the damping parameter intrinsic to cantilever materials. The parameters m_a and c_v represent the added mass per unit length of the cantilever and the added damping due to the fluid motion respectively. The inertia momentum is given by $wt^3/12$ where w is width and t is thickness. The resonant frequency is derived as (Elmer and Dreier 1997):

$$f_n' = \frac{\lambda_n^2}{2\pi} \sqrt{\frac{K}{M_e + m_{ae} + \Delta m}} \qquad (2)$$

where λ_n is the eigenvalue, K is the effective spring constant and M_e is the effective tip mass. When a PEMC sensor is immersed in a liquid medium the surrounding fluid acts as an added mass to the cantilever and results in an additional inertial force. In addition to this added mass effect, a dissipative force proportional to the velocity of the cantilever is included to describe the liquid behavior. This added term is due to the delayed response of the fluid to cantilever motion which leads to phase shifts in the motion of the fluid and the cantilever. On the PEMC sensor the added mass is the sum of the effective mass of the fluid surrounding the cantilever and the target bound to the sensor surface. Using Eq. (3) one can calculate the resonant frequency change as:

$$f_n' - f_n = \frac{1}{2} \frac{f_n \Delta m}{M_{ef}} \qquad (3)$$

where $M_{ef} = M_e + m_{ae'}$ is the effective mass of the cantilever under liquid immersion, Δm is the equivalent mass of the molecules bound to the sensor and f_n and f'_n are the resonant frequencies in the nth mode before and after the binding of molecules. Rearranging Eq. (3) one can get mass-change sensitivity, σ_n:

$$\sigma_n = \frac{\Delta m}{f_n - f_n'} = \frac{2M_{ef}}{f_{nf}} \qquad (4)$$

Resonant-mode micro- and nano-cantilevers actuated in vacuum have shown mass-change sensitivity of attograms or lower (Ilic et al. 2004; Yang et al. 2006). Such a level of sensitivity is far higher than what has been reported for bending-mode cantilever sensors. However, the micro- and nano-scale resonant-mode cantilevers can be used only in vacuum or in air due to viscous damping. The parameter that defines the usefulness of a resonating cantilever in liquid is the Reynolds number (Re) at the operating frequency, and is given as ($2\pi f b^2 \rho / \eta$); b is cantilever width, f is resonant frequency, ρ and η are density and viscosity of fluid (Sader 1998; Van Eysden and Sader 2007). When a sensor is operated at high Re numbers, the dominant forces on the sensor are inertial forces related to mass of the sensor and attached mass. On the other hand, when Re number is low, the dominant forces on the sensor are the viscous forces exerted by the surrounding liquid. Viscous damping not only reduces the quality factor, but also significantly reduces sensitivity to changes in mass (Waggoner et al. 2009). When cantilever width is 1 mm and a high order resonance mode near 1 MHz is used, the sensors do not suffer from viscous damping effects, because the Re number is far greater than 10^6. As a result, they are responsive to mass change and can be monitored continuously both in static and flow conditions.

SENSOR PREPARATION AND EXPERIMENTAL

Sensor Fabrication

Fabrication of PEMC sensors has been reported in detail in one of our earlier publications (Maraldo et al. 2007b). Briefly, a PEMC sensor consists of two layers: a PZT layer (Piezo Systems, Woburn, MA) and a quartz layer (SPI, West Chester, PA), bonded by a non-conductive adhesive. The dimensions of the PZT and glass were 2.5–2.8 × 1 × 0.127 mm, and 2 × 1 × 0.160 mm ($l \times w \times t$), respectively. Wires (30 gauge) were soldered to electrodes of PZT layer and anchored in 6 mm glass tubing. The sensor was spin-coated with polyurethane (Wasser, Auburn, MA) or parylene C (PDS 2010 LABCOTER® 2, SCS) for electrical insulation. Sensor phase and total impedance spectra in air and in deionized (DI) water are shown in Fig. 17.2A. Resonant peak at ~1 MHz is normally used for experiments. A microscopic picture of the fabricated sensor is shown in Fig. 17.2B.

Surface Functionalization

Different chemistries can be applied to the sensor surface for antibody immobilization. Silanization is a widely used method for silica surface functionalization. Organosilanes contain a silicon atom tetrahedrally bound to three similar functional groups, and the fourth is the functional group of interest used in the immobilization reaction. Three silane compounds used in practice are 3-aminopropyl-triethoxysilane, 3-mercaptopropyl-trimethoxysilane and 3-glycidoxypropyl-trimethoxysilane. The hydroxyl group on the glass is converted to one of the functional groups depending

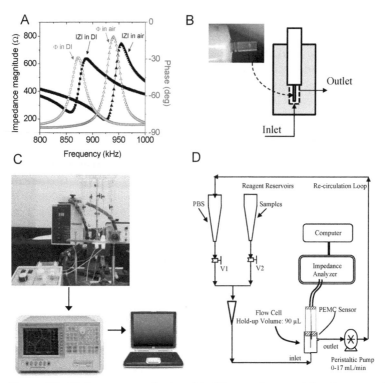

Figure 17.2. (A) Sensor spectra in air and in DI water. Sensor exhibited resonant frequency at 940.71 kHz in air and 873.27 kHz in DI water, which gave a $\Delta f = 67.44$ kHz shift due to the difference in density of the surrounding medium. Resonance is not damped, but persists in liquid because Re number is large. **(B)** Picture of a PEMC sensor and the installed configuration in the flow cell. **(C)** Photographic representation of experimental apparatus arrangement. **(D)** Flow system for detection experiments. Samples containing (oo)cysts at various concentrations are brought into the flow cell in a recirculation mode using a peristaltic pump. Flow rates used were 0–5.0 mL/min.

on the silane used. By activating the amine or carboxyl group on an antibody, one can immobilize the antibody to the sensor surface (Campbell and Mutharasan 2008).

Forming self-assembled monolayers (SAMs) on the sensor surface by chemisorption of thiol compounds (R-SH) onto gold surface is another common method for antibody immobilization (Love et al. 2005). Compounds such as cysteamine have both a functional amine group and a thiol group. It can form a monolayer on gold surface via the thiol group while the exposed amine group is available for activation with glutaraldehyde which can further react with amine groups on the antibody (Xu and Mutharasan 2010b).

Antibody immobilization can also be carried out via Protein G. The Fc region of IgG antibody binds to Protein G leaving the recognition region, Fab, exposed for antigen binding. Immobilization using Protein G offers better opportunity for orienting antibody on the sensor surface (Xu and Mutharasan 2010a).

Flow Configuration and Experimental

Prior to a detection experiment, sensor surface is functionalized with the desired antibodies, using one of the above immobilization methods. Detection experiments were done in the flow apparatus (Fig. 17.2C, D) at various flow rates from 1.0 to 5.0 mL/min. PEMC sensor was installed in the flow cell with a hold-up volume of 90 μL. The flow rate of the flow apparatus was controlled by a peristaltic pump. Five reservoirs were used for injection of various reagents such as antibody, antigen and release buffer. During the course of experiments, only one valve of the reservoir was opened and reagent from that reservoir was pumped into the flow cell for reaction on the sensor surface. Resonant frequency change was monitored in real-time by an impedance analyzer (HP 4192A or Agilent 4294A).

For detection confirmation, three different methods were used. The first method consists of regenerating the sensor surface using lower pH value buffers. With the release of attached oo(cysts) from the sensor surface, resonant frequency will increase due to decrease of attached mass. The second method consists of a second binding antibody for confirmation of the target on the sensor. Attachment of the second antibody to already attached oo(cysts) causes further resonant frequency decrease and is measured. The third method is examination of the sensor surface after detection experiments in a scanning electron microscope (SEM) and counting the oo(cysts) on the sensor surface.

DETECTION OF WATERBORNE PARASITES USING PEMC SENSOR

Typically, the detection experiment was done after the sensor was functionalized with the appropriate antibody. Sensor was installed in the flow cell and samples at various concentrations were pumped through the flow cell. Response of the sensor to attachment of target parasite was measured. In this section, we show several examples of *C. parvum* detection in 25% milk and *G. lamblia* detection in buffer, tap water and river water.

Detection of C. parvum in Buffer and 25% Milk

To test and compare the performance of PEMC sensor in matrixes that contain contaminants, detection experiments in PBS and in milk medium were carried out. Typical responses in PBS and in milk are shown in Fig. 17.3. The sensor was installed in the flow cell and flow rate used was 1 mL/min after functionalizing with IgG anti-*C. parvum*. Various concentrations of *C. parvum* oocysts were introduced into the flow loop and the flow was set in recirculation mode. Resonant frequency responses to sequential additions of 50,250 and 10,000 oocysts in PBS were 150 Hz, 535 Hz and 1,525 Hz, respectively, and is shown in Fig. 17.3A. The sensor showed a rapid resonant frequency decrease in the first ~15 min followed by a slower response reaching steady state in ~30 min. The total frequency response for the cumulative addition of 10,300 *C. parvum* oocysts was 2,210 Hz. Positive controls that expose the same concentrations of oocysts to the blank sensor showed no response and exhibited a noise level of ±10 to ±15 Hz, suggesting that no binding occurred. Negative control with IgG immobilized also showed no response to buffer.

For detection experiments in milk, the sensor was stabilized in PBS at 1 mL/min, and then 1 mL whole milk was introduced and the flow was set in recirculation. As shown in Fig. 17.3B, the resonant frequency decreased by ~380 Hz after a few cycles of mixing due to the density difference between milk and PBS. It is important to recognize that for a given excitation level (100 mV), the volume of surrounding liquid that oscillates with the cantilever is relatively constant for PBS or milk, because viscosities of the

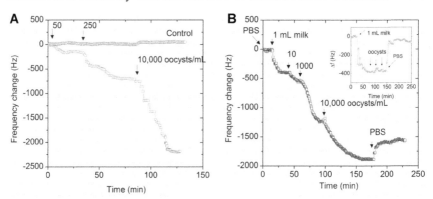

Figure 17.3. Sequential detection of *C. parvum* in PBS **(A)** and in 25% milk medium **(B)**, respectively. For detection in PBS, 1 mL 50, 250 and 10,000 oocysts/mL samples were injected into the flow loop sequentially. For detection in milk, 1 mL 10, 1,000 and 10,000 oocysts/mL samples were injected sequentially. Inset in (B) shows a positive control (bare sensor exposed to oocysts at same concentrations) that shows density response. No response was observed for injections of oocysts in both control experiments. Adapted with permission from Xu, S. and R. Mutharasan. Anal. Chim. Acta, 2010, 669: 81–86. Copyright 2010 Elsevier.

two fluids are similar. When a density change occurs, the oscillating fluid mass increases causing resonant frequency decrease. Given this background, detection of *C. parvum* oocysts at various concentrations were measured. Frequency responses for sequential additions of 10, 1,000 and 10,000 oocysts in 25% milk medium were 145 Hz, 680 Hz and 650 Hz, respectively. We note in Fig. 17.3B, the response was rapid upon sample introduction, followed by a slower change. For the addition of 10,000 oocysts, the response reached a steady state after 50 min, which is much longer than in PBS. The total frequency response for the cumulative *C. parvum* exposure (11,010 oocysts) was 1,745 Hz. After reaching a new steady state, PBS rinse caused a recovery of ~360 Hz, which was approximately the initial density response. The nearly full recovery suggests that no permanent non-specific attachment of milk components occurred on the sensor surface. The control experiment (see inset in Fig. 17.3B) conducted separately confirmed that the initial density response was reversible. Additionally, an unfunctionalized sensor showed no response to various concentrations (10 to 10,000 oocysts/mL) of oocysts injections. By comparing the total resonant frequency changes in PBS and in milk, one notes the response in milk was lower than in PBS. Additionally, the binding in milk also took a longer time, which we attribute to milk constituents impeding or hindering the binding process.

Comparison of various responses indicate that sensor frequency response magnitude was ~45% lower in milk than in PBS for the same *C. parvum* concentration (Xu and Mutharasan 2010a). However, the response of PEMC sensor for 5 oocysts in milk still gave a response. Among the six attempts of detecting 5 oocyst, four gave responses of 140, 50, 90 and 98 Hz, which gave an average of 95 ± 37 Hz (n = 4) with a signal to noise ratio > 4 since the noise level of most of the detection was ~15–20 Hz. Three repeated detection experiment results are shown in Fig. 17.4. As one notes, sensor positive responses to the concentration of 5 oocysts/mL were quite rapid, within 2–3 min. We thus estimate the limit of detection as 5 oocysts/mL.

Detection of *G. lamblia* in Buffer, Tap Water and River Water

The infectious dose of *G. lamblia* cysts has been reported to be as low as 10 to 25 cysts (Huang and White 2006), which means for a biosensor to be of practical use, the limit of detection should be less than 10 cysts. From a series of studies, we found that sensor response to *G. lamblia* concentration is flow rate dependent. That is, higher flow rate yielded higher sensor response (Xu and Mutharasan 2010b). Figure 17.5 shows sensor responses to separate exposures of low concentration samples at 10–500 cysts/mL. Positive control experiment with an unfunctionalized sensor exposed to

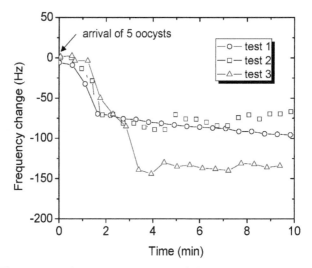

Figure 17.4. Three separate detection experiments with *C. parvum* concentration at 5 oocysts/mL. Sensor was functionalized with anti-*C. parvum* and placed in the flow cell.

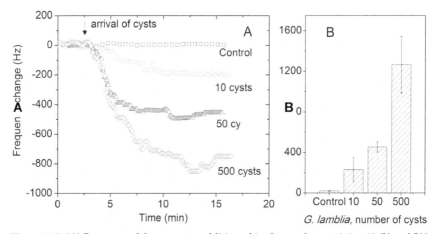

Figure 17.5. (A) Response of the sensor to addition of 1 mL sample containing 10, 50 and 500 cysts/mL *G. lamblia* samples to the flow loop set in recirculation mode at 2.4 mL/min. **(B)** Average frequency responses (n=4) to various concentrations at 2.4 mL/min. Negative control experiments showed responses of 20 ± 5 Hz (n = 4). Reproduced with permission from Xu, S. and R. Mutharasan. Environ Sci Technol. 2010, 44: 1736–1741. Copyright 2010 American Chemical Society.

500 cysts in recirculation showed a small response of 10 ± 6 Hz, which is quantitatively at the noise level. As shown in Fig. 17.5A, for 10, 50 and 500 cysts, the frequency decreases were 190, 450 and 790 Hz, respectively. Responses that were significantly larger than the noise level were observed in the first 5 min. In all of the experiments, step-wise response was not

observed, and we attribute this to sensor dynamics. The signal to noise (S/N) ratio for 10 cysts detection was greater than 10. Average response to repeat experiments (n = 4) is shown in Fig. 17.5B. One notes the sensitivity of the same sensor can be enhanced by flow rate resulting in improved S/N, and therefore a lower limit of detection. S/N ratio is a very important parameter for evaluating the performance and establishing the detection limit of a biosensor. Normally the higher the S/N ratio is, the more reliable are the results.

Further, similar level of sensitivity was examined in practical matrixes, namely tap water and river water. Experiments were designed to examine if the tap and river water samples contained any *Giardia* cysts. Antibody-immobilized sensors exposed to 50 mL of tap/river water in a once-through mode after stabilization in DI water showed no detectable response, which verified that no cysts were present in the tap/river water tested. Negative controls in PBS, tap and river water with antibody-immobilized sensors exposed to the three water samples showed 15 ± 6, 20 ± 5 and 39 ± 9 Hz response, respectively. Positive controls with unfunctionalized sensors exposed to 10,000 cysts spiked in the three water matrixes set in recirculation mode showed 20 ± 5, 18 ± 8 and 45 ± 12 Hz responses, respectively. Sensors for detection experiments were functionalized with anti-*G. lamblia* and installed in flow cell with tap or river water as the running buffer. Flow rate of 2.4 mL/min was used as it gave a good signal-to-noise ratio.

Spiked water samples containing 10 to 10,000 cysts were added to the flow loop following the same procedure as with the PBS samples. The results, summarized in Table 17.1 and Fig. 17.6, show the response in tap water was 4–20 % lower than that obtained in PBS and the response to river water samples was 20–47 % lower at the concentrations tested. The control experiments verified that the positive responses to spiked samples were due to cysts binding to the sensor surface. Lower sensor response in real water samples compared with that in buffer samples is estimated to be due to the interference of both biological and non-biological matters in the sample matrix.

Masking of antigenic sites of immobilized antibodies by organic macromolecules such as humic and fulvic acids in river water has been

Table 17.1. Sensor responses to *G. lamblia* in various media (PBS, tap water and river water).

Conc. (cysts/mL)	PBS		Tap water		River water	
	-Δf (Hz)	S.D. (Hz)	-Δf (Hz)	S.D. (Hz)	-Δf (Hz)	S.D. (Hz)
0	15	6	20	5	39	9
10	231	118	200	42	142	109
100	716	120	691	128	509	371
1,000	1,662	369	1,388	301	852	474
10,000	3,290	1,027	2,442	461	1,746	379

S.D.: standard deviation (n=3).

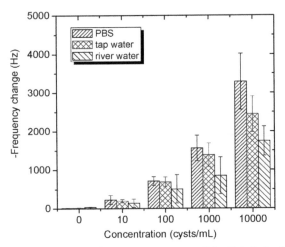

Figure 17.6. Comparison of sensor detection responses at 2.4 mL/min in the three water matrixes: PBS, tap water and river water. In all cases sample size was 1 mL. Both tap and river water samples gave lower sensor response at the same concentration. The decrease of response is attributed to weak binding of materials in water to cysts surface. Reproduced with permission from Xu, S. and R. Mutharasan. Environ Sci Technol. 2010, 44: 1736–1741. Copyright 2010 American Chemical Society.

suggested in the literature (Poitras et al. 2009). It is equally likely that antigenic sites on the cyst may weakly adsorb extraneous matter rendering them unrecognizable by surface-immobilized antibody. The latter is estimated to be the dominant mechanism as previous work on urine and serum samples show that proteinous matter in the sample matrix do not adsorb on PEMC sensors that are in continuous resonance in a flow field (Maraldo and Mutharasan 2007; Rijal and Mutharasan 2007; Xu et al. 2010). Another factor that may also contribute to lower sensor response is the pH at which detection was carried out. The pHs of tap and river waters were 6.7 ± 0.1 and 6.9 ± 0.1, respectively, and are lower than 7.4 at which the PBS experiments were conducted. In any case, the sensor exhibited sensitivity for detecting 10 cysts in tap and river water background.

Assessment of One Liter Samples with 1 Cysts/mL Concentration

An even higher flow rate (5.0 mL/min) was tested in a once-through flow mode for examining if a large one liter sample could be analyzed. The purpose of using higher flow rate is to explore the possibility of large sample processing in a short time without a pre-concentration step for detection. The frequency response of the sensor to 1,000 and 10,000 cysts in one liter sample at 5.0 mL/min is plotted in Fig. 17.7 and compared with the

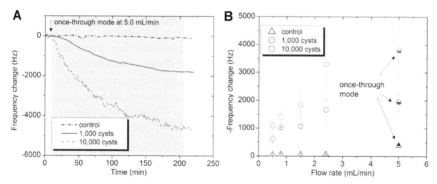

Figure 17.7. **(A)** Detection of 1 cyst/mL and 10 cysts/mL *G. lamblia* in one liter sample at 5.0 mL/min. The one liter sample containing either 1,000 or 10,000 cysts was introduced into the flow loop in a one-through mode followed by PBS rinse. Average responses for 1 and 10 cysts/mL were 1,950 ± 318 Hz and 3,800 ± 895 Hz, respectively. Control responses with 10,000 cysts were 380 ± 42 Hz. **(B)** PEMC sensors responses to 1,000 and 10,000 cysts as a function of flow rate. One-through mode was used in detection at 5.0 mL/min, and the others were on a recirculation mode. Control experiments that exposed an unfunctionalized sensor to 10,000 cysts were carried out at various flow rates. Error bars represent one standard deviation (n = 3 to 5). Adapted with permission from Xu, S. and R. Mutharasan. Environ Sci Technol 2010, 44: 1736–1741. Copyright 2010 American Chemical Society.

response obtained at lower flow rates. On an average, the resonant frequency decreased by 3,800 ± 895 Hz (n=3) when one liter of 10 cysts/mL *G. lamblia* was tested in a once-through fashion. And the response for one liter of 1 cysts/mL *G. lamblia* was 1,950 ± 318 Hz (n = 3). The magnitude of sensor response is larger than the response obtained at lower flow rates. Control experiments were carried out with 10,000 cysts and an unfunctionalized sensor; the responses were 42 ± 24 Hz, 54 ± 36 Hz, 58 ± 24 Hz, 65 ± 21 Hz and 194 ± 50 Hz at flow rate of 0.5, 0.8, 1.5, 2.4 and 5.0 mL/min, respectively. The higher response level at 5.0 mL/min may include some sensor drift as the length of experiment was longer than 4 hr. One notes the sensor responses to 1,000 and 10,000 cysts at 5.0 mL/min are easily distinguishable from the controls. The tests showed detection feasibility of 1 cyst/mL in a one liter sample in 4 hr at 5.0 mL/min. The data in Fig. 17.7A can be normalized by considering the arrival rate of target cysts at the sensor by plotting sensor response against Ct where C is cyst concentration and t is time. Such a plot nearly merges the two profiles into one suggesting that the binding rate when normalized for concentration gave nearly the same binding rate.

The transport of cysts to the sensor surface is the first of a two-step process that leads to sensor response. The second step is the binding reaction to surface-immobilized antibody. The transport step is governed by the flow field in the flow cell. Stokes settling velocity of the cysts (assuming average diameter = 10 μm, density = 1.02 g/cm³) is ~0.000109 cm/s, and the average velocity of the fluid in the flow cell is 0.046–0.46 cm/s for the flow rate range of 0.5 to 5.0 mL/min. Thus the flow velocity is far higher than the settling velocity, and we expect the cyst to follow the fluid flow path through the flow cell. In order to examine the second question, a calculation of the flow field in the flow cell using 2-D Navier-Stokes equation was conducted. Finite element simulation (COMSOL Multiphysics, Boston, MA) at various flow rates from 0.5 to 5.0 mL/min were analyzed (Xu and Mutharasan 2010b). Higher level of interaction of the sample with the sensor surface increases the probability of the cysts binding to it. Higher inlet velocity results in better exposure of *G. lamblia* cysts to the antibody on the sensor surface and accounts for the higher sensor response. Although an even higher flow rate can improve cysts interaction with the sensor surface, a higher noise level in resonant frequency occurs that compromises the measurement. Furthermore, higher flow rate will also cause higher shear stress at the sensor surface that potentially could dislodge bound cysts or impede binding of cysts to antibody.

KEY FACTS OF PEMC SENSOR

Key facts of PEMC sensor are the following:

- PEMC sensor is an electromechanical resonant-mode sensor. It consists of a piezoelectric (PZT) layer and a non-piezoelectric layer. The PZT layer acts both as an actuating and a sensing layer.
- PEMC sensor measures the decrease of resonant frequency as an indicator of target parasite binding to the sensor surface. An impedance analyzer is used for tracking resonant frequency changes in real-time.
- High sensitivity: PEMC sensor has been shown to be highly sensitive both in air and in liquid. Mass change measurement of femtogram (10^{-15} g) has been demonstrated.
- Detection using PEMC sensor is a rapid process. With an antibody functionalized sensor, a positive detection can be obtained in 15 min and the whole detection sequence can be completed in 2 hr.
- PEMC has the unique capability of operating in high liquid flow field. Operation of PEMC sensor at the flow rate as high as 17 mL/min has been shown. PEMC sensor does not suffer from viscous damping effects, since the *Re* number is far greater than 10^6. As a result, they are responsive to mass change and can be monitored continuously both in static and flow conditions.
- PEMC sensor is a vibrating structure, and the vibration amplitude is an intrinsic property for modulation of non-specific adsorption and thus enhances the specificity of the sensor.

APPLICATION TO OTHER AREAS OF HEALTH AND DISEASE

Besides detection of waterborne parasites in buffer and in various water matrixes, PEMC sensor has also been used for detection of a large sample of other pathogens and toxins. Significant examples include *E. coli* O157:H7 (Maraldo et al. 2007b) and *Bacillus anthracis* (Campbell and Mutharasan 2006). A recent study shows that cell viability can be determined within a short time period using PEMC sensor and a dye that accumulates only inside live cells (Xu and Mutharasan 2011). Toxin detection such as staphylococcus enterotoxin B (SEB) in food matrix at as low as 2.5 fg/mL was shown to be feasible (Maraldo and Mutharasan 2007). PEMC sensor has also been used for biomarkers detections such as alpha-methylacyl-CoA racemase (AMACR) in urine samples (Maraldo et al. 2007a).

SUMMARY POINTS

- For detection experiments, PEMC sensor was immobilized with the appropriate antibody and placed in a flow apparatus, and was then exposed to test samples. Resonant frequency of the sensor was monitored using an impedance analyzer. When the (oo)cysts bind to the antibody on the sensor, the resonant frequency of the cantilever sensor decreased and was recorded continuously.
- Detection of *C. parvum* in PBS or 25% milk background is feasible at as low a concentration as 5 oocysts/mL in flow conditions.
- The sensor showed a semi-log relationship between resonant frequency change and *C. parvum* oocysts concentration with a dynamic range of 50–10⁴ oocysts/mL.
- PEMC sensor exhibits sensitive detection of *G. lamblia* cysts in several water matrixes (buffer, tap and river water) in a dynamic range of 10–10⁴ cysts/mL without a pre-concentration step.
- Detecting of as few as 10 cysts per mL was achieved in all three water matrixes tested, and significant sensor response was obtained in 15 min.
- Feasibility of analyzing at a low concentration of 1 cyst/mL in a one liter sample at a high flow rate of 5.0 mL/min was demonstrated.

ACKNOWLEDGEMENT

The authors appreciate the financial support from the Environmental Protection Agency (STAR Grant R833007) during the writing of this chapter.

ABBREVIATIONS

AMACR	:	alpha-methylacyl-CoA racemase
DAPI	:	4,6-diamidino-phenylindole
EPA	:	Environmental protection agency
EIS	:	Electrochemical impedance spectroscopy
IUPAC	:	the International Union of Pure and Applied Chemistry
LOD	:	Limit of detection
PCR	:	Polymerase chain reaction
PEMC sensor	:	Piezoelectric-excited millimeter-sized cantilever sensor
PZT	:	lead zirconate titanate
QCM-D	:	Quartz Crystal Microbalance with dissipation monitoring
SEB	:	Staphylococcus enterotoxin B
SEM	:	Scanning electron microscope
S/N ratio	:	Signal to noise ratio
SPR	:	Surface plasmon resonance

REFERENCES

Adam RD. 2001. Biology of *Giardia lamblia*. Clin Microbiol Rev 14: 447–475.

Campbell GA and R Mutharasan. 2006. Piezoelectric-excited millimeter-sized cantilever (PEMC) sensors detect *Bacillus anthracis* at 300 spores/mL. Biosens Bioelectron 21: 1684–1692.

Campbell GA and R Mutharasan. 2008. Near real-time detection of *Cryptosporidium parvum* oocyst by IgM-functionalized piezoelectric-excited millimeter-sized cantilever biosensor. Biosens Bioelectron 23: 1039–1045.

Elmer FJ and M Dreier. 1997. Eigenfrequencies of a rectangular atomic force microscope cantilever in a medium. J Appl Phys 81: 7709–7714.

Fayer R, U Morgan and SJ Upton. 2000. Epidemiology of *Cryptosporidium*: transmission, detection and identification. Int J Parasitol 30: 1305–1322.

Huang DB and AC White. 2006. An updated review on *Cryptosporidium* and *Giardia*. Gastroenterol. Clin North Am 35: 291–314.

Ilic B, HG Craighead, S Krylov, W Senaratne, C Ober and P Neuzil. 2004. Attogram detection using nanoelectromechanical oscillators. J Appl Phys 95: 3694–3703.

Kang CD, C Cao, J Lee, IS Choi, BW Kim and SJ Sim. 2008. Surface plasmon resonance-based inhibition assay for real-time detection of *Cryptosporidium parvum* oocyst. Water Res 42: 1693–1699.

Lisle JT and JB Rose. 1995. *Cryptosporidium* contamination of water in the USA and UK-A minireview. J Water Supp Res Technol 44: 103–117.

Love JC, LA Estroff, JK Kriebel, RG Nuzzo and GM Whitesides. 2005. Self-assembled monolayers of thiolates on metals as a form of nanotechnology. Chem Rev 105: 1103–1169.

Maraldo D and R Mutharasan. 2007. Detection and confirmation of staphylococcal enterotoxin B in apple juice and milk using piezoelectric-excited millimeter-sized cantilever sensors at 2.5 fg/mL. Anal Chem 79: 7636–7643.

Maraldo D, FU Garcia and R Mutharasan. 2007a. Method for quantification of a prostate cancer biomarker in urine without sample preparation. Anal Chem 79: 7683–7690.

Maraldo D, K Rijal, G Campbell and R Mutharasan. 2007b. Method for label-free detection of femtogram quantities of biologics in flowing liquid samples. Anal Chem 79: 2762–2770.

Mead PS, L Slutsker, V Dietz, LF McCaig, JS Bresee, C Shapiro, PM Griffin and RV Tauxe. 1999. Food-related illness and death in the United States. Emerging Infect Dis 5: 607–625.

Mons C, A Dumetre, S Gosselin, C Galliot and L Moulin. 2009. Monitoring of *Cryptosporidium* and *Giardia* river contamination in Paris area. Water Res 43: 211–217.

Okhuysen PC, CL Chappell, JH Crabb, CR Sterling and HL DuPont. 1999. Virulence of three distinct *Cryptosporidium parvum* isolates for healthy adults. J Infect Dis 180: 1275–1281.

Poitras C, J Fatisson and N Tufenkji. 2009. Real-time microgravimetric quantification of *Cryptosporidium parvum* in the presence of potential interferents. Water Res 43: 2631–2638.

Rijal K and R Mutharasan. 2007. PEMC-based method of measuring DNA hybridization at femtomolar concentration directly in human serum and in the presence of copious noncomplementary strands. Anal Chem 79: 7392–7400.

Sader JE. 1998. Frequency response of cantilever beams immersed in viscous fluids with applications to the atomic force microscope. J Appl Phys 84: 64–76.

Savioli L, H Smith and A Thompson. 2006. *Giardia* and *Cryptosporidium* join the 'Neglected Diseases Initiative'. Trends Parasitol 22: 203–208.

Thevenot DR, K Toth, RA Durst and GS Wilson. 1999. Electrochemical biosensors: Recommended definitions and classification. Pure Appl Chem 71: 2333–2348.

USEPA. 2005. Method 1623: *Cryptosporidium* and *Giardia* in water by filtration/IMS/FA 2005. http://www.epa.gov/microbes/1623de05.pdf.

Van Eysden CA and JE Sader. 2007. Frequency response of cantilever beams immersed in viscous fluids with applications to the atomic force microscope: Arbitrary mode order. J Appl Phys 101: 044908–11.

Waggoner PS, CP Tan, L Bellan and HG Craighead. 2009. High-Q, in-plane modes of nanomechanical resonators operated in air. J Appl Phys 105: 094315–6.

Xu S and R Mutharasan. 2010a. Detection of *Cryptosporidium parvum* in buffer and in complex matrix using PEMC sensors at 5 oocysts mL^{-1}. Anal Chim Acta 669: 81–86.

Xu S and R Mutharasan. 2010b. Rapid and sensitive detection of *Giardia lamblia* Using a piezoelectric cantilever biosensor in finished and source waters. Environ Sci Technol 44: 1736–1741.

Xu S and R Mutharasan. 2011. Sensitive and rapid cell viability measurement using BCECF-AM and cantilever sensor. Anal Chem 83: 1480–1483.

Xu S, H Sharma and R Mutharasan. 2010. Sensitive and Selective Detection of Mycoplasma in Cell Culture Samples Using Cantilever Sensors. Biotechnol Bioeng 105: 1069–1077.

Yang YT, C Callegari, XL Feng, KL Ekinci and ML Roukes. 2006. Zeptogram-scale nanomechanical mass sensing. Nano Lett 6: 583–586.

Zhu L, Q Zhang, HH Feng, S Ang, FS Chauc and WT Liu. 2004. Filter-based microfluidic device as a platform for immunofluorescent assay of microbial cells. Lab Chip 4: 337–341.

18

Bioluminescence-based Biosensor for Food Contamination

Jinping Luo,[1,a] Weiwei Yue[1,b,2] Xiaohong Liu[1,c,2]
Qing Tian[1,d,2] and Xinxia Cai[1,e,2,*]

ABSTRACT

Bacterial contamination in the food industry is an increasing threat to public health, but the traditional methods of testing have several constraints. In order to effectively ensure food quality and safety, rapid and specific detecting systems of bacteria are needed. In this chapter, the design, construction and characterization of three novel biosensors based on adenosine 5′-triphosphate (ATP) bioluminescence have been described. An integrated biosensor for total bacterial count has been first developed by integrating a sampler, relevant reagents and detecting part so that there is no need of extra reagent. For sake of more accuracy, an automatic biosensor for total bacterial count has been

[1]State Key Laboratory of Transducer Technology, Institute of Electronics, Chinese Academy of Sciences, Beijing 100190, P.R. China.
[a]E-mail: jpluo@mail.ie.ac.cn
[b]E-mail: physics_yue@163.com
[c]E-mail: xinqi_0120@163.com
[d]E-mail: tianqing@ustc.edu
[e]E-mail: xxcai@mail.ie.ac.cn
[2]Graduate School of Chinese Academy of Sciences, Beijing 100080, P.R. China.
*Corresponding author

List of abbreviations after the text.

proposed with an automatic injection unit being introduced to control the amount of the relevant reagents. More than 500 real food samples were tested to investigate the characteristics of the two biosensors in detecting bacterial contamination in food. The results show that there are good linear relationships between the bacterial counts detected by the biosensors and that by the conventional method with a correlation coefficient of higher than 0.85. And the statistical results based on food varieties indicate that the total bacterial counts of drinks, jellies and meat products could be detected by the biosensor. Furthermore, a biosensor has been proposed to detect pathogenic bacteria by combining the bioluminescence with immunomagnetic separation. *Escherichia coli* (*E. coli*) was used as the target organism, while other typical enterobacteriaceae bacteria and a typical gram-positive bacterium were used as the contrast bacteria. The bioluminescence intensity connected with the concentration of *E. coli* at the range of $11 \sim 10^6$ CFU/mL. The assay took less than 20 min and showed a favourable specificity to detect *E. coli*. When using corresponding antigens for immunoreaction, the proposed assay could further extend to rapid detection of more pathogens and has potential application in areas such as food hygiene, medical and environmental protection.

INTRODUCTION

The Need for Rapid Detection of Bacterial Contamination

Food safety has been of concern to humankind since the dawn of history, and many of the problems encountered in our supply. In recent decades this concern has grown, because industrialization and growing mass production has increased risks of food contamination, as a result of which foodborne diseases remain one of the most widespread public health problems in the contemporary world (WHO 2010). In many countries, regular inspections of food products and food hygiene are carried out to identify all significant hazards in food processing units for ensuring food safety and quality (Kaferstein 1997). The potential contaminants are the exposure of the food to microorganisms. Microorganisms contaminating foods may render the food unfit for human consumption or cause serious foodborne diseases when consumed (Diane et al. 2010). Consequently, bacterial contamination in food industry is an increasing threat to public health.

In general, the Hazard Analysis and Critical Control Point (HACCP) procedures and product processing applied to food products, which focus on prevention and control and is advocated for every stage in the food chain from primary production to final consumption, are sufficient to protect consumers from the risk of diseases (WHO 1997). In many critical control points, the evaluation of bacterial count is the most frequently performed item to control bacterial contamination. Conventional culture techniques for detecting the bacteria rely on specific microbiological media to isolate and enumerate viable bacterial cells in foods (James and James 1998). These methods are very sensitive, inexpensive and could give both qualitative and quantitative information on the number of the microorganisms present in a food sample. However, the major drawbacks of microbiological methods are their labour-intensiveness due to culture medium preparation, inoculation of plates, colony counting and biochemical identification, and time-consuming as it takes 2–3 d for initial results and even up to 7–10 d for confirmation. Moreover, there is a need for more rapid methods to provide adequate information on the possible presence of pathogens in raw materials and the finished food products, for manufacturing process control and for the monitoring of cleaning and hygiene practices. Therefore, the demand for quick analytical tools for monitoring bacteria contamination in food industry has increased and is still expanding.

Rapid Techniques for Detecting Bacterial Contamination

Alternative and rapid methods can be divided into several categories as follows.

There are several modified conventional methods developed to shorten detection time or simplified detection procedure. Using chromogenic or fluorogenic substrates such as in selective media detection, enumeration and identification can be performed directly on the isolation plate, thus eliminating the use of subculture media and further biochemical tests (Manafi 1996). The hydrophobic grid membrane filter (HGMF) technique consists of capturing the detected bacteria in the HGMF by prefiltering the sample, and the similar incubation and colony counting processes with conventional techniques (Entis 1998). The Petrifilm system (3M) is an alternative to agar-poured plates, which consists of rehydratable nutrients that are embedded into a film along with a gelling agent, soluble in cold water (Gangar et al. 1999). These above methods have improved in plating or colony counting methods compared with the conventional culture techniques, but they still need a suitable incubation time for cultivating microorganisms into visible colonies.

Impedimetry technique is developed on the basis of changes in conductance in a medium where microbial growth and metabolism takes place (Waverla et al. 1998). During the growth of bacteria, the metabolism of such a substance as trimethylamine-N-oxide in the medium results in a large conductance change. When the bacterial numbers reach to above 10^6 CFU/mL, the change of conductance can be detected, and the bacterial counts then obtained according to the interrelation between the change of the conductance in a certain time and the bacterial number of the medium. Several automated systems based on impedimetry are commercially available, and used most commonly to estimate total bacterial counts and for screening of large numbers of samples as this procedure saves substantial time and material (van der Zee and Huis in't Veld 1997). However, the disadvantages of the impedance technique are that the food matrix may influence the analysis and it has little application in testing samples with low numbers of microorganisms.

Flow cytometry is one cell counting technique to detect bacterial contamination without any culture process (Griffiths 1997). It is an optically-based method for analyzing individual cells in complex matrixes. When the microorganisms suspended in a liquid pass a beam of laser light, the light is both scattered and absorbed by the microorganisms, and the extent and the nature of the scattering, may be analyzed by collecting the scattered light with a system of lenses and photocells and can be used to estimate the number, size, and shape of microorganisms. The analysis has a high sensitivity of 10^2–10^3 bacterial cells per mL and can be completed within a few minutes. It is very suitable for detecting low numbers of specific organisms in fluid, but it is unable to distinguish between living and dead cells and interference by the food matrix when used in food microbiology.

Immunology-based methods rely on the specific binding of an antibody to an antigen. Latex agglutination test (Matar et al. 1997), immunological precipitation (Feldsine et al. 1997), enzyme-linked immunosorbent assays (Bennett 2005), enzyme-linked immunomagnetic chemiluminescence (Magliulo et al. 2007) and immunochromatography (ICG) strip test (Shim et al. 2007) are included. In general, these methods consist of the capture of the bacteria cells by relevant antibodies fixed on such solid carriers as latex beads and magnetic particles, further binding with second antibodies labelled with enzyme, addition of the enzymatic substrate to produce a color change in the substrate or chemiluminescence, and final detection of the optical signal. In spite of their very little assay time compared to traditional culture techniques, the immunology-based detection is still lacking the ability to detect microorganisms in "real-time". The problems are probably the low sensitivity of the assays, low affinity of the antibody to the pathogen or other analyte being measured, and potential interference from contaminants.

Bioluminescence is routinely used for the detection of bacterial contamination via detection of the adesosine triphosphate (ATP) that is present in all living cells. This method is based on the reaction between the ATP extrated from the cells and the luciferin-luciferase complex. The total light output of a sample is directly proportional to the amount of ATP present and can be quantitated by luminometers. The bioluminescent method developed by Lundin (Lundin 2000) is known to be in accordance with the conventional microbiological culture methods and can be completed within several minutes. The rapid response time for obtaining results makes this method very suitable for on-line monitoring in HACCP programs (van der Zee and Huis in 't Veld 1997).

Since the biosensor has virtues of high selectivity, fast, simple and other characteristics, there are several novel techniques combined with these above methods with biosensor technology to seek for a practical detection device or system. For example, Madhukar and Li (2007) used an interdigitated array microelectrode based impedance biosensor with magnetic nanoparticle-antibody conjugates for detection of *Escherichia coli (E. coli) O157:H7* with detection limit of 7.4×10^4 colony-forming units per milliliter (CFU/mL). Zhu et al. (2005) demonstrated an integrating waveguide biosensor to detect *E. coli O157:H7* with detection limit of above 10^3 CFU /mL. John et al (2007) and Eum et al (2010) used surface plasmon resonance to detect *E. coli O157:H7* with detection limit of above 10^2 CFU/ mL. All of the above work, however, still required a complicate biosensor or their detection limits were still relatively high.

BIOLUMINESCENCE-BASED BIOSENSOR FOR DETECTING BACTERIAL CONTAMINATION

Recently there have been studies on combining the ATP bioluminescence assay with other technology to detect bacteria. Qiu et al. (2009) used bioluminescence and microfluidics to detect *Salmonella* and *Staphylococcus aureus (S. aureus)*, with detection limit of 10^2 CFU/mL, but needed a multilite luminometer containing the equipment of chemiluminescence and UV detection for bioluminescence detection. Cheng et al. (2009) used ATP bioluminescence combined with biofunctional magnetic nanoparticles to detect *E. coli* and the detection limit is lowered to 20 CFU/mL, but the detection time was about 1 hr. However, all of the above work still required a complicate biosensor or their detection time was still not met with the need of on-line monitoring in HACCP programs. Therefore, a better method is needed to solely detect viable target bacteria in a simpler way with greater sensitivity.

In this chapter, we describe the design, construction and characterization of three novel biosensors based on ATP bioluminescence. An integrated biosensor for total bacterial count was first designed by integrating a sampler, relevant reagents and detecting part so that there is no need of an extra reagent when used. For the sake of more accuracy, it was proposed that an automatic injection unit be introduced to control the adding of relevant reagents and an automatic biosensor for total bacterial count. Several 100s of real food samples were tested to investigate the characteristic of the two biosensors in detecting bacterial contamination in food. Furthermore, a biosensor was proposed to detect pathogenic bacteria by combining the bioluminescence with immunomagnetic separation.

Design and Characterization of the Integrated Biosensor for Total Bacterial Count

Figure 18.1 shows a scheme of the integrated biosensor for detecting total bacterial count. The biosensor contains two parts enveloped in a dark chamber: a home-made sensitive element, and a photomultiplier tube (PMT) used as detector element, which has been previously described (Luo et al. 2009). The former is the key to produce a measurable bioluminescence, and is disposable for the sake of low cost and easy operation. It comprises of a sampler, a cartridge and a microtube which are linked up through a screw thread in a coaxal-suite mode based on the bioluminescence ATP assay combined with a chemical method of ATP extraction. The somatic eliminating reagent, ATP extractant and luminescence reagent are sealed in advance at the swab, the bottom of the cartridge and the microtube, respectively, so that it is no need of an extra reagent when used. Once a sample gets to the sampler through swabbing or dropping, the free ATP or somatic ATP in the sample are hydrolyzed. As the sampler down-spins continuously, the intracellular ATP is released from bacterial cells by the ATP extractant in a few minutes and the residual extractant is subsequently consumed by the neutralizing reagent. When the lowest aluminum foil is

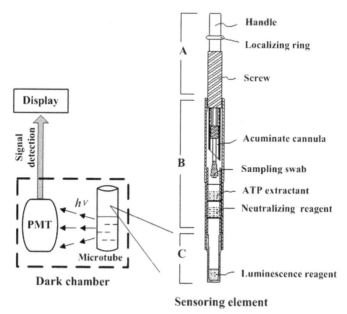

Figure 18.1. Scheme of the integrated biosensor for detecting total bacterial count. The biosensor contains a home-made sensitive element to produce a measurable bioluminescence, and a PMT used as detector element. Adapted with permission from (Luo et al. 2009). Copyright (2009) Elsevier.

torn, most of the solution containing the intracellular ATP drops into the microtube and the ATP quickly reacts with the luminescence reagent to produce bioluminescence. Then, the bioluminescence signal is translated to relevant electrical signal by the PMT and further detected by a handheld luminometer (Zhou et al. 2008). The bacterial count of the sample could be easily quantified by measuring the bioluminescence according to the relationship between the concentration of the intracellular ATP and the bioluminescence intensity.

Considering that real food is a complex sample which includes gram-negative bacteria and gram-positive bacteria, the standard bacterial solutions of *E. coli* mixed with *S. aureus* were taken as the test sample to calibrate the disposable biosensor and obtain the reproducibility. As presented in Fig. 18.2, the disposable biosensor has linear response to the bacterial count of the standard bacterial solutions at a concentration range from 10^3 CFU/mL to 10^8 CFU/mL with a relationship coefficient of 0.933. The total determination time including sampling and ATP extraction was less than 5 min. Moreover, the biosensor exhibits a satisfactory reproducibility for detecting bacterial count with a coefficient variation (CV) of 5.92 % (n = 14). The preliminary results illustrated that the biosensors are capable of rapid detection of the bacterial count in food.

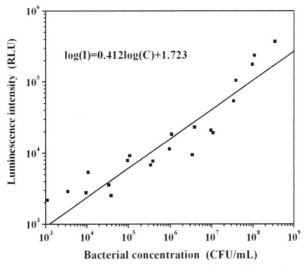

$$\log(I) = 0.412\log(C) + 1.723$$

Figure 18.2. Calibration curve of the integrated biosensor for total bacterial count in which the standard bacterial solutions of *E. coli* mixed with *S. aureus* were taken as the test sample. Where C is the bacterial count obtained by the standard plate count method, and I is the luminescence intensity measured by the biosensor. The bacterial concentrations of the test suspensions were ranging from 10^3 CFU/mL to 10^8 CFU/mL. Adapted with permission from (Luo et al. 2009). Copyright (2009) Elsevier.

More than 470 real food samples including meat products, dairy products, drink, condiment, candy, paddy and so on, were tested by the biosensor and the standard plate count method simultaneously. As shown in Fig. 18.3, there is a good linear relationship between the bacterial counts detected by the two methods with a correlation coefficient of 0.863. To express the offset between bioluminescent results and plate counting results, absolute divergence (Ad) in form of log-log was introduced. The bioluminescent results with Ad≤1 were on the same magnitude with plate count method and these results were thought to be valid. The statistical results based on food varieties are presented in Table 18.1. The incipient results demonstrated that the total bacterial counts of drinks, jellies, meat products and health-care foods could be detected by the biosensor while other kinds of food need to be studied in future work.

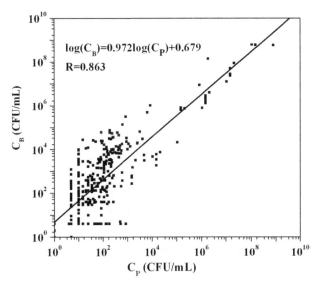

Figure 18.3. Correlation curve of bacterial count detected by the integrated biosensor and by the plate count method. Where C_B and C_P are the bacterial count obtained by the biosensor and that by the standard plate count method, respectively. More than 470 real food samples including meat products, dairy products, drink, condiment, candy, paddy etc., were tested. Raw data is unpublished.

Design and Characterization of the Automatic Biosensor for Total Bacterial Count

Figure 18.4 shows the scheme of an automatic biosensor for detecting total bacterial count. The automatic biosensor consists of three parts: automatic injection unit (AIU), photoelectric detecting unit (PDU) and signal processing and control unit (SPCU). AIU includes three reagent bottles and

Table 18.1. Relationship between absolute divergence and food categories detected by the integrated biosensor. Raw data is unpublished.

Food category	Number of samples (n)	Samples with Ad≤1 (n)	Percent (%)
Drink	58	51	88
Jelly	30	28	93
Meat products	83	66	80
Paddy products	158	65	41
Condiment	39	26	67
Health-care foods	42	38	90
Summation	410	274	67

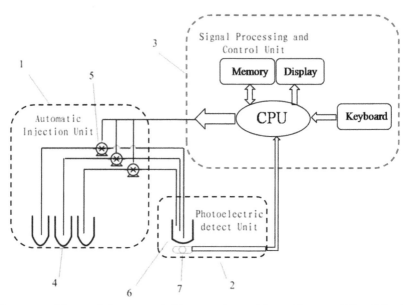

Figure 18.4. Scheme of the automatic biosensor for total bacterial count. 1-Flow Injection Unit; 2-Photoelectric detection Unit; 3-Signal Processing and Control Unit, 4-Reagent Bottle, 5-Peristaltic Pump, 6-Optical bio-sensing cell, 7-PMT. Raw data is unpublished.

three peristaltic pumps which was used to inject the ATP eliminated reagent, ATP extractant and luciferin-luciferase solution respectively into the optical bio-sensing cell under the control of SPCU. PDU contained an optical bio-sensing cell and PMT, which were enveloped in a dark chamber to isolate the light outside. SPCU was comprised of CPU, memory module, display module and other interface modules. SPCU was the key to control the time of adding reagents and reaction time accurately. Since each reagent could be injected into the optical bio-sensing cell punctually under the control of SPCU, each time of measurement was uniform. When 30 μL of food sample were added into the biosensing cell manually, relevant reagents including somatic ATP eliminating reagemt, ATP extractant and the bioluminescent reagent were injected into the cell by the peristaltic pumps, respectively and reacted to produce bioluminescence. PDU collected and converted the bioluminescence signal to relevant electric signal controlled by SPCU and then the result was obtained.

The jelly samples mixed with various standard bacteria suspensions were detected by the automatic biosensor to obtain the calibration equation. As shown in Fig. 18.5, it is obvious that the bioluminescence intensity is directly proportional to the amount of standard bacteria in a linear calibration range from 10^2 CFU/mL to 10^8 CFU/mL with a correlation coefficient higher than 0.97. Though the whole detection time is prolonged to 10 min, the automatic biosensor is more easy-to-operate and has a lower detection limit than the integrated biosensor. More than 500 real food samples including meat products, dairy products, drink, condiment, candy, paddy and so on, were tested by the biosensor and the conventional plate count method simultaneously. As shown in Fig. 18.6, a good linear relationship was obtained between the bacterial counts detected by the two methods with a correlation coefficient of 0.865. The statistical results based on food varieties are presented in Table 18.2. The incipient results demonstrated that surface samples, drink, jelly and meat products could be detected by the bioluminescent technique and other kinds of food need

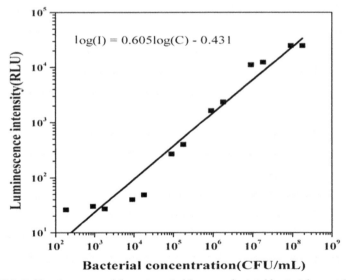

Figure 18.5. Calibration curve of the automatic biosensor for total bacterial count. Where C is the bacterial count obtained by the standard plate count method, and I is the luminescence intensity measured by the automatic biosensor. The bacterial concentrations of the test suspensions were ranged from 10^2 CFU/mL to 10^8 CFU/mL. Raw data is unpublished.

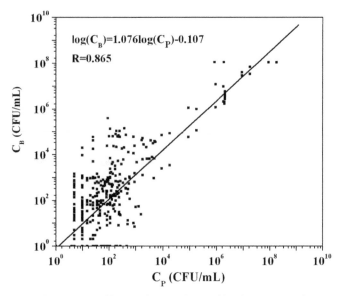

Figure 18.6. Correlation curve of bacterial count detected by the automatic biosensor and by the plant count method. C_B and C_P are the bacterial count obtained by the biosensor and that by the standard plate count method, respectively. More than 510 real food samples including meat products, dairy products, drink, condiment, candy, paddy etc, were tested. Raw data is unpublished.

Table 18.2. Relationship between absolute divergence and food categories detected by the automatic biosensor. Raw data is unpublished.

Food category	Number of samples(n)	Samples with Ad≤1 (n)	Percent (%)
Drink	48	48	100.0
Jelly	46	42	91.3
Meat products	108	93	86.1
Paddy products	177	114	64.4
Fuscous foods	25	10	40.0
Summation	430	333	77.4

to be studied in future work. On the other hand, it was also indicated that fuscous food was not suitable for the bioluminescent technique and our research team is working to solve this program by optical calibration.

Design and Characterization of the Biosensor for Pathogenic Bacteria

By combining the bioluminescence with immunomagnetic separation, a biosensor was also proposed to detect pathogenic bacteria. As shown in Fig. 18.7, the proposed biosensor consists two parts: an immunoenrichment part and an ATP-bioluminescence biosensing part. The immunoenrichment part was designed to capture specific bacteria using immunomagnetic microbeads coated with antibodies by the aid of a magnetic separator, in which the pathogenic bacteria was captured on the immunomagnetic microbeads through indirect antigen-antibody interactions as shown in Fig.18.8. The ATP-bioluminescence biosensing part was designed to detect the bioluminescence as similar to that of the automatic biosensor. When the bioluminescence signal is detected and transformed into a current signal by PMT, it will be transformed into a voltage signal by the signal magnifying part and then processed to be a final result under the control of the CPU. Furthermore, multiplex power supply chips are used to provide energies for different cells in order to lower relevant electro circuit noise and mutual interferences.

A series of *E. coli* suspensions with different concentrations were measured simultaneously by the proposed method and the standard plate count method. As presented in Fig. 18.9, it is obvious that the bioluminescence intensities are connected with the concentration of *E. coli*. Limits of detection of around 11 CFU/mL could be obtained with an upper end cut-off of around 10^6 CFU/mL. The whole procedure for detecting *E. coli* took about 20 min.

To investigate the specificity of the proposed method to detect *E. coli*, such typical enterobacteriaceae as *Klebsiella pneumoiae* (*K. pneumoniae*), *Enterobacter cloacae* (*E. cloacae*), and typical gram-positive bacterium like *S. aureus*, were chosen as the contrast bacteria. The concentrations of the detected bacteria using the conventional culturing method ranged from 10^5 CFU/mL to 10^7 CFU/mL. As seen in Table 18.3, the values of the contrast bacteria detected by the proposed method were far below that of *E. coli* with the same order of magnitude, while close to that of the PBS blank.

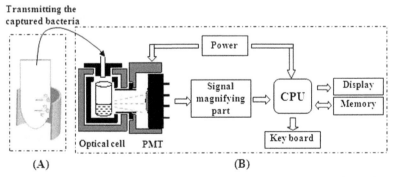

(A) (B)

Figure 18.7. Schematic diagram of the bioluminescence-based biosensor for pathogenic bacteria. It consists two parts: an immunoenrichment part **(A)** to capture specific bacteria using immunomagnetic microbeads coated with antibodies and an ATP-bioluminescence biosensing part **(B)** to detect the bioluminescence. Raw data is unpublished.

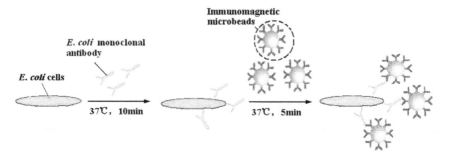

Figure 18.8. Schematic representation of the strategy used to specific capture of pathogenic bacteria. *E. coli* was used as the test example. The *E. coli* was captured on the immunomagnetic microbeads through indirect antigen-antibody interactions. Raw data is unpublished.

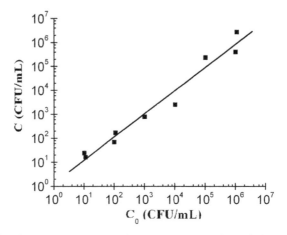

Figure 18.9. Correlation curve between the concentrations of *E. coli* detected by the two methods. Where C and C_0 are the concentration of the *E. coli* suspensions obtained by the proposed method and the conventional culturing method, respectively. Adapted with permission from (Luo et al. 2010).

Table 18.3. Comparison of various bacteria detected by the conventional culture method and the proposed method. Raw data is unpublished.

No.	Sample	Culture method (CFU/mL)	Proposed method (CFU/mL)
1	*E. coli*	1.13×10^5	4.18×10^4
2	*K. pneumoniae*	1.28×10^5	12
3	*E. cloacae*	9.8×10^6	1
4	*S. aureus*	2.46×10^6	1
5	PBS blank	0	0

The results indicated that the assay shows a favourable specificity to detect *E. coli*. It is noted that the results of *E. coli* concentration detected by the proposed method were slightly less than that by the conventional method, due to the loss of bacteria during the washing processes in the experiment. When using corresponding antigens for immunoreaction, the proposed assay could further extend to rapid detection of more pathogens and has potential application in areas such as food hygiene, medical and health, and environmental protection.

CONCLUSION

In this chapter, three novel bioluminescence-based biosensors were described and their characteristics were investigated. The integrated biosensor for total bacterial count was first designed by integrating a sampler, relevant reagents and the detecting part so that there is no need of an extra reagent. The automatic biosensor for total bacterial count was then proposed to improve detection accuracy. Though the entire detection time is prolonged from 5 min to 10 min, the automatic biosensor is more easy-to-operate than the integrated biosensor, and the detection limit is lowered to 10^2 CFU/mL. When 500 real food samples were tested, the results show that there are good linear relationships between the bacterial counts detected by the above biosensors and those of the conventional method with a correlation coefficient of higher than 0.85. The statistical results based on food varieties preliminarily show that the biosensors could be used to detect drinks, jellies and meat products. Furthermore, the biosensor for pathogenic bacteria was proposed by combining the bioluminescence with immunomagnetic separation. The bioluminescence intensity connected with the concentration of *E. coli* at the range of 11~10^6 CFU/mL. The assay took less than 20 min and showed a favourable specificity to detect *E. coli*. The proposed biosensor could further extend to rapid detection of more pathogens when using corresponding antigens for immunoreaction. The three bioluminescence-based biosensors based on ATP bioluminescence have potential applications in on-line monitoring in HACCP programs to detect bacterial contamination in food industry.

KEY FACTS

- Food safety has been of concern to humans since the dawn of history. This includes a number of routines that should be followed to avoid potentially severe health hazards. Food can transmit disease from person to person as well as serve as a growth medium for bacteria that can cause food poisoning.

- Since ATP is a major biological energy source existing in various bacteria with reasonably constant content of approximately 0.47 fg/cell (Lundin et al. 1989), ATP could reflect the existence of living microorganisms.
- Bioluminescence is production of light as a result of a chemical reaction catalyzed by a class of enzymes denominated luciferase, found in multiple living organisms. It relies on the reaction that ATP reacts with luciferin and oxygen to emit a photon of yellow-green light in the presence of the luciferase enzyme, similar to that of a firefly's tail (Lundin 2000). The light intensity is directly proportional to the amount of ATP present.
- The ATP bioluminescence technique for cell detection was first described in the 1960s by National Aeronautics and Space Administration (NASA) scientists who were interested in clinical applications. Since Sharpe first applied the ATP bioluminescence assay to measure microorganism in food during the 1970s (Griffiths 1996), there has been increasing research interest in applications of this technique for evaluating bacterial contamination.
- The HACCP concept was developed in the USA in 1959, further developed in collaboration with NASA and was documented as the HACCP in 1971 (Ioannis SA 2009). It was only in 1985 that its implementation was recommended to all food business operators by the US National Academy of Science. Since then, the system has been further developed worldwide.

DEFINITIONS

- *Adenosine triphosphate (ATP)*: the universal energy donor for metabolism in living organisms and therefore an indicator of life. One molecule of ATP contains three phosphate groups, and it is produced by ATP synthase from inorganic phosphate and adenosine diphosphate or adenosine monophosphate.
- *D-luciferin*: (4S)-2-(6-hydroxy-1,3-benzothiazol-2-yl)-4,5-dihydrothiazole-4-carb-oxylic acid. It is the substrate of the firefly luciferase which becomes oxidized (oxyluciferin) and emits the characteristic yellow light in the reaction.
- *Biosensor*: any device or system capable of detecting a biological entity that combines a biological component with a physicochemical detector component.
- *HACCP*: Hazard Analysis and Critical Control Point. It is a scientific, rational and systematic approach for identification, assessment and control of hazards during production, manufacturing, preparation and use of food to ensure that food is safe when consumed. It

provides a preventive and thus a cost-effective approach to food safety.

- *Total bacterial count*: the total number of bacterial cells. It is the index to evaluate the whole status of food contaminated by bacterial cells.

APPLICATIONS TO AREAS OF HEATH AND DISEASE

Emerging environmental pathogens, such as *Helicobacter pylori* and *Burkholderia pseudomallei*, may well be of significance in some regions, as a result of bacterial contamination of water supplies being an especially significant risk to health (Nicholas 2004). It is important that the quantity of bacterial contamination on surfaces in hospitals such as floors, walls and sinks is measured in tracing routes of infection, identification of human carriers, evaluation of decontamination procedures, bacteriological sureveillance of the institutional environment. These above techniques are expected to be developed in the forthcoming years as a prognostic tool for areas of clinical environment.

SUMMARY POINTS

- The demand for developing rapid analytical tools for monitoring bacteria contamination in food industry has increased and is still expanding.
- These rapid methods including impedimetry, flow cytometry and immunology-based ones, have different disadvantages in application in monitoring bacteria contamination in food industry, thus there is still more work in shortening detection time and improving detection limit.
- The bioluminescence-based biosensor for total bacterial count has been designed, in which a sampler, relevant reagents and detecting part were integrated so that there is no need of extra reagent or large luminometer when used.
- The bioluminescence-based biosensor for total bacterial count has been designed, in which an automatic injection unit was introduced to accurately control the adding of the relevant reagents.
- The biosensor for pathogenic bacteria by combining the bioluminescence with immunomagnetic separation has been proposed.
- Three bioluminescence-based biosensors have potential applications in on-line monitoring in HACCP programs to detect bacterial contamination in food industry.

ACKNOWLEDGEMENTS

This work was sponsored by the Major National Scientific Research Plan (No. 2011CB933202), the China National Funds for Distinguished Young Scientists (No. 61125105), the Strategic Pilot Project for Science and Techlonogy of Chinese Academy of Science (No. XDA06020101), the National Natural Science Foundation of China (No. 61027001 and 61101048), the Hi-Tech R&D Programs of China (No. 2009AA03Z411) and the Chinese Academy of Science Program (No. Y2010015).

ABBREVIATIONS

AIU	:	automatic injection unit
ATP	:	adenosine 5'-triphosphate
CFU	:	colony forming units
CPU	:	central processing unit
EDTA	:	ethylenediaminetetraacetic acid
E. coli	:	*Escherichia coli*
E. cloacae	:	*Enterobacter cloacae*
HACCP	:	Hazard Analysis and Critical Control Point
K. pneumoniae	:	*Klebsiella pneumoiae*
NASA	:	National Aeronautics and Space Administration
PBS	:	phosphate buffered saline
PDU	:	photoelectric detecting unit
PMT	:	photomultiplier tube
PMT	:	photomultiplier tube
S. aureus	:	*Staphylococcus aureus*
SPCU	:	signal processing and control unit
TBC	:	total bacterial count
WHO	:	World Health Organization

REFERENCES

Bennett RW. 2005. Staphylococcal enterotoxin and its rapid identification in foods by enzyme linked immunosorbent assay-based methodology. J Food Prot 68: 1264–1270.

Cheng YX, YJ Liu, JJ Huang, K Li, W Zhang, YZ Xian and LT Jin. 2009. Combining biofunctional magnetic nanoparticles and ATP bioluminescence for rapid detection of *Escherichia coli*, Talanta 77: 1332–1336.

Diane GN, K Marion, V Linda, D Erwin, AK Awa, S Hein, O Marieke, L Merel, T John, S Flemming, van der G Joke and K Hilde. 2010. Food-borne diseases—the challenges of 20 years ago still persist while new ones continue to emerge, Int J Food Microbiol 139: S3–S15.

Entis P. 1998. Direct 24-hour presumptive enumeration of *Escherichia coli* O157:H7 in foods using hydrophobic grid membrane filter followed by serological confirmation: collaborative study. J Assoc Off Anal Chem Int 81: 403–18.

Eum NS, SH Yeom, DH Kwon, HR Kim and SW Kang. 2010. Enhancement of sensitivity using gold nanorods-Antibody conjugator for detection of *E. coli O157:H7* Sens. Actuators B 143: 784–788.

Feldsine PT, AH Lienau, RL Forgey and RD Calhoon. 1997. Visual immunoprecipitate assay (VIP) for Listeria monocytogenes and related Listeria species detection in selected foods: collaborative study. J AOAC Int 80: 791–805.

Gangar V, MS Curiale, K Lindberg and S Gambrel-Lenarz. 1999. Dry rehydratable film method for enumerating confirmed *Escherichia coli* in poultry, meats, and seafood: collaborative study. J Assoc Off Anal Chem Int 82: 73–78.

Griffiths MW. 1996. The role of ATP bioluminescence in the food industry: new light on old problems, Food Technol 6: 62–69.

Griffiths MW. 1997. Rapid Microbiological Methods with Hazard Analysis Critical Control Point. J Assoc Off Anal Chem Int.80: 1143–1150.

Ioannis SA (eds.) 2009. HACCP and ISO 22000: Application to Foods of Animal Origin. Wiley-Blackwell, Oxford, UK.

James PN and BK James. 1998. Diarrheagenic *Escherichia coli*, Clin Microbiol Rev 11: 142–201.

John W, I Joseph and D Chitrita. 2007 Direct detection of *E. coli O157:H7* in selected food systems by a surface plasmon resonance biosensor, LWT 40: 187–192.

Joyce MS and VL Daniel. 2005. Rapid PCR confirmation of *E. coli O157:H7* after evanescent wave fiber optic biosensor detection Biosens Bioelectron 21: 881–887.

Kaferstein FK. 1997. Food safety: a commonly underestimated public health issue, World Health Statistics Quarterly 50: 3–4.

Lundin A. 1989. In: PE. Stanley, BJ McCarthy and R Smither [eds.]. ATP assays in routine microbiology: From visions to realities in the 1980s. Rapid Methods in Microbiology, Blackwell Scientific, Oxford, UK. pp. 11–27.

Lundin A. 2000. Use of Firefly Luciferase in ATP-Related Assays of Biomass, Enzymes and Metabolites. Methods Enzymol. Part C 305: 346–370.

Luo JP, XH Liu, Q Tian, WW Yue, J Zeng, GQ Chen, and XX Cai. 2009. Disposable bioluminescence-based biosensor for detection of bacterial count in food. Anal Biochem 394, 1: 1–6.

Luo JP, RP Liu, Q Tian, XH Liu and XX Cai. 2010. Rapid detection of Escherichia coli using immunomagnetic seperation and bioluminescence ATP assay, Technical Digest of the IMCS-13 2010. The 13th International Meeting on Chemical Sensors, W. Wlodarski, L. Faraone, K. Kalantar-Zadeh and G. Matthews [eds.] ISBN: 978-1-74052-208-3 Perth, Australia.

Madhukar V and YB Li. 2007. Interdigitated array microelectrode based impedance biosensor coupled with magnetic nanoparticle-antibody conjugates for detection of *Escherichia coli O157:H7* in food samples, Biosens. Bioelectron 22: 2408–2414.

Magliulo M, P Simoni, M Guardigli, E Michelini, M Luciani and R Lelli. 2007. A rapid multiplexed chemiluminescent immunoassay for the detection of *Escherichia coli O157: H7, Yersinia enterocolitica, Salmonella typhimurium,* and *Listeria monocytogenes* pathogen bacteria. J Agric Food Chem 55: 4933–4939.

Manafi M. 1996. Fluorogenic and chromogenic substrates in culture media and identification tests. Int J Food Microbiol 31: 45–58.

Matar GM, PS Hayes, WF Bibb, and B Swaminathan. 1997. Listeriolysin O-based latex agglutination test for the rapid detection of *Listeria monocytogenes* in foods. J Food Prot 60: 1038–1040.

Nicholas JA. 2004. Microbial contamination of drinking water and disease outcomes in developing regions, Toxicology 198: 229–238.

Qiu JM, Y Zhou, H Chen and JM Lin. 2009. Immunomagnetic separation and rapid detection of bacteria using bioluminescence and microfluidics. Talanta 79: 787–795.

Shim WB, JG Choi, JY Kim, ZY Yang, KH Lee and MG Kim. 2007. Production of monoclonal antibody against *Listeria monocytogenes* and its application to immunochromato graphy strip test. J Microbiol Biotechnol 17: 1152–61.

van der Zee H and JHJ Huis in't Veld. 1997. Rapid and alternative screening methods for microbiological analysis. J Assoc Off Anal Chem Int 80: 934–940.

Waverla M, H Eisgruber, B Schalch and A Stolle. 1998. Zum Einsatz der Impedanzmessung in de Lebensmittelmikrobiologie. Arch Lebensmittelhyg 19: 76–89.

World Health Organization. 1997. HACCP: introducing the Hazard Analysis and Critical Control System. Food Safety Issues. WHO/FSF/FOS/97.2.

World Health Organization. In: G Rees, K Pond, D Kay, J Bartram and J Santo Domingo [eds.] 2010. Safe Management of Shellfish and Harvest Waters. 1st edn. IWA Publishing, London, UK.

Zhou AY, JP Luo, WW Yue, BS He, QD Yang and XX Cai. 2008. Development of handheld ATP detecting system based on bioluminescence, Chinese J Sens Actuator 21: 543–546.

Zhu PX, RS Daniel, SK Jeffrey, S Appavu, SL Li, A Pete and CM Tang. 2005. Detection of water-borne *E. coli O157* using the integrating waveguide biosensor. Biosens Bioelectron 21: 678–683.

Index

About the Editors

Victor R. Preedy BSc, PhD, DSc, FIBiol, FRCPath, FRSPH is a Professor at King's College London and also at King's College Hospital. He is attached to both the Diabetes and Nutritional Sciences Division and the Department of Nutrition and Dietetics. He is also Director of the Genomics Centre and a member of the School of Medicine. Professor Preedy graduated in 1974 with an Honours Degree in Biology and Physiology with Pharmacology. He gained his University of London PhD in 1981. In 1992, he received his Membership of the Royal College of Pathologists and in 1993 he gained his second doctoral degree, for his outstanding contribution to protein metabolism in health and disease. Professor Preedy was elected as a Fellow to the Institute of Biology in 1995 and to the Royal College of Pathologists in 2000. Since then he has been elected as a Fellow to the Royal Society for the Promotion of Health (2004) and The Royal Institute of Public Health (2004). In 2009, Professor Preedy became a Fellow of the Royal Society for Public Health. In his career Professor Preedy has carried out research at the National Heart Hospital (part of Imperial College London) and the MRC Centre at Northwick Park Hospital. He has collaborated with research groups in Finland, Japan, Australia, USA and Germany. He is a leading expert on the science of health. He has lectured nationally and internationally. To his credit, Professor Preedy has published over 570 articles, which includes 165 peer-reviewed manuscripts based on original research, 90 reviews and over 40 books and volumes.

Dr Vinood B. Patel, PhD is currently a Senior Lecturer in Clinical Biochemistry at the University of Westminster and honorary fellow at King's College London. Dr Patel obtained his degree in Pharmacology from the University of Portsmouth, his PhD in protein metabolism from King's College London in 1997 and carried out post-doctoral research at Wake Forest University School of Medicine, USA where he developed novel biophysical techniques to characterise mitochondrial ribosomes. He presently directs studies on metabolic pathways involved in fatty liver disease, focussing on mitochondrial energy regulation and cell death. Dr Patel has published over 150 articles, including books in the area of nutrition and health prevention.

Color Plate Section

Chapter 1

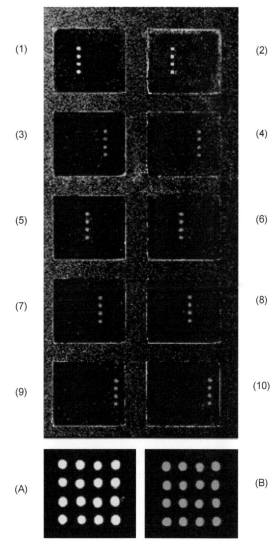

Figure 1.3. Multi-analysis of five different chemicals. Abs addition: In one unit, from left to right: 3% BSA (as the negative control), monoclonal antibodies of CAP, Pap, E_2, NP and Ar were spotted as probes; addition of the 10 units was homogeneous. Antigen conjugates addition: (1) and (2) units: CAP-BSA-Cy3; (3) and (4) units: NP-OVA-Cy3; (5) and (6) units: Pap-OVA-Cy3; (7) and (8) units: E_2-OVA-Cy3; (9) and (10) units: Ar-OVA-Cy3. **(A)** Four mAbs of CAP, NP, E_2 and Ar with their complementary Ag conjugates spotting on slide. **(B)** CAP, NP, E_2 and Ar with the same concentration of 0.01μg/mL mixture were added on the chip (Gao et al. 2009).

Chapter 3

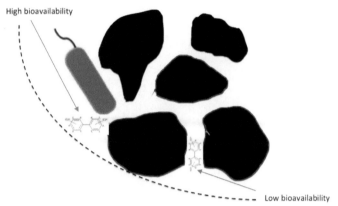

Figure 3.3. Schematization of bioavailability processes in soil environments. Xenobiotics can be trapped within soil particles or within the organic matter, and their biovailability ofrfor ecological receptors such as bacteria is reduced. Unpublished material of the authors.

Chapter 6

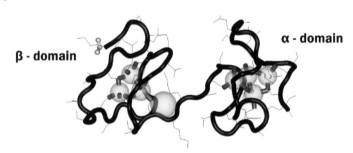

Figure 6.1. Structure of metallothionein. The N-terminal part of the protein is marked as α-domain, which has three binding sites for divalent ions. β-Domain (C-terminal part) has the ability to bind four divalent ions of heavy metals. Due to the property of MT being metal-inducible and, also, due to their high affinity to metal ions, homeostasis of heavy metal levels is probably their most important biological function.

Chapter 8

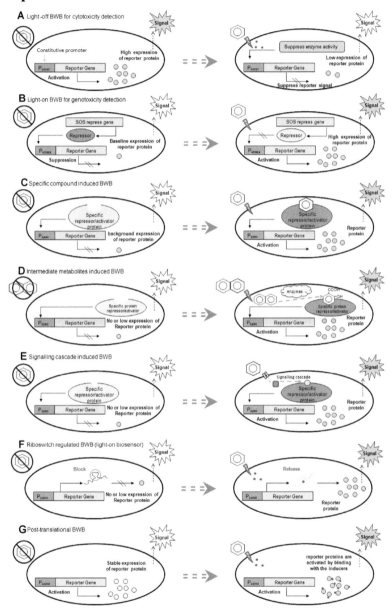

Figure 8.1. Schematic of seven types of bacterial whole cell biosensors (BWBs).

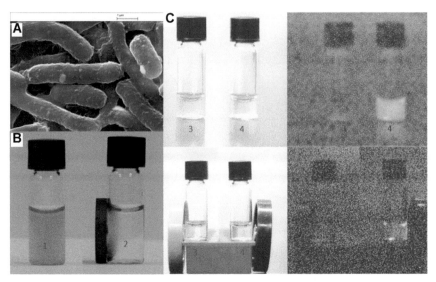

Figure 8.5. Viability and functionality of BWBs (*Acinetobacter* ADPWH_lux) coupled with magnetic nanoparticles (MNPs). **(A)** Scanning Electron Microscope (SEM) image of Acinetobacter ADPWH_lux bioreporters coupled with MNPs. **(B)** Most MNPs coupled cells originally suspended in water (1) were attracted to the side of a permanent magnet (2) after 5 min. **(C)** MNPs coupled cells were present with 0 µM (3) and 100 µM salicylate induction (4). Salicylate induced MNPs coupled cells were attracted to the magnet side (left images were under ordinary light and right in the dark).

Chapter 10

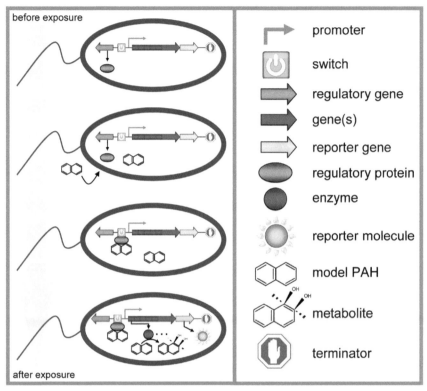

Figure 10.2. Schematic showing idealized/simplified mechanism of response for Class I bacterial bioreporters. Unpublished figure.

Chapter 13

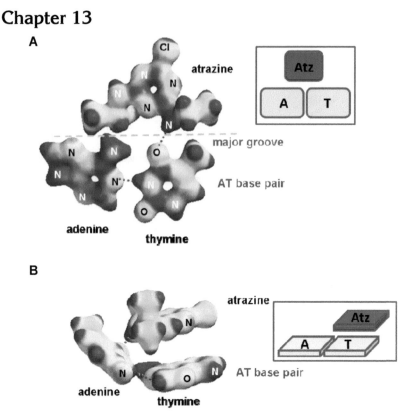

Figure 13.8. Initial stage of the interactions of atrazine with a AT base-pair. (A) In-plane interactions at the major groove, stabilized by hydrogen bonding with T; (B) intercalation of Atz, stabilized by stacking interactions with T and hydrogen bonding with A; atrazine-induced tilt of A with respect to T seen in (B); electron density surfaces ($d = 0.08$ a.u.) with mapped electrostatic potential (color coded: from negative-red to positive-blue). (Unpublished data).

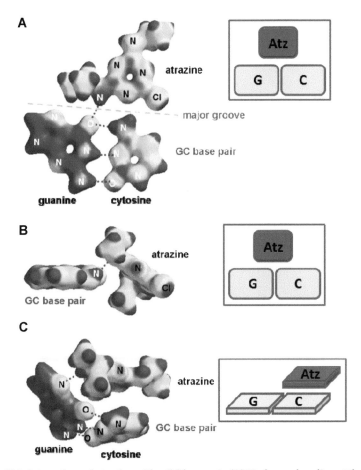

Figure 13.9. Interactions of atrazine with a GC base-pair. **(A)** Hydrogen bonding at the major groove; **(B)** approach towards intercalation, **(C)** intercalation and atrazine-induced tilt of G with respect to C; electron density surfaces ($d = 0.08$ a.u.) with mapped electrostatic potential (color coded: from negative-red to positive-blue). (Unpublished data).

Chapter 16

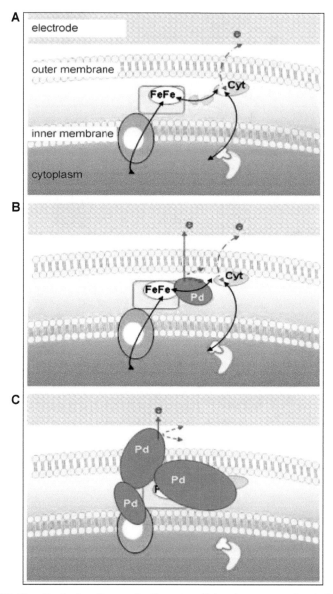

Figure 16.8. Hypothesized pathways for the extracellular electron-transfer chain between the cell and the electrode: **(A)** in the absence of Pd_0 (via periplasmic cytochromes and hydrogenases); **(B)** in the presence of Pd_0 at a low loading; **(C)** in the presence of Pd_0 at a high loading. (FeFe) is an iron-only hydrogenase, Cyt is a c-type cytochrome, and areas shaded gray are the Pd_0 nanoparticles. Adapted from Wu et al. 2010.

T - #0354 - 071024 - C8 - 234/156/17 - PB - 9780367380984 - Gloss Lamination